OXFORD IB STUDY GUIDES

Geoffrey Neuss

Chemistry

FOR THE IB DIPLOMA

OXFORD
UNIVERSITY PRESS

OXFORD
UNIVERSITY PRESS

Great Clarendon Street, Oxford OX2 6DP

Oxford University Press is a department of the University of Oxford.
It furthers the University's objective of excellence in research,
scholarship, and education by publishing worldwide in

Oxford   New York

Auckland   Cape Town   Dar es Salaam   Hong Kong   Karachi
Kuala Lumpur   Madrid   Melbourne   Mexico City   Nairobi
New Delhi   Shanghai   Taipei   Toronto

With offices in

Argentina   Austria   Brazil   Chile   Czech Republic   France   Greece
Guatemala   Hungary   Italy   Japan   Poland   Portugal   Singapore
South Korea   Switzerland   Thailand   Turkey   Ukraine   Vietnam

Oxford is a registered trade mark of Oxford University Press
in the UK and in certain other countries

British Library Cataloguing in Publication Data

Data available

ISBN: 978-0-19-839002-2

10 9 8 7 6 5 4 3

Printed in Great Britain by Bell & Bain Ltd, Glasgow

Paper used in the production of this book is a natural, recyclable product made from wood grown in
sustainable forests. The manufacturing process conforms to the environmental regulations of the country
of origin

Acknowledgments

We are grateful to the following to reproduce the following copyright material.

p82 Thomas Nixon, Karolinska Institutet University Library

Cover photo: Ocean/Corbis

We have tried to trace and contact all copyright holders before publication. If notified the publishers will be
pleased to rectify any errors or omissions at the earliest opportunity.

MIX
Paper from
responsible sources
FSC® C007785

Introduction and acknowledgements

This book replaces the very successful first edition. Like the first edition it is written specifically for students studying Chemistry for the International Baccalaureate Diploma although many students following their own national systems will also find it helpful. It comprehensively covers the new programme that will be examined from 2009 onwards. All the information required for each topic is set out in separate boxes with clear titles that follow faithfully the layout of the syllabus. The first eleven topics cover both the core content needed by all students and the Additional Higher Level material under the main topic headings. The difference between the two levels is clearly distinguished. Both Higher Level and Standard Level students must study at least two of the seven options and each option stands in its own right even if this has meant repeating small amounts of some of the material. Worked examples are included where they are appropriate and at the end of each main topic and option there are practice questions. The majority of these questions are taken from past IB examination papers and I would like to thank the International Baccalaureate for giving me permission to use them. The remaining questions are written to the same IB standard specifically for this book. Answers to the questions are provided. These answers are not necessarily full 'model' answers but they do contain all the information needed to score each possible mark. To help you, the student, gain the highest grade possible the final chapter is devoted to giving you advice on how to study and prepare for the final examination. It also advises you on how to excel at the internally assessed practical component of the course. For those who opt for Chemistry as the subject for their Extended Essay it gives advice and guidance on how to choose the topic and write your Essay. A comprehensive Extended Essay checklist is included to help you gain bonus points towards your IB Diploma.

I have been fortunate at Atlantic College to teach many highly motivated and gifted students who have often challenged me with searching questions. During my association with the International Baccalaureate, the European Baccalaureate and the United World Colleges I have been privileged to meet, work alongside and exchange ideas with many excellent Chemistry teachers who exude a real enthusiasm for their subject. Many of these have influenced me greatly - in particular, John Devonshire, a fellow teacher at Atlantic College and Jacques Furnemont, an Inspector of Chemistry in Belgium. I value greatly their advice, opinions and knowledge. I would also like to pay tribute to Ron Ragsdale and Arden Zipp. The high regard with which IB Chemistry is held today owes much to both of them.

Many IB students and teachers have encouraged me to write this book. I have stuck rigorously to the syllabus to produce a study guide which contains all the necessary subject content required for the examination in one easily accessible and compact format. IB Chemistry is, of course, about much more than the final examination and you are also encouraged to read the IB Course Companion for Chemistry which puts the subject into context and includes wider perspectives such as critical thinking and how it relates to the Theory of Knowledge. Paul Fairbrother and Gareth Hegarty from the IB and Elspeth Boardley and Carolyn Lee from Oxford University Press have been particularly encouraging and helpful. I owe much to Nick Lee at St Clare's Oxford and Chris Talbot from the Anglo-Chinese School in the Republic of Singapore for making constructive suggestions and corrections after spending many hours reading through the drafts. Finally I should like to thank my wife Chris and my friend and colleague John for their patience and unstinting support throughout.

Dr Geoffrey Neuss

CONTENTS

(Italics denote topics which are exclusively Higher Level.)

1 QUANTITATIVE CHEMISTRY

Chemical formulas and the mole concept

ELEMENTS

All substances are made up of one or more elements. An element cannot be broken down by any chemical process into simpler substances. There are just over 100 known elements. The smallest part of an element is called an atom.

Names of the first 20 elements

Atomic Number	Name	Symbol	Relative atomic mass
1	hydrogen	H	1.01
2	helium	He	4.00
3	lithium	Li	6.94
4	beryllium	Be	9.01
5	boron	B	10.81
6	carbon	C	12.01
7	nitrogen	N	14.01
8	oxygen	O	16.00
9	fluorine	F	19.00
10	neon	Ne	20.18
11	sodium	Na	22.99
12	magnesium	Mg	24.31
13	aluminium	Al	26.98
14	silicon	Si	28.09
15	phosphorus	P	30.97
16	sulfur	S	32.06
17	chlorine	Cl	35.45
18	argon	Ar	39.95
19	potassium	K	39.10
20	calcium	Ca	40.08

COMPOUNDS

Some substances are made up of a single element although there may be more than one atom of the element in a particle of the substance. Oxygen is diatomic, that is, a molecule of oxygen contains two oxygen atoms. A compound contains more than one element. For example, a molecule of water contains two hydrogen atoms and one oxygen atom. Water is a compound not an element because it can be broken down chemically into its constituent elements: hydrogen and oxygen.

FORMULAS OF COMPOUNDS

Compounds can be described by different chemical formulas.

Empirical formula (literally the formula obtained by experiment)
This shows the simplest whole number ratio of atoms of each element in a particle of the substance. It can be obtained by either knowing the mass of each element in the compound or from the percentage composition by mass of the compound. The percentage composition can be converted directly into mass by assuming 100 g of the compound are taken.

Example: A compound contains 40.00% carbon, 6.73% hydrogen and 53.27% oxygen by mass, determine the empirical formula.

	Amount /mol	Ratio	
C	40.00/12.01 = 3.33	1	Empirical formula
H	6.73/1.01 = 6.66	2	= CH_2O
O	53.27/16.00 = 3.33	1	

Molecular formula
For molecules this is much more useful as it shows the actual number of atoms of each element in a molecule of the substance. It can be obtained from the empirical formula if the molar mass of the compound is also known.
Methanal CH_2O (M_r =30), ethanoic acid $C_2H_4O_2$ (M_r = 60) and glucose $C_6H_{12}O_6$ (M_r = 180) are different substances with different molecular formulas but all with the same empirical formula CH_2O. Note that subscripts are used to show the number of atoms of each element in the compound.

Structural formula
This shows the arrangement of atoms and bonds within a molecule and is particularly useful in organic chemistry.

The three different formulas can be illustrated using ethene:

CH_2	C_2H_4	H $\diagdown$ $C = C$ $\diagup$ H H $\diagup$ $\diagdown$ H (can also be written $H_2C=CH_2$)
empirical formula	molecular formula	structural formula

MOLE CONCEPT AND AVOGADRO'S CONSTANT

A single atom of an element has an extremely small mass. For example an atom of carbon-12 has a mass of 1.993×10^{-23} g. This is far too small to weigh. A more convenient amount to weigh is 12.00 g. 12.00 g of carbon-12 contains 6.02×10^{23} atoms of carbon-12. This number is known as Avogadro's constant (N_A or L).
Chemists measure amounts of substances in moles. A mole is the amount of substance that contains L particles of that substance. The mass of one mole of **any** substance is known as the **molar mass** and has the symbol M. For example, hydrogen atoms have $\frac{1}{12}$ of the mass of carbon-12 atoms so a mole of hydrogen atoms contains 6.02×10^{23} hydrogen atoms and has a mass of 1.01 g. In reality elements are made up of a mixture of isotopes.
The **relative atomic mass** of an element A_r is the weighted mean of all the naturally occurring isotopes of the element relative to carbon-12. This explains why the relative atomic masses given for the elements above are not whole numbers. The units of molar mass are g mol^{-1} but relative molar masses M_r have no units. For molecules **relative molecular mass** is used. For example, the M_r of glucose, $C_6H_{12}O_6$ = (6 × 12.01) + (12 × 1.01) + (6 × 16.00) = 180.18. For ionic compounds the term **relative formula mass** is used.
Be careful to distinguish between the words **mole** and **molecule**. A molecule of hydrogen gas contains two atoms of hydrogen and has the formula H_2. A mole of hydrogen gas contains 6.02×10^{23} hydrogen molecules made up of two moles (1.20×10^{24}) of hydrogen atoms.

Chemical reactions and equations

PROPERTIES OF CHEMICAL REACTIONS

In all chemical reactions:
- new substances are formed.
- bonds in the reactants are broken and bonds in the products are formed resulting in an energy change between the reacting system and its surroundings.
- there is a fixed relationship between the number of particles of reactants and products resulting in no overall change in mass – this is known as the stoichiometry of the reaction.

CHEMICAL EQUATIONS

Chemical reactions can be represented by chemical equations. Reactants are written on the left hand side and products on the right hand side. The number of moles of each element must be the same on both sides in a balanced chemical equation, e.g. the reaction of nitric acid (one of the acids present in acid rain) with calcium carbonate (the main constituent of marble statues).

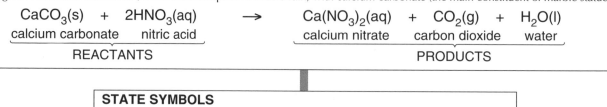

$$CaCO_3(s) \ + \ 2HNO_3(aq) \ \longrightarrow \ Ca(NO_3)_2(aq) \ + \ CO_2(g) \ + \ H_2O(l)$$

calcium carbonate   nitric acid   calcium nitrate   carbon dioxide   water

REACTANTS   PRODUCTS

STATE SYMBOLS

Because the physical state that the reactants and products are in can affect both the rate of the reaction and the overall energy change it is good practice to include the state symbols in the equation.

(s) – solid   (l) – liquid   (g) – gas   (aq) – in aqueous solution

→ OR ⇌

A single arrow → is used if the reaction goes to completion. Sometimes the reaction conditions are written on the arrow:

e.g. $C_2H_4(g) \ + \ H_2(g) \ \xrightarrow{\text{Ni catalyst, 180 °C}} \ C_2H_6(g)$

Reversible arrows are used for reactions where both the reactants and products are present in the equilibrium mixture:

e.g. $3H_2(g) \ + \ N_2(g) \ \underset{250 \text{ atm}}{\overset{\text{Fe(s), 550 °C}}{\rightleftharpoons}} \ 2NH_3(g)$

COEFFICIENTS AND MOLAR RATIO

The coefficient refers to the number in front of each reactant and product in the equation. The coefficients give information on the molar ratio. In the first example above, two moles of nitric acid react with one mole of calcium carbonate to produce one mole of calcium nitrate, one mole of carbon dioxide and one mole of water. In the reaction between hydrogen and nitrogen above, three moles of hydrogen gas react with one mole of nitrogen gas to produce two moles of ammonia gas.

IONIC EQUATIONS

Because ionic compounds are completely dissociated in solution it is sometimes better to use ionic equations. For example, when silver nitrate solution is added to sodium chloride solution a precipitate of silver chloride is formed.

$$Ag^+(aq) + NO_3^-(aq) + Na^+(aq) + Cl^-(aq) \longrightarrow AgCl(s) + Na^+(aq) + NO_3^-(aq)$$

$Na^+(aq)$ and $NO_3^-(aq)$ are spectator ions and do not take part in the reaction. So the ionic equation becomes:

$$Ag^+(aq) + Cl^-(aq) \longrightarrow AgCl(s)$$

From this we can deduce that any soluble silver salt will react with any soluble chloride to form a precipitate of silver chloride.

Mass and gaseous volume relationships

SOLIDS
Normally measured by weighing to obtain the mass.

$$1.000 \text{ kg} = 1000 \text{ g}$$

When weighing a substance the mass should be recorded to show the accuracy of the balance. For example, exactly 16 g of a substance would be recorded as 16.00 g on a balance weighing to + or – 0.01 g but as 16.000 g on a balance weighing to + or – 0.001 g.

MEASUREMENT OF MOLAR QUANTITIES
In the laboratory moles can conveniently be measured using either mass or volume depending on the substances involved.

LIQUIDS
Pure liquids may be weighed or the volume recorded.
The density of the liquid $= \frac{\text{mass}}{\text{volume}}$ and is usually expressed in g cm^{-3}.

GASES
Mass or volume may be used for gases.

SOLUTIONS
Volume is usually used for solutions.

$$1.000 \text{ litre} = 1.000 \text{ dm}^3 = 1000 \text{ cm}^3$$

Concentration is the amount of solute (dissolved substance) in a known volume of solution (solute plus solvent). It is expressed either in g dm^{-3} or, more usually, in mol dm^{-3}. A solution of known concentration is known as a *standard solution*.

To prepare a 1.00 mol dm^{-3} solution of sodium hydroxide dissolve 40.00 g of solid sodium hydroxide in distilled water and then make the total volume up to 1.00 dm^3.

Concentration is often represented by square brackets, e.g.

$$[\text{NaOH(aq)}] = 1.00 \text{ mol dm}^{-3}$$

A 25.0 cm^3 sample of this solution contains

$$1.00 \times \frac{25.0}{1000} = 2.50 \times 10^{-2} \text{ mol of NaOH}$$

CHANGING THE VARIABLES FOR A FIXED MASS OF GAS

$P \propto \frac{1}{V}$ (or PV = constant)
At constant temperature: as the volume decreases the concentration of the particles increases, resulting in more collisions with the container walls. This increase in pressure is inversely proportional to the volume, i.e. doubling the pressure halves the volume.

$P \propto T$ (or $\frac{P}{T}$ = constant)
At constant volume: increasing the temperature increases the average kinetic energy so the force with which the particles collide with the container walls increases. Hence pressure increases and is directly proportional to the absolute temperature, i.e. doubling the absolute temperature doubles the pressure.

$V \propto T$ (or $\frac{V}{T}$ = constant)
At constant pressure: at higher temperatures the particles have a greater average velocity so individual particles will collide with the container walls with greater force. To keep the pressure constant there must be fewer collisions per unit area so the volume of the gas must increase. The increase in volume is directly proportional to the absolute temperature, i.e. doubling the absolute temperature doubles the volume.

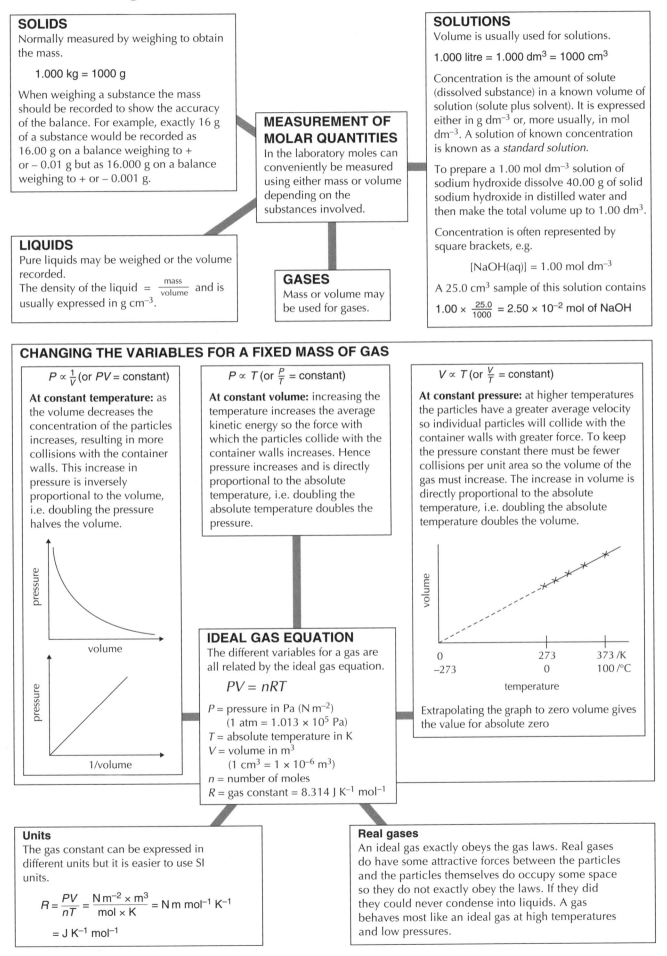

Extrapolating the graph to zero volume gives the value for absolute zero

IDEAL GAS EQUATION
The different variables for a gas are all related by the ideal gas equation.

$$PV = nRT$$

P = pressure in Pa (N m^{-2})
    (1 atm = 1.013 × 10^5 Pa)
T = absolute temperature in K
V = volume in m^3
    (1 cm^3 = 1 × 10^{-6} m^3)
n = number of moles
R = gas constant = 8.314 J K^{-1} mol^{-1}

Units
The gas constant can be expressed in different units but it is easier to use SI units.

$$R = \frac{PV}{nT} = \frac{\text{N m}^{-2} \times \text{m}^3}{\text{mol} \times \text{K}} = \text{N m mol}^{-1} \text{ K}^{-1}$$

$$= \text{J K}^{-1} \text{ mol}^{-1}$$

Real gases
An ideal gas exactly obeys the gas laws. Real gases do have some attractive forces between the particles and the particles themselves do occupy some space so they do not exactly obey the laws. If they did they could never condense into liquids. A gas behaves most like an ideal gas at high temperatures and low pressures.

Molar volume of a gas and calculations

MOLAR VOLUME OF A GAS

The ideal gas equation depends on the amount of gas (number of moles of gas) but not on the nature of the gas. Avogadro's Law states that equal volumes of different gases at the same temperature and pressure contain the same number of moles. From this it follows that one mole of any gas will occupy the same volume at the same temperature and pressure. This is known as the **molar volume of a gas**. At 273K and 1.013×10^5 Pa (1 atm.) pressure this volume is 2.24×10^{-2} m^3 (22.4 dm^3 or 22 400 cm^3).

When the mass of a particular gas is fixed (nR is constant) a useful expression to convert the pressure, temperature and volume under one set of conditions (1) to another set of conditions (2) is:

$$\frac{P_1 V_1}{T_1} = \frac{P_2 V_2}{T_2}$$

In this expression there is no need to convert to SI units as long as the same units for pressure and volume are used on both sides of the equation. However do not forget that T refers to the absolute temperature and must be in kelvin.

CALCULATIONS FROM EQUATIONS

Work methodically.

Step 1. Write down the correct formulas for all the reactants and products.

Step 2. Balance the equation to obtain the correct stoichiometry of the reaction.

Step 3. If the amounts of all reactants are known work out which are in **excess** and which one is the limiting reagent. By knowing the **limiting reagent** the maximum **yield** of any of the products can be determined.

Step 4. Work out the number of moles of the substance required.

Step 5. Convert the number of moles into the mass or volume.

Step 6. Express the answer to the correct number of significant figures and include the appropriate units.

WORKED EXAMPLES

(a) Calculate the volume of hydrogen gas evolved at 273 K and 1 atm pressure when 0.623 g of magnesium reacts with 27.3 cm^3 of 1.25 mol dm^{-3} hydrochloric acid.

Equation:

$Mg(s) + 2HCl(aq) \rightarrow H_2(g) + MgCl_2(aq)$

A_r for Mg = 24.31. Amount of Mg present = $\frac{0.623}{24.31}$ = 2.56×10^{-2} mol

Amount of HCl present = $1.25 \times \frac{27.3}{1000}$ = 3.41×10^{-2} mol

From the equation $2 \times 2.56 \times 10^{-2} = 5.12 \times 10^{-2}$ mol of HCl would be required to react with all of the magnesium.

Therefore the magnesium is in excess and the limiting reagent is the hydrochloric acid.

The maximum amount of hydrogen produced = $\frac{3.41 \times 10^{-2}}{2}$ = 1.705×10^{-2} mol

Volume of hydrogen at 273 K, 1 atm = $1.705 \times 10^{-2} \times 22.4$ = 0.382 dm^3 (or 382 cm^3)

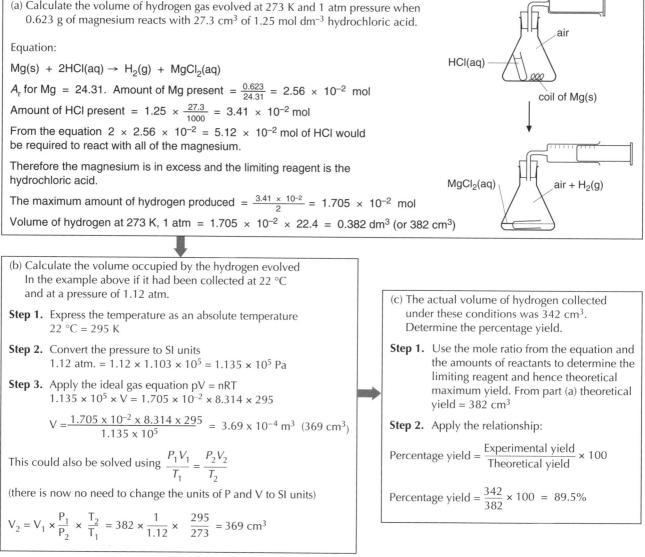

(b) Calculate the volume occupied by the hydrogen evolved
In the example above if it had been collected at 22 °C and at a pressure of 1.12 atm.

Step 1. Express the temperature as an absolute temperature
22 °C = 295 K

Step 2. Convert the pressure to SI units
1.12 atm. = $1.12 \times 1.103 \times 10^5 = 1.135 \times 10^5$ Pa

Step 3. Apply the ideal gas equation pV = nRT
$1.135 \times 10^5 \times V = 1.705 \times 10^{-2} \times 8.314 \times 295$

$V = \frac{1.705 \times 10^{-2} \times 8.314 \times 295}{1.135 \times 10^5}$ = 3.69×10^{-4} m^3 (369 cm^3)

This could also be solved using $\frac{P_1 V_1}{T_1} = \frac{P_2 V_2}{T_2}$

(there is now no need to change the units of P and V to SI units)

$V_2 = V_1 \times \frac{P_1}{P_2} \times \frac{T_2}{T_1} = 382 \times \frac{1}{1.12} \times \frac{295}{273}$ = 369 cm^3

(c) The actual volume of hydrogen collected under these conditions was 342 cm^3. Determine the percentage yield.

Step 1. Use the mole ratio from the equation and the amounts of reactants to determine the limiting reagent and hence theoretical maximum yield. From part (a) theoretical yield = 382 cm^3

Step 2. Apply the relationship:

Percentage yield = $\frac{\text{Experimental yield}}{\text{Theoretical yield}} \times 100$

Percentage yield = $\frac{342}{382} \times 100$ = 89.5%

4 Quantitative chemistry

IB QUESTIONS – QUANTITATIVE CHEMISTRY

1. What is the mass in grams of one molecule of ethanoic acid CH_3COOH?

   A. 0.1   **B.** 3.6×10^{25} **C.** 1×10^{-22} **D.** 60

2. Which is not a true statement?

   A. One mole of methane contains four moles of hydrogen atoms

   B. One mole of ^{12}C has a mass of 12.00 g

   C. One mole of hydrogen gas contains 6.02×10^{23} atoms of hydrogen

   D. One mole of methane contains 75% of carbon by mass

3. A pure compound contains 24 g of carbon, 4 g of hydrogen and 32 g of oxygen.

   No other elements are present. What is the empirical formula of the compound?

   A. $C_2H_4O_2$ **B.** CH_2O **C.** CH_4O **D.** CHO

4. Which one of the following statements about SO_2 is/are correct?

   I. One mole of SO_2 contains 1.8×10^{24} atoms

   II. One mole of SO_2 has a mass of 64 g

   A. Both I and II   **B.** Neither I nor II   **C.** I only   **D.** II only

5. What is the empirical formula for the compound $C_6H_5(OH)_2$?

   A. C_6H_6O **B.** $C_6H_5O_2H_2$ **C.** C_6H_7O **D.** $C_6H_7O_2$

6. Phosphorus burns in oxygen to produce phosphorus pentoxide P_4O_{10}.

   What is the sum of the coefficients in the balanced equation?

   $$\_P_4(s) + \_O_2(g) \rightarrow \_P_4O_{10}(s)$$

   A. 3   **B.** 5   **C.** 6   **D.** 7

7. Magnesium reacts with hydrochloric acid according to the following equation:

   $$Mg(s) + 2HCl(aq) \rightarrow MgCl_2(aq) + H_2(g)$$

   What mass of hydrogen will be obtained if 100 cm^3 of 2.00 mol dm^{-3} HCl are added to 4.86 g of magnesium?

   A. 0.2g   **B.** 0.4g   **C.** 0.8g   **D.** 2.0g

8. Butane burns in oxygen according to the equation below.

   $$2C_4H_{10}(g) + 13O_2(g) \rightarrow 8CO_2(g) + 10H_2O(l)$$

   If 11.6 g of butane is burned in 11.6 g of oxygen which is the limiting reagent?

   A. Butane   **C.** Neither

   B. Oxygen   **D.** Oxygen and butane

9. When 250 cm^3 of 3.00 mol dm^{-3} HCl(aq) is added to 350 cm^3 of 2.00 mol dm^{-3} HCl(aq) the concentration of the solution of hydrochloric acid obtained in mol dm^{-3} is:

   A. 2.42   **B.** 1.45   **C.** 2.90   **D.** 2.50

10. Sulfuric acid and sodium hydroxide react together according to the equation:

    $$H_2SO_4(aq) + 2NaOH(aq) \rightarrow Na_2SO_4(aq) + 2H_2O(l)$$

    What volume of 0.250 mol dm^{-3} NaOH is required to neutralise exactly 25.0 cm^3 of 0.125 mol dm^{-3} H_2SO_4?

    A. 25.0 cm^3 **B.** 12.5 cm^3 **C.** 50 cm^3 **D.** 6.25 cm^3

11. Separate samples of two gases, each containing a pure substance, are found to have the same density under the same conditions of temperature and pressure. Which statement about these two samples **must** be correct?

    A. They have the same volume

    B. They have the same relative molecular mass

    C. There are equal numbers of moles of gas in the two samples

    D. They condense at the same temperature

12. Which expression represents the density of a gas sample of relative molar mass, M_r, at temperature T, and pressure, P?

    A. $\dfrac{PM_r}{T}$ **C.** $\dfrac{PM_r}{RT}$

    B. $\dfrac{RT}{PM_r}$ **D.** $\dfrac{RM_r}{PT}$

13. A 250 cm^3 sample of an unknown gas has a mass of 1.42 g at 35 °C and 0.85 atmospheres. Which expression gives its molar mass, M_r? (R = 82.05 cm^3 atm K^{-1} mol^{-1})

    A. $\dfrac{1.42 \times 82.05 \times 35}{0.25 \times 0.85}$ **C.** $\dfrac{1.42 \times 250 \times 0.85}{82.05 \times 308}$

    B. $\dfrac{1.42 \times 82.05 \times 308}{0.25 \times 0.85}$ **D.** $\dfrac{1.42 \times 82.05 \times 308}{250 \times 0.85}$

14. Aspirin, $C_9H_8O_4$, is made by reacting ethanoic anhydride, $C_4H_6O_3$ (M_r = 102.1), with 2-hydroxybenzoic acid (M_r = 138.1), according to the equation:

    $$2C_7H_6O_3 + C_4H_6O_3 \rightarrow 2C_9H_8O_4 + H_2O$$

    (a) If 15.0 g 2-hydroxybenzoic acid is reacted with 15.0 g ethanoic anhydride, determine the limiting reagent in this reaction.

    (b) Calculate the maximum mass of aspirin that could be obtained in this reaction.

    (c) If the mass obtained in this experiment was 13.7 g, calculate the percentage yield of aspirin.

15. 14.48 g of a metal sulfate with the formula M_2SO_4 were dissolved in water. Excess barium nitrate solution was added in order to precipitate all the sulfate ions in the form of barium sulfate. 9.336 g of precipitate was obtained.

    (a) Calculate the amount of barium sulfate $BaSO_4$ precipitated.

    (b) Calculate the amount of sulfate ions present in the 14.48 g of M_2SO_4.

    (c) What is the relative molar mass of M_2SO_4?

    (d) Calculate the relative atomic mass of M and hence identify the metal.

2 ATOMIC STRUCTURE

The atom

COMPOSITION OF ATOMS
The smallest part of an element is an atom. It used to be thought that atoms are indivisible but they can be broken down into many different sub-atomic particles. All atoms, with the exception of hydrogen, are made up of three fundamental sub-atomic particles – protons, neutrons, and electrons.

The hydrogen atom, the simplest atom of all, contains just one proton and one electron. The actual mass of a proton is 1.672×10^{-24} g but it is assigned a relative value of 1. The mass of a neutron is virtually identical and also has a relative mass of 1. Compared to a proton and a neutron an electron has negligible mass with a relative mass of only $\frac{1}{2000}$. Neutrons are neutral particles. An electron has a charge of 1.602×10^{-19} coulombs which is assigned a relative value of –1. A proton carries the same charge as an electron but of an opposite sign so has a relative value of +1. All atoms are neutral so must contain equal numbers of protons and electrons.

SUMMARY OF RELATIVE MASS AND CHARGE

Particle	Relative mass	Relative charge
proton	1	+1
neutron	1	0
electron	5×10^{-4}	–1

SIZE AND STRUCTURE OF ATOMS
Atoms have a radius in the order of 10^{-10} m. Almost all of the mass of an atom is concentrated in the nucleus which has a very small radius in the order of 10^{-14} m. All the protons and neutrons (collectively called nucleons) are located in the nucleus. The electrons are to be found in energy levels or shells surrounding the nucleus. Much of the atom is empty space.

MASS NUMBER A
Equal to the number of protons and neutrons in the nucleus.

ATOMIC NUMBER Z
Equal to the number of protons in the nucleus and to the number of electrons in the atom. The atomic number defines which element the atom belongs to and consequently its position in the Periodic Table.

SHORTHAND NOTATION FOR AN ATOM OR ION

$$^{A}_{Z}X^{n+/n-}$$

CHARGE
Atoms have no charge so n = 0 and this is left blank. However by losing one or more electrons atoms become positive ions, or by gaining one or more electrons atoms form negative ions.

EXAMPLES

Symbol	Atomic number	Mass number	Number of protons	Number of neutrons	Number of electrons
$^{9}_{4}Be$	4	9	4	5	4
$^{40}_{20}Ca^{2+}$	20	40	20	20	18
$^{37}_{17}Cl^{-}$	17	37	17	20	18

ISOTOPES
All atoms of the same element must contain the same number of protons, however they may contain a different number of neutrons. Such atoms are known as isotopes. Chemical properties are related to the number of electrons so isotopes of the same element have identical chemical properties. Since their mass is different their physical properties such as density and boiling point are different.

Examples of isotopes: $^{1}_{1}H$ $^{2}_{1}H$ $^{3}_{1}H$ $^{12}_{6}C$ $^{14}_{6}C$ $^{35}_{17}Cl$ $^{37}_{17}Cl$.

RELATIVE ATOMIC MASS
The two isotopes of chlorine occur in the ratio of 3:1. That is, naturally occurring chlorine contains 75% $^{35}_{17}Cl$ and 25% $^{37}_{17}Cl$. The weighted mean molar mass is thus:

$$\frac{(75 \times 35) + (25 \times 37)}{100} = 35.5 \text{ g mol}^{-1}$$

and the relative atomic mass is 35.5. Accurate values to 2 d.p. for all the relative atomic masses of the elements are given in Table 5 of the IB data booklet. These are the values which must be used when performing calculations in the examinations.

Mass spectrometer and relative atomic mass

MASS SPECTROMETER

Relative atomic masses can be determined using a mass spectrometer. A *vaporized* sample is injected into the instrument. Atoms of the element are *ionized* by being bombarded with a stream of high energy electrons in the ionization chamber. In practice the instrument is set so that only ions with a single positive charge are formed. The resulting unipositive ions pass through holes in parallel plates under the influence of an electric field where they are *accelerated*. The ions are then *deflected* by an external magnetic field.

The amount of deflection depends both on the mass of the ion and its charge. The smaller the mass and the higher the charge the greater the deflection. Ions with a particular mass/charge ratio are then recorded on a *detector* which measures both the mass and the relative amounts of all the ions present.

DIAGRAM OF A MASS SPECTROMETER

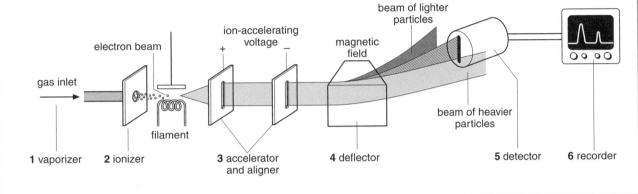

THE MASS SPECTRUM OF NATURALLY OCCURRING LEAD

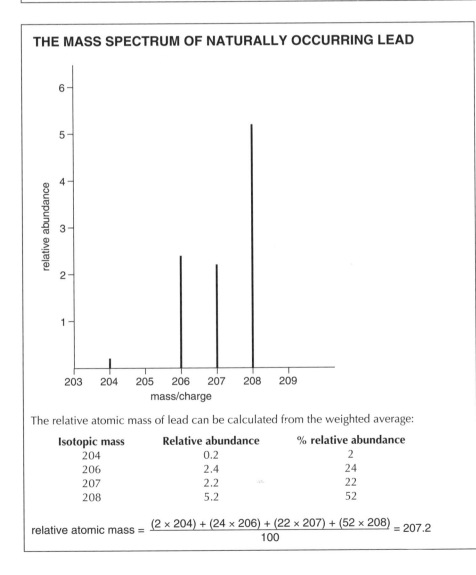

The relative atomic mass of lead can be calculated from the weighted average:

Isotopic mass	Relative abundance	% relative abundance
204	0.2	2
206	2.4	24
207	2.2	22
208	5.2	52

$$\text{relative atomic mass} = \frac{(2 \times 204) + (24 \times 206) + (22 \times 207) + (52 \times 208)}{100} = 207.2$$

USES OF RADIOACTIVE ISOTOPES

Isotopes have many uses in chemistry and beyond. Many, but by no means all, isotopes of elements are radioactive as the nuclei of these atoms break down spontaneously. When they break down these radioisotopes emit radiation which is dangerous to living things. There are three different forms of radiation. Gamma (γ) radiation is highly penetrating whereas alpha (α) radiation, can be stopped by a few centimetres of air and beta (β) radiation by a thin sheet of aluminium. Radioisotopes can occur naturally or be created artificially. Their uses include nuclear power generation, the sterilization of surgical instruments in hospitals, crime detection, finding cracks and stresses in metals and the preservation of food. $^{14}_{6}C$ is used for carbon dating, $^{60}_{27}Co$ is used in radiotherapy and $^{131}_{53}I$ and $^{125}_{53}I$ are used as tracers in medicine for treating and diagnosing illness.

Emission spectra

THE ELECTROMAGNETIC SPECTRUM

Electromagnetic waves can travel through space and, depending on the wavelength, also through matter. The velocity of travel c is related to its wavelength λ and its frequency f. Velocity is measured in m s^{-1}, wavelength in m and frequency in s^{-1} so it is easy to remember the relationship between them:

$$c = \lambda \times f$$

$$(\text{m s}^{-1}) \quad (\text{m}) \quad (\text{s}^{-1})$$

Electromagnetic radiation is a form of energy. The smaller the wavelength and thus the higher the frequency the more energy the wave possesses. Electromagnetic waves have a wide range of wavelengths ranging from low energy radio waves to high energy γ-radiation. Visible light occupies a very narrow part of the spectrum.

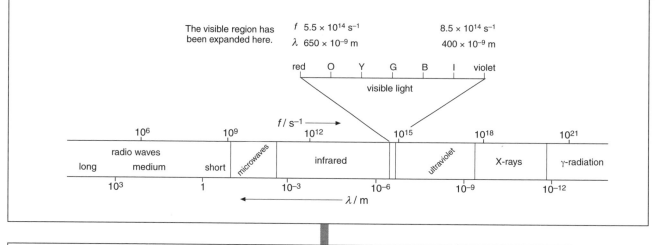

ATOMIC EMISSION SPECTRA

White light is made up of all the colours of the spectrum. When it is passed through a prism a **continuous spectrum** of all the colours can be obtained.

When energy is supplied to individual elements they emit a spectrum which only contains emissions at particular wavelengths. Each element has its own characteristic spectrum known as a **line spectrum** as it is not continuous.

The visible hydrogen spectrum

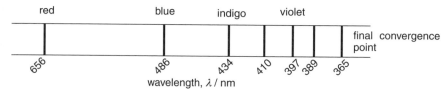

Note that the spectrum consists of discrete lines and that the lines converge towards the high energy (violet) end of the spectrum. A similar series of lines at even higher energy also occurs in the ultraviolet region of the spectrum and several other series of lines at lower energy can be found in the infrared region of the spectrum.

EXPLANATION OF EMISSION SPECTRA

When energy is supplied to an atom electrons are excited (gain energy) from their lowest (ground) state to an excited state. Electrons can only exist in certain fixed energy levels. When electrons drop from a higher level to a lower level they emit energy. This energy corresponds to a particular wavelength and shows up as a line in the spectrum. When electrons return to the first level ($n = 1$) the series of lines occurs in the ultraviolet region as this involves the largest energy change. The visible region spectrum is formed by electrons dropping back to the $n = 2$ level and the first series in the infrared is due to electrons falling to the $n = 3$ level. The lines in the spectrum converge because the energy levels themselves converge.

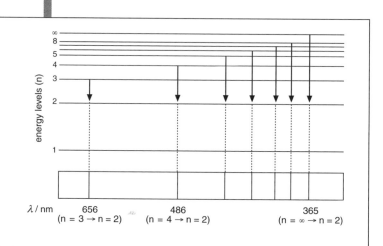

Electron arrangement

EVIDENCE FROM IONIZATION ENERGIES

The first ionization energy of an element is defined as the energy required to remove one electron from an atom in its gaseous state. It is measured in kJ mol^{-1}.

$$X(g) \rightarrow X^+(g) + e^-$$

A graph of first ionization energies plotted against atomic number shows a repeating pattern.

It can be seen that the highest value is for helium, an atom that contains two protons and two electrons. The two electrons are in the lowest level and are held tightly by the two protons. For lithium it is relatively easy to remove an electron, which suggests that the third electron in lithium is in a higher energy level than the first two. The value then generally increases until element 10, neon, is reached before it drops sharply for sodium. This graph provides evidence that the levels can contain different numbers of electrons before they become full.

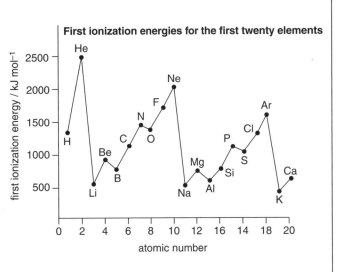

First ionization energies for the first twenty elements

Level	Maximum number of electrons
1 (K shell)	2
2 (L shell)	8
3 (M shell)	8 (or 18)

ELECTRON ARRANGEMENT

The arrangement of electrons in an atom is known as its electronic configuration. Each energy level or shell is separated by a dot (or a comma). The electrons in the highest main energy level (outermost level) are known as the **valence electrons**.

Element	Electron configuration	Element	Electron configuration
H	1	Na	2.8.1
He	2 (first level full)	Mg	2.8.2
Li	2.1	Al	2.8.3
Be	2.2	Si	2.8.4
B	2.3	P	2.8.5
C	2.4	S	2.8.6
N	2.5	Cl	2.8.7
O	2.6	Ar	2.8.8 (third level full)
F	2.7	K	2.8.8.1
Ne	2.8 (second level full)	Ca	2.8.8.2

EVIDENCE FOR SUB-LEVELS

The graph already shown above was for the first ionization energy for the first twenty elements. Successive ionization energies for the same element can also be measured, e.g. the second ionization energy is given by:

$$X^+(g) \rightarrow X^{2+}(g) + e^-$$

As more electrons are removed the pull of the protons holds the remaining electrons more tightly so increasingly more energy is required to remove them, hence a logarithmic scale is usually used. A graph of the successive ionization energies for potassium also provides evidence of the number of electrons in each main level.

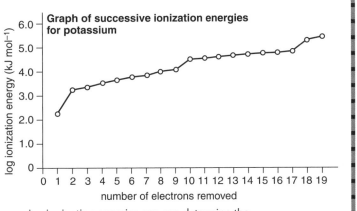

Graph of successive ionization energies for potassium

By looking to see where the first 'large jump' occurs in successive ionization energies one can determine the number of valence electrons (and hence the group in the Periodic Table to which the element belongs).

If the graph for first ionization energies is examined more closely then it can be seen that the graph does not increase regularly. This provides evidence that the main levels are split into sub-levels.

TYPES OF ORBITAL

Electrons are found in orbitals. Each orbital can contain a maximum of two electrons each with opposite spins. The first level contains just one orbital, called an s orbital. The second level contains one s orbital and three p orbitals. The 2p orbitals are all of equal energy but the sub-level made up of these three 2p orbitals is slightly higher in energy than the 2s orbital. This explains why the first ionization energy of B is lower than Be as a higher energy 2p electron is being removed from the B compared with a lower energy 2s electron from Be.

Principal level (shell)	Number of each type of orbital				Maximum number of electrons in level
	s	p	d	f	
1	1	–	–	–	2
2	1	3	–	–	8
3	1	3	5	–	18
4	1	3	5	7	32

The relative position of all the sub-levels for the first four main energy levels is shown.

Note that the 4s sub-level is below the 3d sub-level. This explains why the third level is sometimes stated to hold 8 or 18 electrons.

Electrons with opposite spins tend to repel each other. When orbitals of the same energy (degenerate) are filled the electrons will go singly into each orbital first before they pair up to minimize repulsion. This explains why there is a regular increase in the first ionization energies going from B to N as the three 2p orbitals each gain one electron. Then there is a slight decrease between N and O as one of the 2p orbitals gains a second electron before a regular increase again.

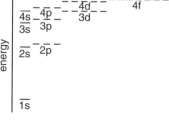

Relative energies of sub-levels within an atom

SHAPES OF ORBITALS

An electron has the properties of both a particle and a wave. Heisenberg's uncertainty principle states that it is impossible to know the exact position of an electron at a precise moment in time. An orbital describes the three-dimensional shape where there is a high probability that the electron will be located.

s orbitals are spherical and the three p orbitals are orthogonal (at right angles) to each other.

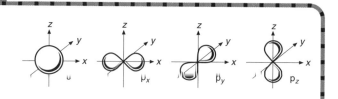

AUFBAU PRINCIPLE

The electronic configuration can be determined by following the aufbau (building up) principle. The orbitals with the lowest energy are filled first. Each orbital can contain a maximum of two electrons. Orbitals within the same sub-shell are filled singly first – this is known as Hund's rule,

e.g.   F $1s^2 2s^2 2p^5$
V $1s^2 2s^2 2p^6 3s^2 3p^6 4s^2 3d^3$.

To save writing out all the lower levels the configuration may be shortened by building on the last noble gas configuration, e.g. V is more usually written:

   [Ar] $4s^2 3d^3$.

(When writing electronic configurations check that for a neutral atom the sum of the superscripts adds up to the atomic number of the element.)

Sometimes boxes are used to represent orbitals so the number of unpaired electrons can easily be seen,

e.g.

C $1s^2 2s^2 2p^2$

ELECTRONIC CONFIGURATION AND THE PERIODIC TABLE

An element's position in the Periodic Table is related to its valence electrons so the electronic configuration of any element can be deduced from the Table, e.g. iodine ($Z = 53$) is a p block element. It is in group 7 so its configuration will contain $ns^2 np^5$. If one takes H and He as being the first period then iodine is in the fifth period so n = 5. The full configuration for iodine will therefore be:

$1s^2 2s^2 2p^6 3s^2 3p^6 4s^2 3d^{10} 4p^6 5s^2 4d^{10} 5p^5$ or [Kr] $5s^2 4d^{10} 5p^5$

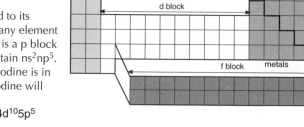

IB QUESTIONS – ATOMIC STRUCTURE

1. Which of the following particles contain more electrons than **neutrons**?

   I. ^1_1H II. $^{35}_{17}Cl^-$ III. $^{39}_{19}K^+$

   A. I only   C. I and II only

   B. II only   D. II and III only

2. The atom with the same number of neutrons as ^{54}Cr is

   A. ^{50}Ti B. ^{51}V C. ^{53}Fe D. ^{55}Mn

3. All isotopes of tin (Sn) have the same

   I. number of protons

   II. number of neutrons

   III. mass number

   A. I only   C. III only

   B. II only   D. I and III only

4. Which one of the following sets represents a pair of isotopes?

   A. $^{14}_6C$ and $^{14}_7N$ C. $^{32}_{16}S$ and $^{32}_{16}S^{2-}$

   B. O_2 and O_3 D. $^{206}_{82}Pb$ and $^{208}_{82}Pb$

5. The atomic and mass numbers for four different nuclei are given in the table below. Which two are isotopes?

	atomic number	mass number
I.	101	258
II.	102	258
III.	102	260
IV.	103	259

   A. I and II   C. II and IV

   B. II and III   D. III and IV

6. Which species contains 16 protons, 17 neutrons and 18 electrons?

   A. $^{32}S^-$ B. $^{33}S^{2-}$ C. $^{34}S^-$ D. $^{35}S^{2-}$

7. Spectra have been used to study the arrangements of electrons in atoms. An emission spectrum consists of a series of bright lines that converge at high frequencies. Such emission spectra provide evidence that electrons are moving from

   A. lower to higher energy levels with the higher energy levels being closer together.

   B. lower to higher energy levels with the lower energy levels being closer together.

   C. higher to lower energy levels with the lower energy levels being closer together.

   D. higher to lower energy levels with the higher energy levels being closer together.

8. Which electron transition in a hydrogen atom releases the most energy?

   A. $n = 2 \rightarrow n = 1$ C. $n = 6 \rightarrow n = 3$

   B. $n = 4 \rightarrow n = 2$ D. $n = 7 \rightarrow n = 6$

9. An element has the electronic configuration 2.7. What would be the electronic configuration of an element with similar chemical properties?

   A. 2.6   B. 2.8   C. 2.7.1   D. 2.8.7

10. An element with the symbol Z has the electron configuration 2.8.6. Which species is this element most likely to form?

    A. The ion Z^{2+} C. The compound H_2Z

    B. The ion Z^{6+} D. The compound Z_6F

HL

The following diagram should be used to answer question 11.

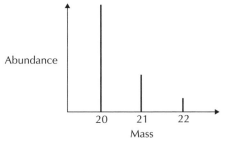

11. According to the mass spectrum above, the relative atomic mass of the element shown is best expressed as

    A. 20.0.   C. 21.0.

    B. between 20.0 and 21.0.   D. between 21.0 and 22.0.

12. The first four ionization energies (kJ mol^{-1}) for a particular element are 550, 1064, 4210 and 5500 respectively. This element should be placed in the same Group as

    A. Li   B. Be   C. B   D. C

13. Which ionization requires the most energy?

    A. $Na(g) \rightarrow Na^+(g) + e^-$

    B. $Na^+(g) \rightarrow Na^{2+}(g) + e^-$

    C. $Mg(g) \rightarrow Mg^+(g) + e^-$

    D. $Mg^+(g) \rightarrow Mg^{2+}(g) + e^-$

14. Which one of the following atoms in its ground state has the greatest number of unpaired electrons?

    A. Al   B. Si   C. P   D. S

15. All of the following factors affect the value of the ionization energy of an atom **except** the

    A. mass of the atom.

    B. charge on the nucleus.

    C. size of the atom.

    D. main energy level from which the electron is removed.

The Periodic Table and physical properties (1)

THE PERIODIC TABLE

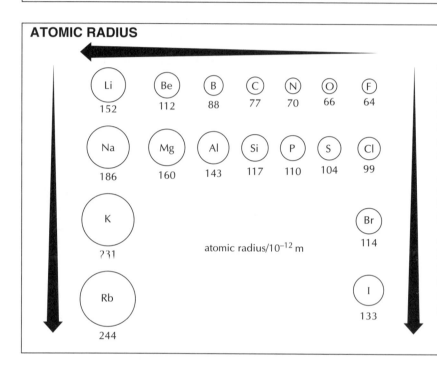

In the Periodic Table elements are placed in order of increasing atomic number. Elements with the same number of valence electrons are placed vertically in the same **group**. The groups are numbered from 1 to 8 (or 0). Some groups have their own name:

- Group 1 – alkali metals
- Group 7 – halogens
- Group 8 or 0 – noble gases (sometimes also called rare gases or inert gases).

Elements with the same outer shell of valence electrons are placed horizontally in the same **period**. The transition elements are located between groups 2 and 3.

ATOMIC RADIUS

Li 152	Be 112	B 88	C 77	N 70	O 66	F 64
Na 186	Mg 160	Al 143	Si 117	P 110	S 104	Cl 99
K ?31						Br 114
Rb 244			atomic radius/10^{-12} m			I 133

The atomic radius is the distance from the nucleus to the outermost electron. Since the position of the outermost electron can never be known precisely, the atomic radius is usually defined as half the distance between the nuclei of two bonded atoms of the same element.

As a group is descended the outermost electron is in a higher energy level, which is further from the nucleus, so the radius increases.

Across a period electrons are being added to the same energy level, but the number of protons in the nucleus increases. This attracts the energy level closer to the nucleus and the atomic radius decreases across a period.

IONIC RADIUS

It is important to distinguish between positive ions (**cations**) and negative ions (**anions**). Both cations and anions increase in size down a group as the outer level gets further from the nucleus.

Cations contain fewer electrons than protons so the electrostatic attraction between the nucleus and the outermost electron is greater and the ion is smaller than the parent atom. It is also smaller because the number of electron shells has decreased by one. Across the period the ions contain the same number of electrons (**isoelectronic**), but an increasing number of protons, so the ionic radius decreases.

Anions contain more electrons than protons so are larger than the parent atom. Across a period the size decreases because the number of electrons remains the same but the number of protons increases.

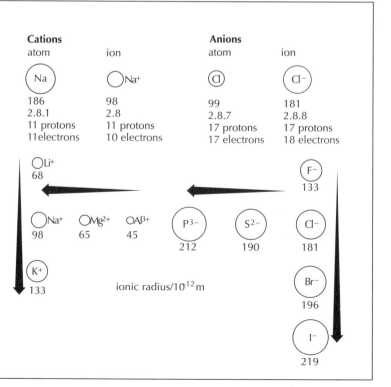

Cations		**Anions**	
atom	ion	atom	ion
Na 186 2.8.1 11 protons 11 electrons	Na⁺ 98 2.8 11 protons 10 electrons	Cl 99 2.8.7 17 protons 17 electrons	Cl⁻ 181 2.8.8 17 protons 18 electrons

Li⁺ 68
Na⁺ 98 Mg²⁺ 65 Al³⁺ 45 P³⁻ 212 S²⁻ 190 Cl⁻ 181
K⁺ 133

F⁻ 133
Cl⁻ 181
Br⁻ 196
I⁻ 219

ionic radius/10^{-12} m

The Periodic Table and physical properties (2)

PERIODICITY
Elements in the same group tend to have similar chemical and physical properties. There is a change in chemical and physical properties across a period. The repeating pattern of physical and chemical properties shown by the different periods is known as **periodicity**.

These periodic trends can clearly be seen in atomic radii, ionic radii, ionization energies, electronegativities and melting points.

MELTING POINTS
Melting points depend both on the structure of the element and on the type of attractive forces holding the atoms together. Using period 3 as an example:

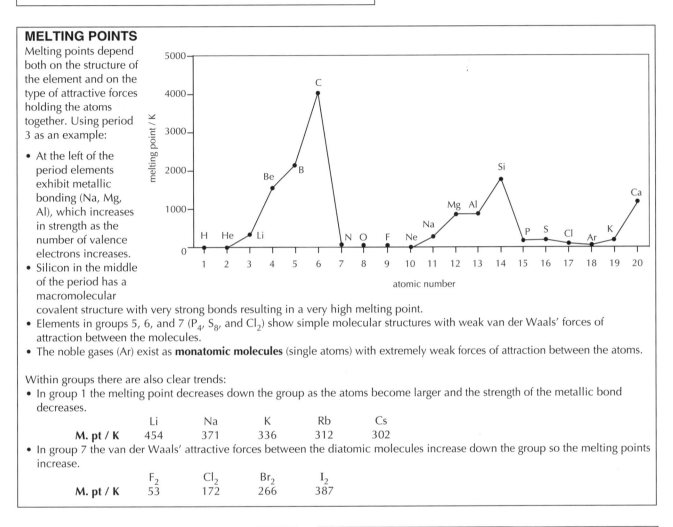

- At the left of the period elements exhibit metallic bonding (Na, Mg, Al), which increases in strength as the number of valence electrons increases.
- Silicon in the middle of the period has a macromolecular covalent structure with very strong bonds resulting in a very high melting point.
- Elements in groups 5, 6, and 7 (P_4, S_8, and Cl_2) show simple molecular structures with weak van der Waals' forces of attraction between the molecules.
- The noble gases (Ar) exist as **monatomic molecules** (single atoms) with extremely weak forces of attraction between the atoms.

Within groups there are also clear trends:
- In group 1 the melting point decreases down the group as the atoms become larger and the strength of the metallic bond decreases.

	Li	Na	K	Rb	Cs
M. pt / K	454	371	336	312	302

- In group 7 the van der Waals' attractive forces between the diatomic molecules increase down the group so the melting points increase.

	F_2	Cl_2	Br_2	I_2
M. pt / K	53	172	266	387

ELECTRONEGATIVITY
Electronegativity is a relative measure of the attraction that an atom has for a shared pair of electrons when it is covalently bonded to another atom. As the size of the atom decreases the electronegativity increases, so the value increases across a period and decreases down a Group. The three most electronegative elements are F, N, and O.

```
H
2.1
  ────────────────────────────►

Li   Be   B    C    N    O    F
1.0  1.5  2.0  2.5  3.0  3.5  4.0

Na                            Cl
0.9                           3.0

K                             Br
0.8                           2.8

                              I
                              2.5
```

FIRST IONIZATION ENERGY
The definition of first ionization energy has been given on page 9. The values decrease down each group as the outer electron is further from the nucleus and therefore less energy is required to remove it, e.g. for the group 1 elements, Li, Na and K.

Element:	Li	Na	K
Electron arrangement	2.1	2.8.1	2.8.8.1
First ionization energy (kJ mol^{-1})	519	494	418

Generally the values increase across a period. This is because the extra electrons are filling the same energy level and the extra protons in the nucleus attract this energy level closer making it harder to remove an electron, e.g. for the third period.

Element	Na	Mg	Al	Si	P	S	Cl	Ar
Number of protons	11	12	13	14	15	16	17	18
Electron arrangement	2.8.1	2.8.2	2.8.3	2.8.4	2.8.5	2.8.6	2.8.7	2.8.8
First ionization energy (kJ mol^{-1})	494	736	577	786	1060	1000	1260	1520

The Periodic Table and chemical properties

CHEMICAL PROPERTIES OF ELEMENTS IN THE SAME GROUP

Group 1 – the alkali metals

Lithium, sodium, and potassium all contain one electron in their outer shell. They are all reactive metals and are stored under liquid paraffin to prevent them reacting with air. They react by losing their outer electron to form the metal ion. Because they can readily lose an electron they are good reducing agents. The reactivity increases down the group as the outer electron is in successively higher energy levels and less energy is required to remove it.

They are called alkali metals because they all react with water to form an alkali solution of the metal hydroxide and hydrogen gas. Lithium floats and reacts quietly, sodium melts into a ball which darts around on the surface, and the heat generated from the reaction with potassium ignites the hydrogen.

$$2Li(s) + 2H_2O(l) \rightarrow 2Li^+(aq) + 2OH^-(aq) + H_2(g)$$

$$2Na(s) + 2H_2O(l) \rightarrow 2Na^+(aq) + 2OH^-(aq) + H_2(g)$$

$$2K(s) + 2H_2O(l) \rightarrow 2K^+(aq) + 2OH^-(aq) + H_2(g)$$

They all also react readily with chlorine, bromine and iodine to form ionic salts, e.g.

$$2Na(s) + Cl_2(g) \rightarrow 2Na^+Cl^-(s)$$

$$2K(s) + Br_2(l) \rightarrow 2K^+Br^-(s)$$

$$2Li(s) + I_2(g) \rightarrow 2Li^+I^-(s)$$

Group 7 – the halogens

The halogens react by gaining one more electron to form halide ions. They are good oxidizing agents. The reactivity decreases down the group as the outer shell is increasingly at higher energy levels and further from the nucleus. This, together with the fact that there are more electrons between the nucleus and the outer shell, decreases the attraction for an extra electron.

Chlorine is a stronger oxidizing agent than bromine, so can remove the electron from bromide ions in solution to form chloride ions and bromine. Similarly both chlorine and bromine can oxidize iodide ions to form iodine.

$$Cl_2(aq) + 2Br^-(aq) \rightarrow 2Cl^-(aq) + Br_2(aq)$$

$$Cl_2(aq) + 2I^-(aq) \rightarrow 2Cl^-(aq) + I_2(aq)$$

$$Br_2(aq) + 2I^-(aq) \rightarrow 2Br^-(aq) + I_2(aq)$$

Test for halide ions

The presence of halide ions in solution can be detected by adding silver nitrate solution. The silver ions react with the halide ions to form a precipitate of the silver halide. The silver halides can be distinguished by their colour. These silver halides react with light to form silver metal. This is the basis of photography.

$$Ag^+(aq) + X^-(aq) \rightarrow AgX(s)$$

where X = Cl, Br, or I

AgCl white

AgBr cream

AgI yellow

light

$$Ag(s) + \tfrac{1}{2}X_2$$

CHANGE FROM METALLIC TO NON-METALLIC NATURE OF THE ELEMENTS ACROSS PERIOD 3

Metals tend to be shiny and are good conductors of heat and electricity. Sodium, magnesium, and aluminium all conduct electricity well. Silicon is a semi-conductor and is called a **metalloid** as it possesses some of the properties of a metal and some of a non-metal. Phosphorus, sulfur, chlorine, and argon are non-metals and do not conduct electricity. Metals can also be distinguished from non-metals by their chemical properties. Metal oxides tend to be basic, whereas non-metal oxides tend to be acidic.

Sodium oxide and magnesium oxide are both basic and react with water to form hydroxides,

e.g. $Na_2O(s) + H_2O(l) \rightarrow 2NaOH(aq)$ $MgO(s) + H_2O(l) \rightarrow Mg(OH)_2$

Aluminium is a metal but its oxide is amphoteric, that is, it can be either basic or acidic depending on whether it is reacting with an acid or a base.

The remaining elements in period 3 have acidic oxides. For example, sulfur trioxide reacts with water to form sulfuric acid, and phosphorus pentoxide reacts with water to form phosphoric(V) acid.

$$SO_3(g) + H_2O(l) \rightarrow H_2SO_4(aq)$$ $$P_4O_{10}(s) + 6H_2O(l) \rightarrow 4H_3PO_4(aq)$$

Oxides of the third period (sodium → argon)

OXIDES OF PERIOD 3 ELEMENTS

The oxides of sodium, magnesium, and aluminium are all ionic. This accounts for their high melting points and electrical conductivity when molten. Silicon dioxide has a diamond-like macromolecular structure with a high boiling point. At the other end of the period the difference in electronegativities between the element and oxygen is small, resulting in simple covalent molecular structures with low melting and boiling points.

The acid–base properties of the oxides are also linked to their structure. The oxides of the electropositive elements are very basic and form solutions that are alkaline.

$$Na_2O(s) + H_2O(l) → 2Na^+(aq) + 2OH^-(aq)$$

$$MgO(s) + H_2O(l) → Mg(OH)_2$$

The amphoteric nature of aluminium oxide can be seen from its reactions with hydrochloric acid and sodium hydroxide.

Acting as a base: $Al_2O_3(s) + 6HCl(aq) → 2AlCl_3(aq) + 3H_2O(l)$

Acting as an acid: $Al_2O_3(s) + 2NaOH(aq) + 3H_2O(l) → 2NaAl(OH)_4(aq)$
                                    sodium aluminate

Silicon dioxide behaves as a weak acid. It does not react with water but will form sodium silicate with sodium hydroxide.

$$SiO_2(s) + 2NaOH(aq) → Na_2SiO_3(aq) + H_2O(aq)$$

The oxides of phosphorus, sulfur, and chlorine are all strongly acidic.

$$SO_2(g) + H_2O(l) → H_2SO_3 \text{ sulfurous acid}$$

$$P_4O_{10}(s) + 6H_2O(l) → 4H_3PO_4(aq) \text{ phosphoric acid}$$

$$Cl_2O_7(l) + H_2O(l) → 2HClO_4(aq) \text{ perchloric acid}$$

Oxides of period 3 elements

Formula	Na_2O	MgO	Al_2O_3	SiO_2	P_4O_{10} (P_4O_6)	SO_3 (SO_2)	Cl_2O_7 (Cl_2O)
State at 25 °C	Solid	Solid	Solid	Solid	Solid (Solid)	Liquid (Gas)	Liquid (Gas)
Melting point / °C	1275	2852	2027	1610	24	17	−92
Boiling point / °C	–	3600	2980	2230	175	45	80
Electrical conductivity in molten state	Good	Good	Good	Very poor	None	None	None
Structure	Ionic			Covalent macromolecular	Simple covalent molecular		
Reaction with water	Forms NaOH(aq), an alkaline solution	Forms $Mg(OH)_2$, weakly alkaline	Does not react	Does not react	P_4O_{10} forms H_3PO_4, an acidic solution	SO_3 forms H_2SO_4, a strong acid	Cl_2O_7 forms $HClO_4$, an acidic solution
Nature of oxide	Basic		Amphoteric	Acidic			

CHLORIDES OF PERIOD 3 ELEMENTS

The physical properties of the chlorides are related to the structure in the same way as the oxides. Sodium chloride and magnesium chloride are ionic – they conduct electricity when molten and have high melting points. Aluminium chloride is covalent and is a poor conductor. Unlike silicon dioxide, silicon tetrachloride has a simple molecular structure as do the remaining chlorides in the period. These molecules are held together by weak van der Waals' forces, which results in low melting and boiling points.

Sodium chloride dissolves in water to give a neutral solution, magnesium chloride gives a slightly acidic solution with water. All the other chlorides including aluminium chloride react vigorously with water to produce acidic solutions of hydrochloric acid together with fumes of hydrogen chloride.

$$2AlCl_3(s) + 3H_2O(l) \rightarrow Al_2O_3(s) + 6HCl(aq)$$

$$SiCl_4(l) + 4H_2O(l) \rightarrow Si(OH)_4(aq) + 4HCl(aq)$$

$$PCl_3(l) + 3H_2O(l) \rightarrow H_3PO_3(aq) + 3HCl(aq)$$

Chlorine itself reacts with water to some extent to form an acidic solution.

$$Cl_2(aq) + H_2O(l) \rightleftharpoons HCl(aq) + HClO(aq)$$

Chlorides of period 3 elements

Formula	NaCl	MgCl$_2$	Al$_2$Cl$_6$	SiCl$_4$	PCl$_3$ (PCl$_5$)	(S$_2$Cl$_2$)	Cl$_2$
State at 25 °C	Solid	Solid	Solid	Liquid	Liquid (Solid)	Liquid	Gas
Melting point / °C	801	714	178 (sublimes)	−70	−112	−80	−101
Boiling point / °C	1413	1412	–	58	76	136	−35
Electrical conductivity in molten state	Good	Good	Poor	None	None	None	None
Structure	Ionic		Simple covalent molecular				
Reaction with water	Dissolve easily		Fumes of HCl produced				Some reaction with water
Nature of solution	Neutral	Weakly acidic	Acidic				

HL d-block elements (first row)

THE FIRST ROW TRANSITION ELEMENTS

Element	(Sc)	Ti	V	Cr	Mn	Fe	Co	Ni	Cu	(Zn)
Electron configuration [Ar]	$4s^23d^1$	$4s^23d^2$	$4s^23d^3$	$4s^13d^5$	$4s^23d^5$	$4s^23d^6$	$4s^23d^7$	$4s^23d^8$	$4s^13d^{10}$	$4s^23d^{10}$

A transition element is defined as an element that possesses an incomplete d sub-level in one or more of its oxidation states. Scandium is not a typical transition metal as its common ion Sc^{3+} has no d electrons. Zinc is not a transition metal as it contains a full d sub-level in all its oxidation states. (Note: for Cr and Cu it is more energetically favourable to half-fill and completely fill the d sub-level respectively so they contain only one 4s electron).

Variable oxidation states

The 3d and 4s sub-levels are very similar in energy. When transition metals lose electrons they lose the 4s electrons first. All transition metals can show an oxidation state of +2. Some of the transition metals can form the +3 or +4 ion (e.g. Fe^{3+}, Mn^{4+}) as the ionization energies are such that up to two d electrons can also be lost. The M^{4+} ion is rare and in the higher oxidation states the element is usually found not as the free metal ion but either covalently bonded or as the oxyanion, such as MnO_4^-. Some common examples of variable oxidation states in addition to +2 are:

Cr(+3)	$CrCl_3$	chromium(III) chloride
Cr(+6)	$Cr_2O_7^{2-}$	dichromate(VI) ion
Mn(+4)	MnO_2	manganese(IV) oxide
Mn(+7)	MnO_4^-	manganate(VII) ion
Fe(+3)	Fe_2O_3	iron(III) oxide
Cu(+1)	Cu_2O	copper(I) oxide

CHARACTERISTIC PROPERTIES OF TRANSITION ELEMENTS

Formation of complex ions

Because of their small size d-block ions attract species that are rich in electrons. Such species are known as **ligands**. Ligands are neutral molecules or anions which contain a non-bonding pair of electrons. These electron pairs can form co-ordinate covalent bonds with the metal ion to form **complex ions**.

A common ligand is water and most (but not all) transition metal ions exist as hexahydrated complex ions in aqueous solution, e.g. $[Fe(H_2O)_6]^{3+}$. Ligands can be replaced by other ligands. A typical example is the addition of ammonia to an aqueous solution of copper(II) sulfate to give the deep blue colour of the tetraaminecopper(II) ion. Similarly if concentrated hydrochloric acid is added to a solution of Cu^{2+}(aq) the yellow tetrachlorocopper(II) anion is formed. Note: in this ion the overall charge on the ion is –2 as the four ligands each have a charge of –1.

$$[CuCl_4]^{2-} \underset{H_2O}{\overset{Cl^-}{\rightleftharpoons}} [Cu(H_2O)_4]^{2+} \underset{H_2O}{\overset{NH_3}{\rightleftharpoons}} [Cu(NH_3)_4]^{2+}$$

The number of lone pairs bonded to the metal ion is known as the **co-ordination number**. Compounds with a co-ordination number of six are octahedral in shape, those with a co-ordination number of four are tetrahedral or square planar, whereas those with a co-ordination number of two are usually linear.

Co-ordination number	6	4	2
Examples	$[Fe(CN)_6]^{3-}$	$[CuCl_4]^{2-}$	$[Ag(NH_3)_2]^+$
	$[Fe(OH)_3(H_2O)_3]$	$[Cu(NH_3)_4]^{2+}$	

Coloured complexes

In the free ion the five d orbitals are degenerate (of equal energy). However, in complexes the d orbitals are split into two distinct levels. The energy difference between the levels corresponds to a particular wavelength or frequency in the visible region of the spectrum. When light falls on the complex, energy of a particular wavelength is absorbed and electrons are excited from the lower level to the higher level.

Cu^{2+}(aq) appears blue because it is the complementary colour to the wavelengths that have been absorbed. The amount the orbitals are split depends on the nature of the transition metal, the oxidation state, the shape of the complex, and the nature of the ligand, which explains why different complexes have different colours. If the d orbital is completely empty, as in Sc^{3+}, or completely full, as in Cu^+ or Zn^{2+}, no transitions within the d level can take place and the complexes are colourless.

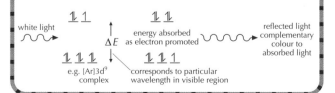

white light → energy absorbed as electron promoted → reflected light complementary colour to absorbed light

ΔE corresponds to particular wavelength in visible region

e.g. [Ar]3d⁹ complex

Catalytic behaviour

Many transition elements and their compounds are very efficient catalysts, that is, they increase the rate of chemical reactions. This helps to make industrial processes, such as the production of ammonia and sulfuric acid, more efficient and economic. Platinum and palladium are used in catalytic converters fitted to cars. In the body, iron is found in haem and cobalt is found in vitamin B_{12}. Other common examples include:

Iron in the Haber process
$$3H_2(g) + N_2 \overset{Fe(s)}{\rightleftharpoons} 2NH_3(g)$$

Vanadium(V) oxide in the Contact process
$$2SO_2(g) + O_2(g) \overset{V_2O_5(s)}{\rightleftharpoons} 2SO_3(g)$$

Nickel in hydrogenation reactions
$$C_2H_4(g) + H_2(g) \overset{Ni(s)}{\longrightarrow} C_2H_6(g)$$

Manganese(IV) oxide with hydrogen peroxide
$$2H_2O_2(aq) \overset{MnO_2(s)}{\longrightarrow} 2H_2O(l) + O_2(g)$$

IB QUESTIONS – PERIODICITY

1. In the periodic table, elements are arranged in order of increasing

    A. atomic number. **C.** number of valence electrons.

    B. atomic mass. **D.** electronegativity.

2. Which one of the following series is arranged in order of increasing value?

    A. The first ionisation energies of: oxygen, fluorine, neon.

    B. The radii of: H^- ion, H atom, H^+ ion.

    C. The electronegativities of: chlorine, bromine, iodine.

    D. The boiling points of: iodine, bromine, chlorine.

3. Which property increases with increasing atomic number for both the alkali metals and the halogens?

    A. Ionisation energies **C.** Electronegativities

    B. Melting points **D.** Atomic radii

4. Which set of reactants below is expected to produce the most vigorous reaction?

    A. $Na(s) + Cl_2(g)$ **C.** $K(s) + Cl_2(g)$

    B. $Na(s) + Br_2(g)$ **D.** $K(s) + Br_2(g)$

5. Which one of the following statements about the halogen group is correct?

    A. First ionisation energies increase from F to I.

    B. Fluorine has the smallest tendency to be reduced.

    C. Cl_2 will oxidise $I^-(aq)$.

    D. I_2 is a stronger oxidising agent than F_2.

6. Strontium is an element in Group 2 of the Periodic Table with atomic number 38. Which of the following statements about strontium is NOT correct?

    A. Its first ionisation energy is lower than that of calcium.

    B. It has two electrons in its outermost energy level.

    C. Its atomic radius is smaller than magnesium.

    D. It forms a chloride with the formula $SrCl_2$.

7. Which one of the following elements has the lowest first ionisation energy?

    A. Li **C.** Mg

    B. Na **D.** Al

8. Which element is most similar chemically to the element with 14 electrons?

    A. Al **C.** Ge

    B. As **D.** P

9. 0.01 mole samples of the following oxides were added to separate 1 dm^3 portions of water. Which will produce the most acidic solution?

    A. $Al_2O_3(s)$ **C.** $Na_2O(s)$

    B. $SiO_2(s)$ **D.** $SO_3(g)$

10. Which reaction occurs readily?

    I. $Br_2(aq) + 2I^-(aq) \rightarrow I_2(aq) + 2Br^-(aq)$

    II. $Br_2(aq) + 2Cl^-(aq) \rightarrow Cl_2(aq) + 2Br^-(aq)$

    A. I only **C.** Both I and II

    B. II only **D.** Neither I nor II

(HL) ———————————————————————————————

11. In which region of the Periodic Table would the element with the electronic structure below be located?

    $1s^2 2s^2 2p^6 3s^2 3p^6 3d^{10} 4s^2 4p^6 4d^6 5s^2$

    A. group 6 **C.** s block

    B. noble gases **D.** d block

12. A certain element has the electronic configuration $1s^2 2s^2 2p^6 3s^2 3p^6 4s^2 3d^3$. Which oxidation state(s) would this element most likely show?

    A. +2 only **C.** +2 and +5 only

    B. +3 only **D.** +2, +3, +4, +5

13. Which ion is colourless?

    A. $[Cr(H_2O)_6]^{3+}$ **C.** $[Cu(NH_3)_4]^{2+}$

    B. $[Fe(CN)_6]^{4-}$ **D.** $[Zn(H_2O)_4]^{2+}$

14. Which of the following chlorides give neutral solutions when added to water?

    I. NaCl

    II. Al_2Cl_6

    III. PCl_3

    A. I only **C.** II and III only

    B. I and II only **D.** I, II and III

15. Based on melting points, the dividing line between ionic and covalent chlorides of the elements Mg to S lies between

    A. Mg and Al. **C.** Si and P.

    B. Al and Si. **D.** P and S.

16. The colours of the compounds of d-block elements are due to electron transitions

    A. between different d orbitals.

    B. between d orbitals and s orbitals.

    C. among the attached ligands.

    D. from the metal to the attached ligands.

Ionic bonding

IONIC BOND

When atoms combine they do so by trying to achieve an inert gas configuration. Ionic compounds are formed when electrons are transferred from one atom to another to form ions with complete outer shells of electrons. In an ionic compound the positive and negative ions are attracted to each other by strong electrostatic forces, and build up into a strong lattice. Ionic compounds have high melting points as considerable energy is required to overcome these forces of attraction.

The classic example of an ionic compound is sodium chloride Na^+Cl^-, formed when sodium metal burns in chlorine. Chlorine is a covalent molecule, so each atom already has an inert gas configuration. However, the energy given out when the ionic lattice is formed is sufficient to break the bond in the chlorine molecule to give atoms of chlorine. Each sodium atom then transfers one electron to a chlorine atom to form the ions.

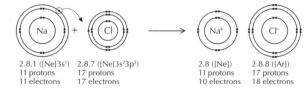

2.8.1 ([Ne]3s¹)	2.8.7 ([Ne]3s²3p⁵)	2.8 ([Ne])	2.8.8 ([Ar])

2.8.1 ([Ne]$3s^1$)
11 protons
11 electrons

2.8.7 ([Ne]$3s^2 3p^5$)
17 protons
17 electrons

2.8 ([Ne])
11 protons
10 electrons

2.8.8 ([Ar])
17 protons
18 electrons

The charge carried by an ion depends on the number of electrons the atom needed to lose or gain to achieve a full outer shell.

	Cations				Anions		
Group 1	**Group 2**	**Group 3**		**Group 5**	**Group 6**	**Group 7**
+1	+2	+3		−3	−2	−1
Li^+ Na^+ K^+	Mg^{2+} Ca^{2+}	Al^{3+}		N^{3-} P^{3-}	O^{2-} S^{2-}	F^- Cl^- Br^-

Thus in magnesium chloride two chlorine atoms each gain one electron from a magnesium atom to form $Mg^{2+}Cl^-_2$. In magnesium oxide two electrons are transferred from magnesium to oxygen to give $Mg^{2+}O^{2-}$. Transition metals can form more than one ion. For example, iron can form Fe^{2+} and Fe^{3+} and copper can form Cu^+ and Cu^{2+}.

FORMULAS OF IONIC COMPOUNDS

It is easy to obtain the correct formula as the overall charge of the compound must be zero.

lithium fluoride Li^+F^- magnesium chloride $Mg^{2+}Cl^-_2$ aluminium bromide $Al^{3+}Br^-_3$
sodium oxide $Na^+_2O^{2-}$ calcium sulfide $Ca^{2+}S^{2-}$ iron(III) oxide $Fe^{3+}_2O^{2-}_3$
potassium nitride $K^+_3N^{3-}$ calcium phosphide $Ca^{2+}_3P^{3-}_2$ iron(II) oxide $Fe^{2+}O^{2-}$

Note: the formulas above have been written to show the charges carried by the ions. Unless asked specifically to do this it is common practice to omit the charges and simply write LiF, $MgCl_2$, etc.

IONS CONTAINING MORE THAN ONE ELEMENT

In ions formed from more than one element the charge is often spread (delocalized) over the whole ion. An example of a positive ion is the ammonium ion NH_4^+, in which all four N–H bonds are identical. Negative ions are sometimes known as acid radicals as they are formed when an acid loses one or more H^+ ions.

hydroxide OH^-
nitrate NO_3^- (from nitric acid, HNO_3)
sulfate SO_4^{2-} } from sulfuric acid, H_2SO_4
hydrogensulfate HSO_4^-

carbonate CO_3^{2-} } from carbonic acid, H_2CO_3
hydrogencarbonate HCO_3^-
ethanoate CH_3COO^- (from ethanoic acid, CH_3COOH)

The formulas of the ionic compounds are obtained in exactly the same way. Note: brackets are used to show that the subscript covers all the elements in the ion.

sodium nitrate $Na^+NO_3^-$ calcium carbonate $Ca^{2+}CO_3^{2-}$ aluminium hydroxide $Al^{3+}(OH^-)_3$
ammonium sulphate $(NH_4^+)_2SO_4^{2-}$ magnesium ethanoate $Mg^{2+}(CH_3COO^-)_2$

IONIC OR COVALENT?

Ionic compounds are formed between metals on the left of the Periodic Table and non-metals on the right of the Periodic Table; that is, between elements in groups 1, 2, and 3 with a low electronegativity (electropositive elements) and elements with a high electronegativity in groups 5, 6, and 7. Generally the difference between the electronegativity values needs to be greater than about 1.8 for ionic bonding to occur.

	Al	F	Al	O	Al	Cl	Al	Br
Electronegativity	1.5	4.0	1.5	3.5	1.5	3.0	1.5	2.8
Difference in electronegativity		2.5		2.0		1.5		1.3
Formula		AlF_3		Al_2O_3		Al_2Cl_6		Al_2Br_6
Type of bonding		ionic		ionic		intermediate between ionic and covalent		covalent
M. pt / °C		1265		2050		Sublimes at 180		97

Covalent bonding

SINGLE COVALENT BONDS

Covalent bonding involves the sharing of one or more pairs of electrons so that each atom in the molecule achieves an inert gas configuration. The simplest covalent molecule is hydrogen. Each hydrogen atom has one electron in its outer shell. The two electrons are shared and attracted electrostatically by both positive nuclei resulting in a directional bond between the two atoms to form a molecule. When one pair of electrons is shared the resulting bond is known as a single covalent bond. Another example of a diatomic molecule with a single covalent bond is chlorine, Cl_2.

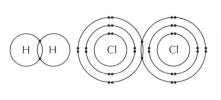

LEWIS STRUCTURES

In the Lewis structure (also known as electron dot structure) all the valence electrons are shown. There are various different methods of depicting the electrons. The simplest method involves using a line to represent one pair of electrons. It is also acceptable to represent single electrons by dots, crosses or a combination of the two. The four methods below are all correct ways of showing the Lewis structure of fluorine.

$$|\overline{F}-\overline{F}| \qquad \overset{xx\ xx}{\underset{xx\ xx}{\overset{x}{\times}F\overset{x}{\times}F\overset{x}{\times}}} \qquad :\overset{\bullet\bullet}{\underset{\bullet\bullet}{F}}:\overset{\bullet\bullet}{\underset{\bullet\bullet}{F}}: \qquad \overset{xx\ \bullet\bullet}{\underset{xx\ \bullet\bullet}{\overset{x}{\times}F\overset{\bullet}{:}F\overset{\bullet}{:}}}$$

Sometimes just the shared pairs of electrons are shown, e.g. F–F. This gives information about the bonding in the molecule, but it is not the Lewis structure as it does not show all the valence electrons.

SINGLE COVALENT BONDS

$$\underset{methane}{\overset{H}{\underset{H}{H-\overset{|}{\underset{|}{C}}-H}}} \qquad \underset{\substack{tetrafluoro-\\methane}}{\overset{|\overline{F}|}{\underset{|\overline{F}|}{\overline{F}-\overset{|}{\underset{|}{C}}-\overline{F}|}}} \qquad \underset{ammonia}{\overset{}{\underset{H}{H-\overline{N}-H}}} \qquad \underset{water}{\overset{}{\underset{H}{H-\overline{O}|}}} \qquad \underset{\substack{hydrogen\\fluoride}}{H-\overline{F}|}$$

The carbon atom (electronic configuration 2.4) has four electrons in its outer shell and requires a share in four more electrons. It forms four single bonds with elements that only require a share in one more electron, such as hydrogen or chlorine. Nitrogen (2.5) forms three single bonds with hydrogen in ammonia leaving one non-bonded pair of electrons (also known as a lone pair). In water there are two non-bonded pairs and in hydrogen fluoride three non-bonded pairs.

MULTIPLE COVALENT BONDS

In some compounds atoms can share more than one pair of electrons to achieve an inert gas configuration.

$$\underset{oxygen}{\langle O{=}O\rangle} \qquad \underset{nitrogen}{|N{\equiv}N|} \qquad \underset{\substack{carbon\ dioxide}}{\langle O{=}C{=}O\rangle} \qquad \underset{ethene}{\overset{H}{\underset{H}{}}{>}C{=}C{<}\overset{H}{\underset{H}{}}} \qquad \underset{ethyne}{H{-}C{\equiv}C{-}H}$$

CO-ORDINATE (DATIVE) BONDS

The electrons in the shared pair may originate from the same atom. This is known as a coordinate covalent bond.

$$\underset{Carbon\ monoxide}{|C{\equiv}O|} \qquad \left[\overset{H}{\underset{H}{H-\overset{|}{\underset{|}{N}}-H}}\right]^{+} \qquad \left[\overset{H}{\underset{H}{|\overset{|}{\underset{|}{O}}-H}}\right]^{+} \qquad \left(\overset{\longleftarrow}{\underset{bond}{co\text{-}ordinate}}\right)$$

ammonium ion   hydroxonium ion

Sulfur dioxide and sulfur trioxide are both sometimes shown as having a coordinate bond between sulfur and oxygen or they are shown as having double bonds between the sulfur and the oxygen. Both are acceptable.

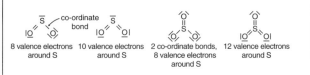

8 valence electrons around S   10 valence electrons around S   2 co-ordinate bonds, 8 valence electrons around S   12 valence electrons around S

BOND LENGTH AND BOND STRENGTH

The strength of attraction that the two nuclei have for the shared electrons affects both the length and strength of the bond. Although there is considerable variation in the bond lengths and strengths of single bonds in different compounds, double bonds are generally much stronger and shorter than single bonds. The strongest covalent bonds are shown by triple bonds.

		Length / nm	Strength / kJ mol^{-1}
Single bonds	Cl–Cl	0.199	242
	C–C	0.154	348
Double bonds	C=C	0.134	612
	O=O	0.121	496
Triple bonds	C≡C	0.120	837
	N≡N	0.110	944

e.g. ethanoic acid:

double bond between C and O shorter and stronger than single bond

BOND POLARITY

In diatomic molecules containing the same element (e.g. H_2 or Cl_2) the electron pair will be shared equally, as both atoms exert an identical attraction. However, when the atoms are different the more electronegative atom exerts a greater attraction for the electron pair. One end of the molecule will thus be more electron rich than the other end, resulting in a polar bond. This relatively small difference in charge is represented by δ+ and δ–. The bigger the difference in electronegativities the more polar the bond.

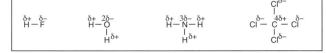

Shapes of simple molecules and ions

VSEPR THEORY

The shapes of simple molecules and ions can be determined by using the **valence shell electron pair repulsion (VSEPR)** theory. This states that pairs of electrons arrange themselves around the central atom so that they are as far apart from each other as possible. There will be greater repulsion between non-bonded pairs of electrons than between bonded pairs. Since all the electrons in a multiple bond must lie in the same direction, double and triple bonds count as one pair of electrons. Strictly speaking the theory refers to negative charge centres, but for most molecules this equates to pairs of electrons.

This results in five basic shapes depending on the number of pairs.

No. of charge centres	Shape	Name of shape	Bond angle(s)
2	O—O—O	linear	180°
3		trigonal planar	120°
4		tetrahedral	109.5°
5		trigonal bipyramidal	90°, 120°, 180°
6		octahedral	90°, 180°

WORKING OUT THE ACTUAL SHAPE

To work out the actual shape of a molecule calculate the number of pairs of electrons around the *central* atom, then work out how many are bonding pairs and how many are non-bonding pairs. (For ions the number of electrons which equate to the charge on the ion must also be included when calculating the total number of electrons.)

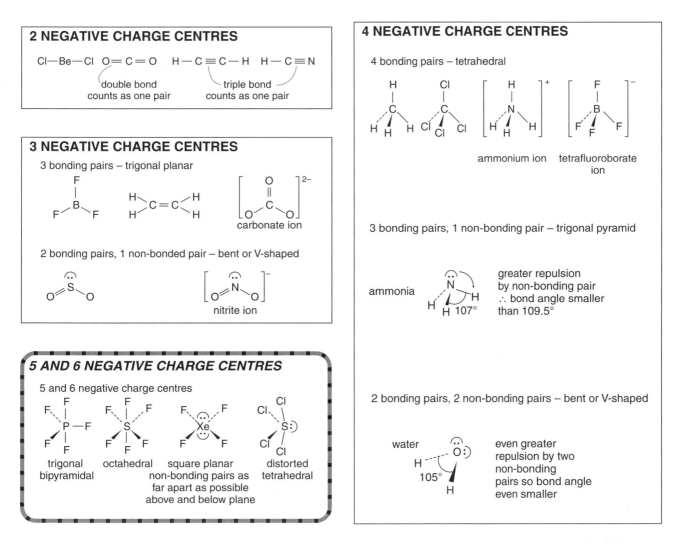

2 NEGATIVE CHARGE CENTRES

Cl—Be—Cl   O=C=O   H—C≡C—H   H—C≡N

double bond counts as one pair

triple bond counts as one pair

3 NEGATIVE CHARGE CENTRES

3 bonding pairs – trigonal planar

carbonate ion

2 bonding pairs, 1 non-bonded pair – bent or V-shaped

nitrite ion

5 AND 6 NEGATIVE CHARGE CENTRES

5 and 6 negative charge centres

trigonal bipyramidal   octahedral   square planar non-bonding pairs as far apart as possible above and below plane   distorted tetrahedral

4 NEGATIVE CHARGE CENTRES

4 bonding pairs – tetrahedral

ammonium ion   tetrafluoroborate ion

3 bonding pairs, 1 non-bonding pair – trigonal pyramid

ammonia   greater repulsion by non-bonding pair ∴ bond angle smaller than 109.5°

H H 107°

2 bonding pairs, 2 non-bonding pairs – bent or V-shaped

water   even greater repulsion by two non-bonding pairs so bond angle even smaller

105°

Intermolecular forces and allotropes of carbon

MOLECULAR POLARITY

Whether a molecule is polar, or not, depends both on the relative electronegativities of the atoms in the molecule and on its shape. If the individual bonds are polar then it does not necessarily follow that the molecule will be polar as the resultant dipole may cancel out all the individual dipoles.

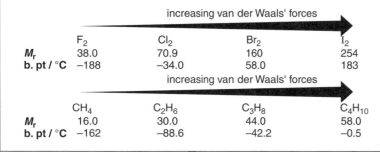

Van der Waals' forces

Even in non-polar molecules the electrons can at any one moment be unevenly spread. This produces temporary instantaneous dipoles. An instantaneous dipole can induce another dipole in a neighbouring particle resulting in a weak attraction between the two particles. Van der Waals' forces increase with increasing mass.

increasing van der Waals' forces

	F_2	Cl_2	Br_2	I_2
M_r	38.0	70.9	160	254
b. pt / °C	−188	−34.0	58.0	183

increasing van der Waals' forces

	CH_4	C_2H_6	C_3H_8	C_4H_{10}
M_r	16.0	30.0	44.0	58.0
b. pt / °C	−162	−88.6	−42.2	−0.5

Dipole:dipole forces

Polar molecules are attracted to each other by electrostatic forces. Although still relatively weak the attraction is stronger than van der Waals' forces.

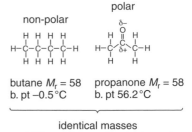

butane $M_r = 58$ propanone $M_r = 58$
b. pt −0.5 °C   b. pt 56.2 °C

identical masses
(different intermolecular forces)

INTERMOLECULAR FORCES

The covalent bonds between the atoms *within* a molecule are very strong. The forces of attraction *between* the molecules are much weaker. These intermolecular forces depend on the polarity of the molecules.

ALLOTROPES OF CARBON

Allotropes occur when an element can exists in different crystalline forms. In diamond each carbon atom is covalently bonded to four other carbon atoms to form a giant covalent structure. All the bonds are equally strong and there is no plane of weakness in the molecule so diamond is exceptionally hard and because all the electrons are localized it does not conduct electricity. Both silicon and silicon dioxide, $SiO2$, form similar giant tetrahedral structures.

diamond

In graphite each carbon atom has very strong bonds to three other carbon atoms to give layers of hexagonal rings. There are only very weak bonds between the layers. The layers can slide over each other so graphite is an excellent lubricant and because the electrons are delocalized between the layers it is a good conductor of electricity.

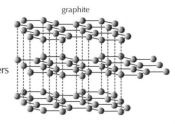
graphite

A third allotrope of carbon is buckminsterfullerene. Sixty carbon atoms are arranged in hexagons and pentagons to give a geodesic spherical structure similar to a football. Following the initial discovery of buckminsterfullerene many other similar carbon molecules have been isolated. This has led to a new branch of science called nanotechnology.

Hydrogen bonding

Hydrogen bonding occurs when hydrogen is bonded directly to a small highly electronegative element, such as fluorine, oxygen, or nitrogen. As the electron pair is drawn away from the hydrogen atom by the electronegative element, all that remains is the proton in the nucleus as there are no inner electrons. The proton attracts a non-bonding pair of electrons from the F, N, or O resulting in a much stronger dipole:dipole attraction. Water has a much higher boiling point than the other group 6 hydrides as the hydrogen bonding between water molecules is much stronger than the dipole:dipole bonding in the remaining hydrides. A similar trend is seen in the hydrides of group 5 and group 7. Hydrogen bonds between the molecules in ice result in a very open structure. When ice melts the molecules can move closer to each other so that water has its maximum density at 4 °C.

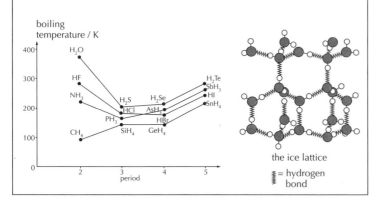

the ice lattice

≷ = hydrogen bond

Metallic bonding and physical properties related to bonding type

METALLIC BONDING

The valence electrons in metals become detached from the individual atoms so that metals consist of a close packed lattice of positive ions in a sea of delocalized electrons. A metallic bond is the attraction that two neighbouring positive ions have for the delocalized electrons between them. Metals are malleable, that is, they can be bent and reshaped under pressure. They are also ductile, which means they can be drawn out into a wire.

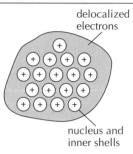

delocalized electrons

nucleus and inner shells

Metals are malleable and ductile because the close-packed layers of positive ions can slide over each other without breaking more bonds than are made.

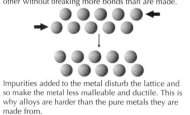

Impurities added to the metal disturb the lattice and so make the metal less malleable and ductile. This is why alloys are harder than the pure metals they are made from.

TYPE OF BONDING AND PHYSICAL PROPERTIES

Melting and boiling points

When a liquid turns into a gas the attractive forces between the particles are completely broken so boiling point is a good indication of the strength of intermolecular forces. When solids melt the crystal structure is broken down, but there are still some attractive forces between the particles. Melting points are affected by impurities. These weaken the structure and result in lower melting points.

Covalent macromolecular structures have extremely high melting and boiling points. Metals and ionic compounds also tend to have relatively high boiling points due to ionic attractions. Hydrogen bonds are in the order of $\frac{1}{10}$ th the strength of a covalent bond whereas van der Waals' forces are in the order of less than $\frac{1}{100}$ of a covalent bond. The weaker the attractive forces the more volatile the substance.

diamond lattice

NaCl

Diamond (melting point over 4000 °C)
All bonds in the macromolecular structure covalent

Sodium chloride (melting point 801 °C)
Ions held strongly in ionic lattice

Compound	propane	ethanal	ethanol
M_r	44	44	46
M. pt / °C	−42.2	20.8	78.5
Polarity	non-polar	polar	polar
Bonding type	van der Waals'	dipole:dipole	hydrogen bonding

Solubility

'Like tends to dissolve like'. Polar substances tend to dissolve in polar solvents, such as water, whereas non-polar substances tend to dissolve in non-polar solvents, such as heptane or tetrachloromethane. Organic molecules often contain a polar head and a non-polar carbon chain tail. As the non-polar carbon chain length increases in an homologous series the molecules become less soluble in water. Ethanol itself is a good solvent for other substances as it contains both polar and non-polar ends.

CH_3OH
C_2H_5OH
C_3H_7OH
C_4H_9OH

decreasing solubility in water

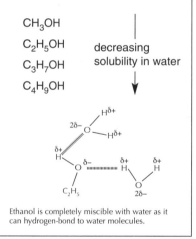

Ethanol is completely miscible with water as it can hydrogen-bond to water molecules.

Conductivity

For conductivity to occur the substance must possess electrons or ions that are free to move. Metals (and graphite) contain delocalized electrons and are excellent conductors. Molten ionic salts also conduct electricity, but are chemically decomposed in the process. Where all the electrons are held in fixed positions, such as diamond or in simple molecules, no electrical conductivity occurs.

When a potential gradient is applied to the metal, the delocalized electrons can move towards the positive end of the gradient carrying charge.

When an ionic compound melts, the ions are free to move to oppositely charged electrodes. Note: in molten ionic compounds it is the ions that carry the charge, not free electrons.

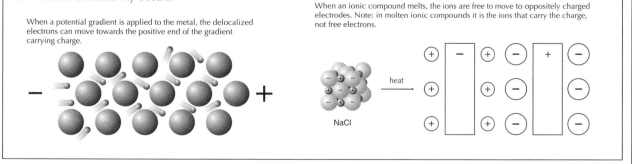

NaCl

heat

COMBINATION OF ATOMIC ORBITALS TO FORM MOLECULAR ORBITALS

Although the Lewis representation is a useful model to represent covalent bonds it does make the false assumption that all the valence electrons are the same. A more advanced model of bonding considers the combination of atomic orbitals to form molecular orbitals.

σ bonds

A σ (sigma) bond is formed when two atomic orbitals on different atoms overlap along a line drawn through the two nuclei. This occurs when two s orbitals overlap, an s orbital overlaps with a p orbital, or when two p orbitals overlap 'head on'.

π bonds

A π (pi) bond is formed when two p orbitals overlap 'sideways on'. The overlap now occurs above and below the line drawn through the two nuclei. A π bond is made up of two regions of electron density.

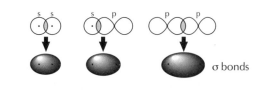

HYBRIDIZATION (1)

sp³ hybridization

Methane provides a good example of sp³ hybridization. Methane contains four equal C–H bonds pointing towards the corners of a tetrahedron with bond angles of 109.5°. A free carbon atom has the configuration $1s^2 2s^2 2p^2$. It cannot retain this configuration in methane. Not only are there only two unpaired electrons, but the p orbitals are at 90° to each other and will not give bond angles of 109.5° when they overlap with the s orbitals on the hydrogen atoms.

When the carbon bonds in methane one of its 2s electrons is promoted to a 2p orbital and then the 2s and three 2p orbitals *hybridize* to form four new hybrid orbitals. These four new orbitals arrange themselves to be as mutually repulsive as possible, i.e. tetrahedrally. Four equal σ bonds can then be formed with the hydrogen atoms.

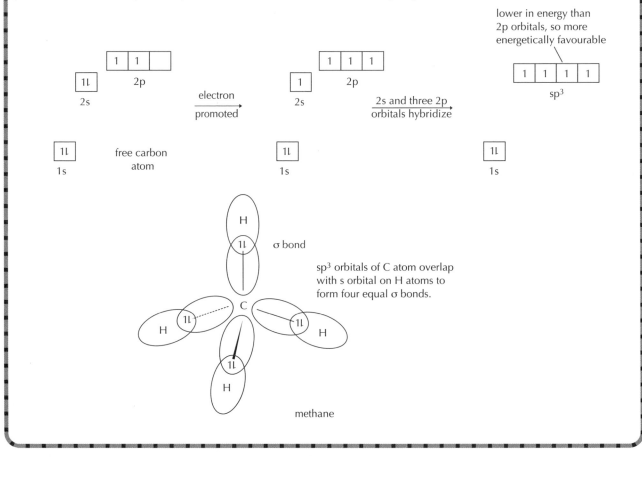

methane

HYBRIDIZATION (2)

sp² hybridization

sp² hybridization occurs in ethene. After a 2s electron on the carbon atom is promoted the 2s orbital hybridizes with two of the 2p orbitals to form three new planar hybrid orbitals with a bond angle of 120° between them. These can form σ bonds with the hydrogen atoms and also a σ bond between the two carbon atoms. Each carbon atom now has one electron remaining in a 2p orbital. These can overlap to form a π bond. Ethene is thus a planar molecule with a region of electron density above and below the plane.

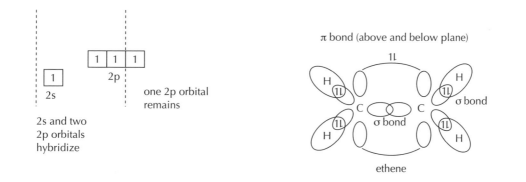

sp hybridization

sp hybridization occurs when the 2s orbital hybridizes with just one of the 2p orbitals to form two new linear sp hybrid orbitals with an angle of 180° between them. The remaining two p orbitals on each carbon atom then overlap to form two π bonds. An example is ethyne.

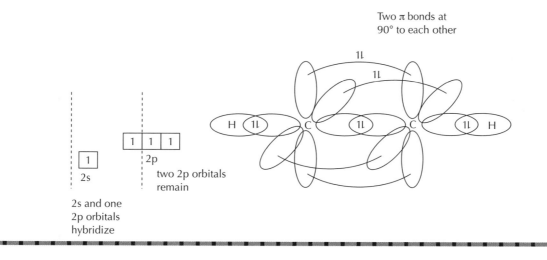

RELATIONSHIP BETWEEN TYPE OF HYBRIDIZATION, LEWIS STRUCTURE, AND MOLECULAR SHAPES

Molecular shapes can be arrived at either by using the VSEPR theory or by knowing the type of hybridization. Hybridization can take place between any s and p orbital in the same energy level and is not just restricted to carbon compounds. If the shape and bond angles are known from using Lewis structures then the type of hybridization can be deduced. Similarly if the type of hybridization is known the shape and bond angles can be deduced.

Hybridization	Regular bond angle	Examples
sp³	109.5°	
sp²	120°	
sp	180°	

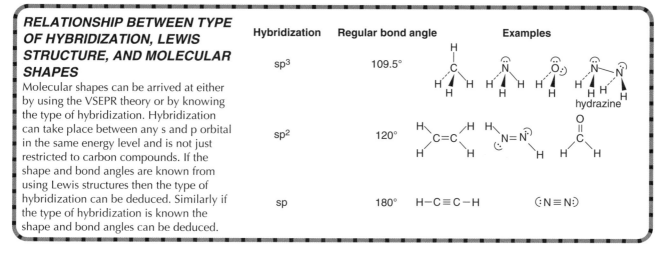

RESONANCE STRUCTURES

When writing the Lewis structures for some molecules it is possible to write more than one correct structure. For example, ozone can be written:

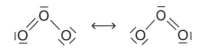

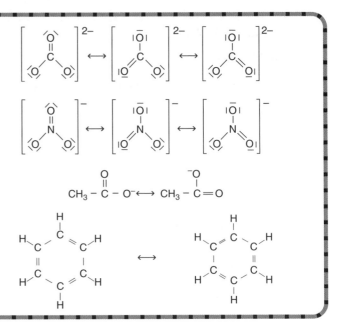

These two structures are known as resonance hybrids. They are extreme forms of the true structure, which lies somewhere between the two. Evidence that this is true comes from bond lengths, as the bond lengths between the oxygen atoms in ozone are both the same and are intermediate between an O=O double bond and an O–O single bond. Resonance structures are usually shown with a double headed arrow between them. Other common compounds which can be written using resonance structures are shown here.

DELOCALIZATION OF ELECTRONS

Resonance structures can also be explained by the delocalization of electrons. For example, in the ethanoate ion the carbon atom and the two oxygen atoms each have a p orbital containing one electron after the σ bonds have been formed. Instead of forming just one double bond between the carbon atom and one of the oxygen atoms the electrons can delocalize over all three atoms. This is energetically more favourable than forming just one double bond.

Delocalization can occur whenever alternate double and single bonds occur between carbon atoms. The delocalization energy in benzene is about 150 kJ mol^{-1}, which explains why the benzene ring is so resistant to addition reactions.

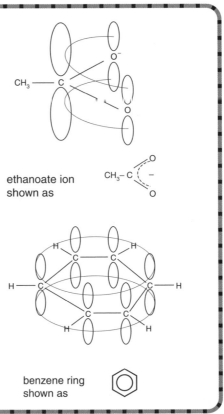

ethanoate ion shown as

benzene ring shown as

IB QUESTIONS – BONDING

1. Which compound contains both covalent and ionic bonds?

   A. sodium carbonate, Na_2CO_3

   B. magnesium bromide, $MgBr_2$

   C. dichloromethane, CH_2Cl_2

   D. ethanoic acid, CH_3COOH

2. Which pair of elements is most likely to form a covalently bonded compound?

   A. Li and Cl   **C.** Ca and S

   B. P and O   **D.** Zn and Br

3. Given the following electronegativities,

   H: 2.2   N: 3.0   O: 3.5   F: 4.0

   which bond would be the most polar?

   A. O–H in H_2O **C.** N–O in NO_2

   B. N–F in NF_2 **D.** N–H in NH_3

4. What is the correct Lewis structure for methanal?

   A. H:C:::O:H

   B.   H
  H:C::O:

   C.   H
  C::O:
  H

   D. :C:O:H
    H

5. When CH_4, NH_3, H_2O, are arranged in order of **increasing** bond angle, what is the correct order?

   A. CH_4, NH_3, H_2O **C.** NH_3, CH_4, H_2O

   B. NH_3, H_2O, CH_4 **D.** H_2O, NH_3, CH_4

6. When the H–N–H bond angles in the species NH_2^-, NH_3, NH_4^+ are arranged in order of increasing bond angle (smallest bond angle first), which order is correct?

   A. $NH_2^- < NH_3 < NH_4^+$ **C.** $NH_3 < NH_2^- < NH_4^+$

   B. $NH_4^+ < NH_3 < NH_2^-$ **D.** $NH_2^- < NH_4^+ < NH_3$

7. In which of the following pairs does the second substance have the lower boiling point?

   A. F_2, Cl_2 **C.** C_2H_6, C_3H_8

   B. H_2O, H_2S **D.** CH_3OCH_3, CH_3CH_2OH

8. In which of the following substances would hydrogen bonding be expected to occur?

   I. CH_4

   II. CH_3COOH

   III. CH_3OCH_3

   A. II only   **C.** I and III only

   B. I and III only   **D.** I, II and III

9. Which one of the following statements is correct?

   A. The energy absorbed when liquid ammonia boils is used to overcome the covalent bonds **within** the ammonia molecule.

   B. The energy absorbed when solid phosphorus (P_4) melts is used to overcome the ionic bonds **between** the phosphorus molecules.

   C. The energy absorbed when sodium chloride dissolves in water is used to form ions.

   D. The energy absorbed when copper metal melts is used to overcome the non-directional metallic bonds between the copper atoms.

10. A solid has a melting point of 1440 °C. It conducts heat and electricity. It does not dissolve in water or in organic solvents. The bond between the particles is most likely to be

   A. covalent.   **C.** ionic.

   B. dipole:dipole.   **D.** metallic.

(HL)

---

11. What are the types of hybridization of the carbon atoms in the compound

   $H_2ClC–CH_2–COOH$?
        1   2   3

	1	*2*	*3*		*1*	*2*	*3*
A.	sp^2	sp^2	sp^2	**C.**	sp^3	sp^3	sp^2
B.	sp^3	sp^2	sp	**D.**	sp^3	sp^3	sp

12. Which molecule or ion does not have a tetrahedral shape?

   A. XeF_4 **C.** BF_4^-

   B. $SiCl_4$ **D.** NH_4^+

13. When the substances below are arranged in order of increasing carbon–carbon bond length (shortest bond first), what is the correct order?

   I. H_2CCH_2 II. H_3CCH_3 III. ⬡

   A. I < II < III   **C.** II < I < III

   B. I < III < II   **D.** III < II < I

14. Which of the following species is considered to involve sp^3 hybridization?

   I. BCl_3 II. CH_4 III. NH_3

   A. I only   **C.** I and III only

   B. II only   **D.** II and III only

15. How many π bonds are present in CO_2?

   A. One   **C.** Three

   B. Two   **D.** Four

16. When the following substances are arranged in order of increasing melting point (lowest melting point first) the correct order is

   A. $CH_3CH_2CH_3$, CH_3COCH_3, $CH_3CH_2CH_2OH$

   B. $CH_3CH_2CH_3$, $CH_3CH_2CH_2OH$, CH_3COCH_3

   C. CH_3COCH_3, $CH_3CH_2CH_2OH$, $CH_3CH_2CH_3$

   D. $CH_3CH_2CH_2OH$, $CH_3CH_2CH_3$, CH_3COCH_3

Enthalpy changes

EXOTHERMIC AND ENDOTHERMIC REACTIONS

Energy is defined as the ability to do work, that is, move a force through a distance. It is measured in joules.

Energy = force × distance
(J)   (N × m)

In a chemical reaction energy is required to break the bonds in the reactants, and energy is given out when new bonds are formed in the products. The most important type of energy in chemistry is heat. If the bonds in the products are stronger than the bonds in the reactants then the reaction is said to be **exothermic**, as heat is given out to the surroundings. Examples of exothermic processes include combustion and neutralization. In **endothermic** reactions heat is absorbed from the surroundings because the bonds in the reactants are stronger than the bonds in the products.

The internal energy stored in the reactants is known as its **enthalpy**, H. The absolute value of the enthalpy of the reactants cannot be known, nor can the enthalpy of the products, but what can be measured is the difference between them, ΔH. By convention ΔH has a negative value for exothermic reactions and a positive value for endothermic reactions. It is normally measured under standard conditions of 1 atm pressure at a temperature of 298 K. The **standard enthalpy change of a reaction** is denoted by $\Delta H^{\ominus}$.

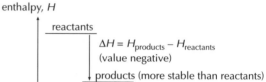

enthalpy, H

$\Delta H = H_{products} - H_{reactants}$
(value negative)

reactants

products (more stable than reactants)

Representation of exothermic reaction using an enthalpy diagram.

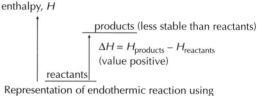

enthalpy, H

products (less stable than reactants)

$\Delta H = H_{products} - H_{reactants}$
(value positive)

reactants

Representation of endothermic reaction using an enthalpy diagram.

TEMPERATURE AND HEAT

It is important to be able to distinguish between heat and temperature as the terms are often used loosely.

- Heat is a measure of the total energy in a given amount of substance and therefore depends on the amount of substance present.
- Temperature is a measure of the 'hotness' of a substance. It represents the average kinetic energy of the substance, but is independent of the amount of substance present.

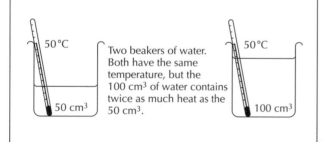

50°C

50 cm³

Two beakers of water. Both have the same temperature, but the 100 cm³ of water contains twice as much heat as the 50 cm³.

50°C

100 cm³

CALORIMETRY

The enthalpy change for a reaction can be measured experimentally by using a calorimeter. In a simple calorimeter all the heat evolved in an exothermic reaction is used to raise the temperature of a known mass of water. For endothermic reactions the heat transferred from the water to the reaction can be calculated by measuring the lowering of temperature of a known mass of water.

To compensate for heat lost by the water in exothermic reactions to the surroundings as the reaction proceeds a plot of temperature against time can be drawn. By extrapolating the graph, the temperature rise that would have taken place had the reaction been instantaneous can be calculated.

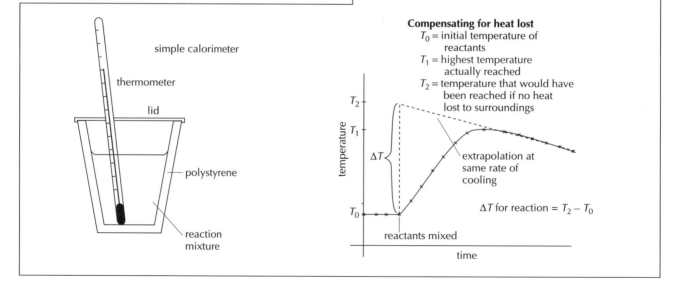

simple calorimeter

thermometer

lid

polystyrene

reaction mixture

Compensating for heat lost
T_0 = initial temperature of reactants
T_1 = highest temperature actually reached
T_2 = temperature that would have been reached if no heat lost to surroundings

T_2
T_1
ΔT
T_0

temperature

extrapolation at same rate of cooling

ΔT for reaction = $T_2 - T_0$

reactants mixed

time

ΔH calculations

CALCULATION OF ENTHALPY CHANGES

The heat involved in changing the temperature of any substance can be calculated from the equation:

Heat energy = mass (m) × specific heat capacity (c) × temperature change (ΔT)

The specific heat capacity of water is 4.18 kJ kg^{-1} K^{-1}. That is, it requires 4.18 kilojoules of energy to raise the temperature of one kilogram of water by one kelvin.

Enthalpy changes are normally quoted in kJ mol^{-1}, for either a reactant or a product, so it is also necessary to work out the number of moles involved in the reaction which produces the heat change in the water.

Worked example 1

50.0 cm^3 of 1.00 mol dm^{-3} hydrochloric acid solution was added to 50.0 cm^3 of 1.00 mol dm^{-3} sodium hydroxide solution in a polystyrene beaker. The initial temperature of both solutions was 16.7 °C. After stirring and accounting for heat loss the highest temperature reached was 23.5 °C. Calculate the enthalpy change for this reaction.

Step 1. Write equation for reaction

$HCl(aq) + NaOH(aq) \rightarrow NaCl(aq) + H_2O(l)$

Step 2. Calculate molar quantities

Amount of HCl = $\frac{50.0}{1000} \times 1.00 = 5.00 \times 10^{-2}$ mol

Amount of NaOH = $\frac{50.0}{1000} \times 1.00 = 5.00 \times 10^{-2}$ mol

Therefore the heat evolved will be for 5.00×10^{-2} mol

Step 3. Calculate heat evolved

Total volume of solution = 50.0 + 50.0 = 100 cm^3
Assume the solution has the same density and specific heat capacity as water then
mass of 'water' = 100 g = 0.100 kg
Temperature change = 23.5 − 16.7 = 6.8 °C = 6.8 K
Heat evolved in reaction = 0.100 × 4.18 × 6.8 = 2.84 kJ
$\quad\quad$ = 2.84 kJ (for 5.00×10^{-2} mol)

ΔH for reaction $= -2.84 \times \dfrac{1}{5.00 \times 10^{-2}} = -56.8$ kJ mol^{-1}

(negative value as the reaction is exothermic)

Worked example 2

A student used a simple calorimeter to determine the enthalpy change for the combustion of ethanol.

$$C_2H_5OH(l) + 3O_2(g) \rightarrow 2CO_2(g) + 3H_2O(l)$$

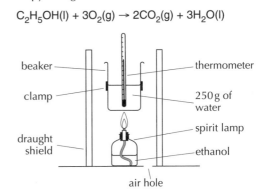

When 0.690 g (0.015 mol) of ethanol was burned it produced a temperature rise of 13.2 K in 250 g of water. Calculate ΔH for the reaction.

$$\frac{\text{Heat evolved}}{\text{by 0.015 mol}} = \frac{250}{1000} \times 4.18 \times 13.2 = 13.79 \text{ kJ}$$

$$\Delta H = -13.79 \times \frac{1}{0.015} = -920 \text{ kJ mol}^{-1}$$

Note: the Data Book value is −1371 kJ mol^{-1}. Reasons for the discrepancy include the fact that not all the heat produced is transferred to the water, the water loses some heat to the surroundings, and there is incomplete combustion of the ethanol.

Worked example 3

50.0 cm^3 of 0.200 mol dm^{-3} copper(II) sulfate solution was placed in a polystyrene cup. After two minutes 1.20 g of powdered zinc was added. The temperature was taken every 30 seconds and the following graph obtained. Calculate the enthalpy change for the reaction taking place.

Step 1. Write the equation for the reaction

$Cu^{2+}(aq) + Zn(s) \rightarrow Cu(s) + Zn^{2+}(aq)$

Step 2. Determine the limiting reagent

Amount of $Cu^{2+}(aq)$ = 50.0/1000 × 0.200 = 0.0100 mol

Amount of Zn(s) = 1.20/65.37 = 0.0184 mol

∴ $Cu^{2+}(aq)$ is the limiting reagent

Step 3. Extrapolate the graph (*already done*) to compensate for heat loss and determine ΔT

$\Delta T = 10.4$ °C

Step 4. Calculate the heat evolved in the experiment for 0.0100 mol of reactants

Heat evolved = 50.0/1000 × 4.18 × 10.4 °C = 2.17 kJ

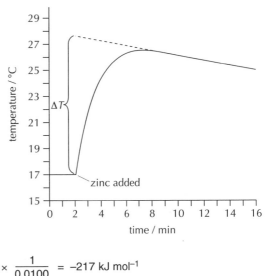

Step 5. Express this as the enthalpy change for the reaction $\Delta H = -2.17 \times \dfrac{1}{0.0100} = -217$ kJ mol^{-1}

Bond enthalpies and Hess' law

BOND ENTHALPIES
Enthalpy changes can also be calculated directly from bond enthalpies.
For a diatomic molecule the bond enthalpy is defined as the enthalpy change for the process

$$X–Y(g) \rightleftharpoons X(g) + Y(g) \text{ Note the gaseous state.}$$

For bond formation the value is negative and for bond breaking the value is positive. If the bond enthalpy values are known for all the bonds in the reactants and products then the overall enthalpy change can be calculated.

Some average bond enthalpies
All values in kJ mol^{-1}

H–H	436	C=C	612	C≡C	837
C–C	348	O=O	496	N≡N	944
C–H	412				
O–H	463				
N–H	388				
N–N	163				

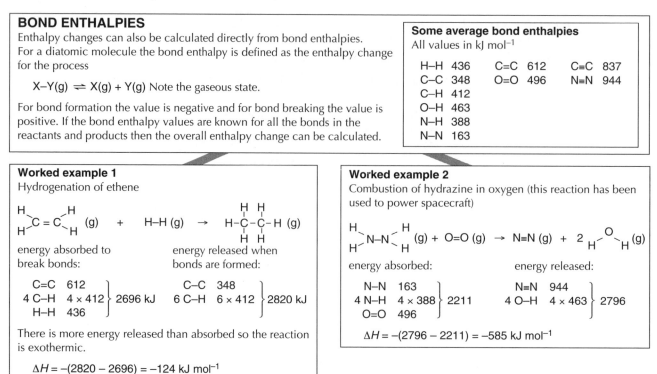

Worked example 1
Hydrogenation of ethene

energy absorbed to break bonds:

$$\left.\begin{array}{ll} \text{C=C} & 612 \\ \text{4 C–H} & 4 \times 412 \\ \text{H–H} & 436 \end{array}\right\} 2696 \text{ kJ}$$

energy released when bonds are formed:

$$\left.\begin{array}{ll} \text{C–C} & 348 \\ \text{6 C–H} & 6 \times 412 \end{array}\right\} 2820 \text{ kJ}$$

There is more energy released than absorbed so the reaction is exothermic.

$$\Delta H = -(2820 - 2696) = -124 \text{ kJ mol}^{-1}$$

Worked example 2
Combustion of hydrazine in oxygen (this reaction has been used to power spacecraft)

energy absorbed:

$$\left.\begin{array}{ll} \text{N–N} & 163 \\ \text{4 N–H} & 4 \times 388 \\ \text{O=O} & 496 \end{array}\right\} 2211$$

energy released:

$$\left.\begin{array}{ll} \text{N≡N} & 944 \\ \text{4 O–H} & 4 \times 463 \end{array}\right\} 2796$$

$$\Delta H = -(2796 - 2211) = -585 \text{ kJ mol}^{-1}$$

LIMITATIONS OF USING AVERAGE BOND ENTHALPIES
Average bond enthalpies can only be used if all the reactants and products are in the gaseous state. If water were a liquid product in the above example then even more heat would be evolved since the enthalpy change of vaporization of water would also need to be included in the calculation.

In the above calculations **average** bond enthalpies have been used. These have been obtained by considering a number of similar compounds. In practice the energy of a particular bond will vary slightly in different compounds, as it will be affected by neighbouring atoms. So ΔH values obtained from using average bond enthalpies will not necessarily be very accurate.

HESS' LAW
Hess' law states that the enthalpy change for a reaction depends only on the difference between the enthalpy of the products and the enthalpy of the reactants. It is independent of the reaction pathway.

The enthalpy change going from A to B is the same whether the reaction proceeds directly to A or whether it goes via an intermediate.

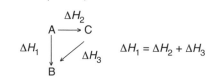

$$\Delta H_1 = \Delta H_2 + \Delta H_3$$

This law is a statement of the law of conservation of energy. It can be used to determine enthalpy changes, which cannot be measured directly. For example, the enthalpy of combustion of both carbon and carbon monoxide to form carbon dioxide can easily be measured directly, but the combustion of carbon to carbon monoxide cannot. This can be represented by an energy cycle.

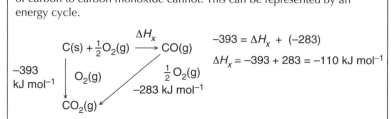

$$-393 = \Delta H_x + (-283)$$
$$\Delta H_x = -393 + 283 = -110 \text{ kJ mol}^{-1}$$

Worked example 3
Calculate the standard enthalpy change when one mole of methane is formed from its elements in their standard states. The standard enthalpies of combustion $\Delta H_c^\ominus$ of carbon, hydrogen, and methane are -393, -286, and -890 kJ mol^{-1} respectively.

Step 1. Write the equation for the enthalpy change with the unknown $\Delta H^\ominus$ value. Call this value $\Delta H_x^\ominus$.

$$C(s) + 2H_2(g) \xrightarrow{\Delta H_x^\ominus} CH_4(g)$$

Step 2. Construct an energy cycle showing the different routes to the products (in this case the products of combustion)

$$
\begin{array}{ccc}
C(s) \quad + \quad 2H_2(g) & \xrightarrow{\Delta H_x^\ominus} & CH_4(g) \\
\downarrow O_2(g) \quad \downarrow O_2(g) & & \swarrow 2O_2(g) \\
CO_2(g) \; + \; 2H_2O(l) & &
\end{array}
$$

Step 3. Use Hess' law to equate the energy changes for the two different routes

$$\underbrace{\Delta H_c^\ominus(C) + 2\Delta H_c^\ominus(H_2)}_{\text{direct route}} = \underbrace{\Delta H_x^\ominus + \Delta H_c^\ominus(CH_4)}_{\text{route via methane}}$$

Step 4. Rearrange the equation and substitute the values to give the answer

$$\Delta H_x^\ominus = \Delta H_c^\ominus(C) + 2\Delta H_c^\ominus(H_2) - \Delta H_c^\ominus(CH_4)$$
$$= -393 + (2 \times -286) - (-890) \text{ kJ mol}^{-1}$$
$$= -75 \text{ kJ mol}^{-1}$$

Energy cycles

STANDARD ENTHALPY CHANGE OF FORMATION $\Delta H_f^{\ominus}$

The standard enthalpy change of formation of a compound is the enthalpy change when one mole of the compound is formed from its elements in their standard states at 298 K and 1 atm pressure. From this it follows that $\Delta H_f^{\ominus}$ for an element in its standard state will be zero. The standard enthalpy change of combution, $\Delta H_c^{\ominus}$, is the enthalpy change when one mole of a substance is completely combusted in oxygen under standard conditions, An accurate value for the standard enthalpy change of formation of ethanol can be determined from the following cycle.

$$2C(s) \quad + \quad 3H_2(g) \quad + \quad \tfrac{1}{2}O_2(g) \xrightarrow{\Delta H_f^{\ominus}\ (C_2H_5OH)} C_2H_5OH(l)$$

$2 \times \Delta H_f^{\ominus}(CO_2)\big\downarrow\ 2O_2(g) \quad 3 \times \Delta H_f^{\ominus}(H_2O)\ \big\downarrow 1\tfrac{1}{2}O_2(g) \qquad\qquad 3O_2(g)$

$$2CO_2(g) \quad + \quad 3H_2O(l) \xleftarrow{\Delta H_c^{\ominus}\ (C_2H_5OH)}$$

By Hess' law: $\Delta H_f^{\ominus}(C_2H_5OH) = 2 \times \Delta H_f^{\ominus}(CO_2) + 3 \times \Delta H_f^{\ominus}(H_2O) - \Delta H_c^{\ominus}(C_2H_5OH)$

Substituting the relevant values $\Delta H_f^{\ominus}(C_2H_5OH) = (2 \times -393.5) + (3 \times -285.8) - (-1371) = -273.4$ kJ mol^{-1}

BORN–HABER CYCLES

Born–Haber cycles are simply energy cycles for the formation of ionic compounds. The enthalpy change of formation of sodium chloride can be considered to occur through a number of separate steps.

$$Na(s) \quad + \quad \tfrac{1}{2}Cl_2 \xrightarrow{\Delta H_f^{\ominus}(NaCl)} Na^+Cl^-(s)$$

$\big\downarrow \Delta H_{at}^{\ominus}(Na) \qquad\qquad \big\downarrow \Delta H_{at}^{\ominus}(Cl)$

$Na(g) \qquad\qquad\quad Cl(g)$

$\big\downarrow \Delta H_{IE}^{\ominus}(Na) \qquad\quad \big\downarrow \Delta H_{EA}^{\ominus}(Cl) \quad\nearrow \Delta H_{latt}^{\ominus}(NaCl)$

$Na^+(g) \quad + \quad Cl^-(g)$

Using Hess' law:

$\Delta H_f^{\ominus}(NaCl) = \Delta H_{at}^{\ominus}(Na) + \Delta H_{IE}^{\ominus}(Na) + \Delta H_{at}^{\ominus}(Cl) + \Delta H_{EA}^{\ominus}(Cl) + \Delta H_{latt}^{\ominus}(NaCl)$

Substituting the relevant values:
$\Delta H_f^{\ominus}(NaCl) = +108 + 494 + 121 - 364 - 771 = -412$ kJ mol^{-1}

Note: it is the large lattice enthalpy that mainly compensates for the endothermic processes and leads to the enthalpy of formation of ionic compounds having a negative value.

LATTICE ENTHALPY $\Delta H_{latt}^{\ominus}$

The lattice enthalpy relates either to the endothermic process of turning a crystalline solid into its gaseous ions or to the exothermic process of turning gaseous ions into a crystalline solid.

$$MX(s) \rightleftharpoons M^+(g) + X^-(g)$$

The sign of the lattice enthalpy indicates whether the lattice is being formed (−) or broken (+).

The size of the lattice enthalpy depends both on the size of the ions and on the charge carried by the ions.

cation size increasing →			anion size increasing →		
LiCl	NaCl	KCl	NaCl	NaBr	NaI
Lattice enthalpy / kJ mol⁻¹					
846	771	701	771	733	684

charge on cation increasing →		charge on anion increasing →	
NaCl	MgCl$_2$	MgCl$_2$	MgO
Lattice enthalpy / kJ mol⁻¹			
771	2493	2493	3889

ENTHALPY OF ATOMIZATION $\Delta H_{at}^{\ominus}$

The standard enthalpy of atomization is the standard enthalpy change when one mole of gaseous atoms is formed from the element in its standard state under standard conditions. For diatomic molecules this is equal to half the bond dissociation enthalpy.

$$\tfrac{1}{2}Cl_2(g) \rightarrow Cl(g) \qquad \Delta H_{at}^{\ominus} = +121 \text{ kJ mol}^{-1}$$

ELECTRON AFFINITY $\Delta H_{EA}^{\ominus}$

The electron affinity is the enthalpy change when an electron is added to an isolated atom in the gaseous state, i.e.

$$X(g) + e^- \rightarrow X^-(g)$$

Atoms 'want' an extra electron so electron affinity values are negative for the first electron. However, when oxygen forms the O^{2-} ion the overall process is endothermic:

$$O(g) + e^- \rightarrow O^-(g) \qquad \Delta H^{\ominus} = -142 \text{ kJ mol}^{-1}$$

$$O^-(g) + e^- \rightarrow O^{2-}(g) \qquad \Delta H^{\ominus} = +844 \text{ kJ mol}^{-1}$$

overall $\overline{O(g) + 2e^- \rightarrow O^{2-}(g) \qquad \Delta H^{\ominus} = +702 \text{ kJ mol}^{-1}}$

USE OF BORN–HABER CYCLES

Like any energy cycle Born–Haber cycles can be used to find the value of an unknown. They can also be used to assess how ionic a substance is. The lattice enthalpy can be calculated theoretically by considering the charge and size of the constituent ions. It can also be obtained indirectly from the Born–Haber cycle. If there is good agreement between the two values then it is reasonable to assume that there is a high degree of ionic character, e.g. NaCl. However, if there is a big difference between the two values then it is because the compounds possesses a considerable degree of covalent character, e.g. AgCl.

	NaCl	AgCl
Theoretical value / kJ mol⁻¹	766	770
Experimental value / kJ mol⁻¹	771	905

DISORDER

In nature systems naturally tend towards disorder. An increase in disorder can result from:

- mixing different types of particles, e.g. the dissolving of sugar in water
- a change in state where the distance between the particles increases, e.g. liquid water → steam
- the increased movement of particles, e.g. heating a liquid or gas
- increasing the number of particles, e.g.
  $2H_2O_2(l) \rightarrow 2H_2O(l) + O_2(g)$.

The greatest increase in disorder is usually found where the number of particles in the gaseous state increases.

The change in the disorder of a system is known as the entropy change, ΔS. The more disordered the system becomes the more positive the value of ΔS becomes. Systems which become more ordered will have negative ΔS values.

$$NH_3(g) + HCl(g) \quad \rightarrow \quad NH_4Cl(s) \quad \Delta S = -284 \text{ J K}^{-1} \text{ mol}^{-1}$$

(two moles   (one mole
of gas)   of solid)

ABSOLUTE ENTROPY VALUES

The standard entropy of a substance is the entropy change per mole that results from heating the substance from 0 K to the standard temperature of 298 K. Unlike enthalpy, absolute values of entropy can be measured. The standard entropy change for a reaction can then be determined by calculating the difference between the entropy of the products and the reactants.

$$\Delta S^{\ominus} = S^{\ominus} \text{ (products)} - S^{\ominus} \text{ (reactants)}$$

e.g. for the formation of ammonia

$$3H_2(g) + N_2(g) \rightleftharpoons 2NH_3(g)$$

the standard entropies of hydrogen, nitrogen, and ammonia are respectively 131, 192, and 192 J K^{-1} mol^{-1}.

Therefore per mole of reaction

$$\Delta S^{\ominus} = 2 \times 192 - [(3 \times 131) + 192] = -201 \text{ J K}^{-1}\text{mol}^{-1}$$

(or per mole of ammonia $\Delta S^{\ominus} = \dfrac{-201}{2} = -101$ J K^{-1}mol^{-1})

SPONTANEITY

A reaction is said to be spontaneous if it causes a system to move from a less stable to a more stable state. This will depend both upon the enthalpy change and the entropy change. These two factors can be combined and expressed as the Gibbs energy change ΔG, often known as the 'free energy change'.

The standard free energy change $\Delta G^{\ominus}$ is defined as:

$$\Delta G^{\ominus} = \Delta H^{\ominus} - T\Delta S^{\ominus}$$

Where all the values are measured under standard conditions. For a reaction to be spontaneous it must be able to do work, that is $\Delta G^{\ominus}$ must have a negative value.

Note: the fact that a reaction is spontaneous does not necessarily mean that it will proceed without any input of energy. For example, the combustion of coal is a spontaneous reaction and yet coal is stable in air. It will only burn on its own accord after it has received some initial energy so that some of the molecules have the necessary activation energy for the reaction to occur.

Spontaneity of a reaction

POSSIBLE COMBINATIONS FOR FREE ENERGY CHANGES

Some reactions will always be spontaneous. If $\Delta H^{\ominus}$ is negative or zero and $\Delta S^{\ominus}$ is positive then $\Delta G^{\ominus}$ must always have a negative value. Conversely if $\Delta H^{\ominus}$ is positive or zero and $\Delta S^{\ominus}$ is negative then $\Delta G^{\ominus}$ must always be positive and the reaction will never be spontaneous.

For some reactions whether or not they will be spontaneous depends upon the temperature. If $\Delta H^{\ominus}$ is positive or zero and $\Delta S^{\ominus}$ is positive, then $\Delta G^{\ominus}$ will only become negative at high temperatures when the value of $T\Delta S^{\ominus}$ exceeds the value of $\Delta H^{\ominus}$.

Type	$\Delta H^{\ominus}$	$\Delta S^{\ominus}$	$T\Delta S^{\ominus}$	$\Delta H^{\ominus} - T\Delta S^{\ominus}$	$\Delta G^{\ominus}$
1	0	+	+	(0) – (+)	–
2	0	–	–	(0) – (–)	+
3	–	+	+	(–) – (+)	–
4	+	–	–	(+) – (–)	+
5	+	+	+	(+) – (+)	– or +
6	–	–	–	(–) – (–)	+ or –

Type 1. Mixing two gases. $\Delta G^{\ominus}$ is negative so gases will mix of their own accord. Gases do not unmix of their own accord (Type 2) as $\Delta G^{\ominus}$ is positive.

Type 3. $(NH_4)_2Cr_2O_7(s) \rightarrow N_2(g) + Cr_2O_3(s) + 4H_2O(g)$

The decomposition of ammonium dichromate is spontaneous at all temperatures.

Type 4. $N_2(g) + 2H_2(g) \rightarrow N_2H_4(g)$

The formation of hydrazine from its elements will never be spontaneous.

Type 5. $CaCO_3(s) \rightarrow CaO(s) + CO_2(g)$

The decomposition of calcium carbonate is only spontaneous at high temperatures.

Type 6. $C_2H_4(g) + H_2(g) \rightarrow C_2H_6(g)$

Above a certain temperature this reaction will cease to be spontaneous.

DETERMINING THE VALUE OF $\Delta G^{\ominus}$

The precise value of $\Delta G^{\ominus}$ for a reaction can be determined from $\Delta G_f^{\ominus}$ values using an energy cycle, e.g. to find the standard free energy of combustion of methane given the standard free energies of formation of methane, carbon dioxide, water, and oxygen.

$$\Delta G_x^{\ominus}$$
$$CH_4(g) + 2O_2 \rightarrow CO_2(g) + 2H_2O(l)$$
$$C(s) + 2O_2(g) + 2H_2(g)$$

By Hess' law

$\Delta G_x^{\ominus} = [\Delta G_x^{\ominus}(CO_2) + 2\Delta G_f^{\ominus}(H_2O)] - [\Delta G_f^{\ominus}(CH_4) + 2\Delta G_f^{\ominus}(O_2)]$

Substituting the actual values

$\Delta G_x^{\ominus} = [-394 + 2 \times (-237)] - [-50 + 2 \times 0] = -818$ kJ mol^{-1}

$\Delta G^{\ominus}$ values can also be calculated from using the equation $\Delta G^{\ominus} = \Delta H^{\ominus} - T\Delta S^{\ominus}$. For example, in Type 5 in the adjacent list the values for $\Delta H^{\ominus}$ and $\Delta S^{\ominus}$ for the thermal decomposition of calcium carbonate are +178 kJ mol^{-1} and +165.3 J K^{-1} mol^{-1} respectively. Note that the units of $\Delta S^{\ominus}$ are different to those of $\Delta H^{\ominus}$.

At 25 °C (298 K) the value for $\Delta G^{\ominus} = 178 - 298 \times \dfrac{165.3}{1000}$

$$= +129 \text{ kJ mol}^{-1}$$

which means that the reaction is not spontaneous.

The reaction will become spontaneous when $T\Delta S^{\ominus} > \Delta H^{\ominus}$.

$T\Delta S^{\ominus} = \Delta H^{\ominus}$ when $T = \dfrac{\Delta H^{\ominus}}{\Delta S^{\ominus}} = \dfrac{178}{165.3/1000} = 1077$ K (804 °C)

Therefore above 804 °C the reaction will be spontaneous.

Note: this calculation assumes that the entropy value is independent of temperature, which is not strictly true.

IB QUESTIONS – ENERGETICS

1. Which statement about this reaction is correct?

   $2Fe(s) + 3CO_2(g) \rightarrow Fe_2O_3(s) + 3CO(g) \quad \Delta H^{\ominus} = +26.6$ kJ

   A. 26.6 kJ of energy are released for every mole of Fe reacted

   B. 26.6 kJ of energy are absorbed for every mole of Fe reacted

   C. 53.2 kJ of energy are released for every mole of Fe reacted

   D. 13.3 kJ of energy are absorbed for every mole of Fe reacted

2. When solutions of HCl and NaOH are mixed the temperature increases. The reaction:

   $H^+(aq) + OH^-(aq) \rightarrow H_2O(l)$

   A. is endothermic with a positive $\Delta H^{\ominus}$.

   B. is endothermic with a negative $\Delta H^{\ominus}$.

   C. is exothermic with a positive $\Delta H^{\ominus}$.

   D. is exothermic with a negative $\Delta H^{\ominus}$.

3.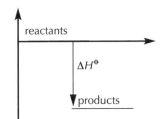

   What can be deduced about the relative stability of the reactants and products and the sign of $\Delta H^{\ominus}$, from the enthalpy level diagram above?

	Relative stability	Sign of $\Delta H^{\ominus}$
A.	Products more stable	–
B.	Products more stable	+
C.	Reactants more stable	–
D.	Reactants more stable	+

4. For the reaction:

$$2C(s) + 2H_2(g) \rightarrow C_2H_4(g) \qquad \Delta H_1^\ominus = +52.3 \text{ kJ}$$

If $\Delta H_2^\ominus = -174.4$ kJ for the reaction:

$$C_2H_2(g) + H_2(g) \rightarrow C_2H_4(g)$$

what can be said about the value of $\Delta H_3^\ominus$ for the reaction below?

$$2C(s) + H_2(g) \rightarrow C_2H_2(g)$$

A. $\Delta H_3^\ominus$ must be negative.

B. $\Delta H_3^\ominus$ must be a positive number smaller than 52.3 kJ.

C. $\Delta H_3^\ominus$ must be a positive number larger than 52.3 kJ.

D. No conclusion can be made about $\Delta H_3^\ominus$ without the value of H for $H_2(g)$.

5. The enthalpy changes for two different hydrogenation reactions of C_2H_2 are:

$$C_2H_2 + H_2 \rightarrow C_2H_4 \quad \Delta H_1^\ominus$$

$$C_2H_2 + 2H_2 \rightarrow C_2H_6 \quad \Delta H_2^\ominus$$

Which expression represents the enthalpy change for the reaction below?

$$C_2H_4 + H_2 \rightarrow C_2H_6 \quad \Delta H^\ominus = ?$$

A. $\Delta H_1^\ominus + \Delta H_2^\ominus$

B. $\Delta H_1^\ominus - \Delta H_2^\ominus$

C. $\Delta H_2^\ominus - \Delta H_1^\ominus$

D. $-\Delta H_1^\ominus - \Delta H_2^\ominus$

6.

$$2KHCO_3(s) \xrightarrow{\Delta H^\ominus} K_2CO_3(s) + CO_2(g) + H_2O(l)$$
$$+2HCl(aq) \qquad \Delta H_2^\ominus \qquad +2HCl(aq)$$
$$\Delta H_1^\ominus$$
$$2KCl(aq) + 2CO_2(g) + 2H_2O(l)$$

This cycle may be used to determine $\Delta H^\ominus$ for the decomposition of potassium hydrogencarbonate. Which expression can be used to calculate $\Delta H^\ominus$?

A. $\Delta H^\ominus = \Delta H_1^\ominus + \Delta H_2^\ominus$ **C.** $\Delta H^\ominus = \frac{1}{2}\Delta H_1^\ominus - \Delta H_2^\ominus$

B. $\Delta H^\ominus = \Delta H_1^\ominus - \Delta H_2^\ominus$ **D.** $\Delta H^\ominus = \Delta H_2^\ominus - \Delta H_1^\ominus$

7. Use the bond energies for H–H (436 kJ mol^{-1}), Br–Br (193 kJ mol^{-1}) and H–Br (366 kJ mol^{-1}) to calculate $\Delta H^\ominus$ (in kJ mol^{-1}) for the reaction:

$$H_2(g) + Br_2(g) \rightarrow 2HBr(g)$$

A. 263   **C.** –103

B. 103   **D.** –263

8. The average bond enthalpy for the C–H bond is 412 kJ mol^{-1}. Which process has an enthalpy change closest to this value?

A. $CH_4(g) \rightarrow C(s) + 2H_2(g)$

B. $CH_4(g) \rightarrow C(g) + 2H_2(g)$

C. $CH_4(g) \rightarrow C(s) + 4H(g)$

D. $CH_4(g) \rightarrow CH_3(g) + H(g)$

9. Which of the changes below occurs with the greatest increase in entropy?

A. $Na_2O(s) + H_2O(l) \rightarrow 2Na^+(aq) + 2OH^-(aq)$

B. $NH_3(g) + HCl(g) \rightarrow NH_4Cl(s)$

C. $H_2(g) + I_2(g) \rightarrow 2HI(g)$

D. $C(s) + CO_2(g) \rightarrow 2CO(g)$

10. How would this reaction at 298 K be described in thermodynamic terms?

$$2H_2O(g) \rightarrow 2H_2(g) + O_2(g)$$

A. Endothermic with a significant increase in entropy

B. Endothermic with a significant decrease in entropy

C. Exothermic with a significant increase in entropy

D. Exothermic with a significant decrease in entropy

11. The enthalpy change, $\Delta H^\ominus$, for a chemical reaction is –10 kJ mol^{-1} and the entropy change, $\Delta S^\ominus$, is –10 JK^{-1} mol^{-1} at 27 °C. What is the value of $\Delta G^\ominus$ (in J) for this reaction?

A. –260

B. –7000

C. –9730

D. –13000

12. At 0 °C, the mixture formed when the following reaction reaches equilibrium consists mostly of $N_2O_4(g)$.

$$2NO_2(g) \rightleftharpoons N_2O_4(g)$$

What are the signs of ΔG, ΔH, ΔS at this temperature?

	ΔG	ΔH	ΔS
A.	+	+	+
B.	–	–	–
C.	–	+	+
D.	+	+	–

13. Which substance has the largest lattice energy?

A. NaF   **C.** MgO

B. KCl   **D.** CaS

14. Which factor(s) will cause the lattice enthalpy of ionic compounds to increase in magnitude?

I. an increase in the charge on the ions

II. an increase in the size of ions

A. I only   **C.** Both I and II

B. II only   **D.** Neither I nor II

15. The Born–Haber cycle for the formation of potassium chloride includes the steps below:

I. $K(g) \rightarrow K^+(g) + e^-$ III. $Cl(g) + e^- \rightarrow Cl^-(g)$

II. $\frac{1}{2}Cl_2(g) \rightarrow Cl(g)$ IV. $K^+(g) + Cl^-(g) \rightarrow KCl(s)$

Which of these steps are exothermic?

A. I and II only   **C.** I, II and II only

B. III and IV only   **D.** I, III and IV only

Rates of reaction and collision theory

RATE OF REACTION

Chemical kinetics is the study of the factors affecting the rate of a chemical reaction. The rate of a chemical reaction can be defined either as the increase in the concentration of one of the products per unit time or as the decrease in the concentration of one of the reactants per unit time. It is measured in mol dm^{-3} s^{-1}.

The change in concentration can be measured by using any property that differs between the reactants and the products. Common methods include mass or volume changes when a gas is evolved, absorption using a spectrometer when there is a colour change, pH changes when there is a change in acidity, and electrical conductivity when there is a change in the ionic concentrations. Data loggers could be used for all these methods. A graph of concentration against time is then usually plotted. The rate at any stated point in time is then the gradient of the graph at that time. Rates of reaction usually decrease with time as the reactants are used up.

The reaction of hydrochloric acid with calcium carbonate can be used to illustrate three typical curves that could be obtained depending on whether the concentration of reactant, the volume of the product or the loss in mass due to the carbon dioxide escaping is followed.

$$CaCO_3(s) + 2HCl(aq) \rightarrow CaCl_2(aq) + H_2O(l) + CO_2(g)$$

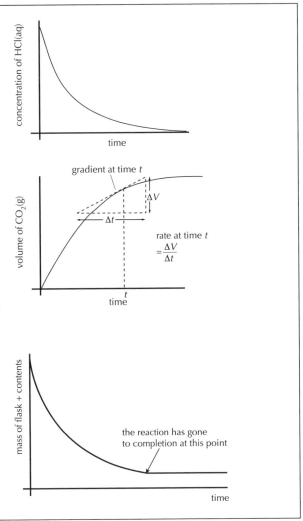

MAXWELL–BOLTZMANN DISTRIBUTION

The moving particles in a gas or liquid do not all travel with the same velocity. Some are moving very fast and others much slower. The faster they move the more kinetic energy they possess. The distribution of kinetic energies is shown by a Maxwell–Boltzmann curve.

As the temperature increases the area under the curve does not change as the total number of particles remains constant. More particles have a very high velocity resulting in an increase in the average kinetic energy, which leads to a broadening of the curve.

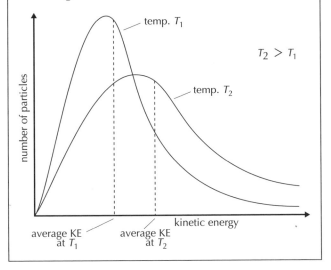

COLLISION THEORY

For a reaction between two particles to occur three conditions must be met.
- The particles must collide.
- They must collide with the appropriate geometry or orientation so that the reactive parts of the particles come into contact with each other.
- They must collide with sufficient energy to bring about the reaction.

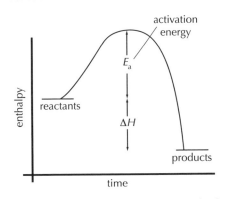

This minimum amount of energy required is known as the **activation energy**. Any factor that either increases the frequency of the collisions or increases the energy with which they collide will make the reaction go faster.

Factors affecting the rate of reaction

TEMPERATURE

As the temperature increases the particles will move faster so there will be more collisions per second. However, the main reason why an increase in temperature increases the rate is that more of the colliding particles will possess the necessary activation energy resulting in more successful collisions. As a rough rule of thumb an increase of 10 °C doubles the rate of a chemical reaction.

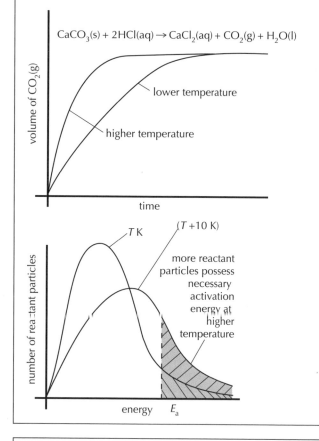

$$CaCO_3(s) + 2HCl(aq) \rightarrow CaCl_2(aq) + CO_2(g) + H_2O(l)$$

SURFACE AREA

In a solid substance only the particles on the surface can come into contact with a surrounding reactant. If the solid is in powdered form then the surface area increases dramatically and the rate increases correspondingly.

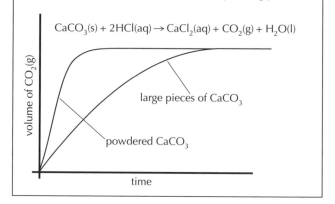

$$CaCO_3(s) + 2HCl(aq) \rightarrow CaCl_2(aq) + CO_2(g) + H_2O(l)$$

CONCENTRATION

The more concentrated the reactants the more collisions there will be per second per unit volume. As the reactants get used up their concentration decreases. This explains why the rate of most reactions gets slower as the reaction proceeds. (Some exothermic reactions do initially speed up if the heat that is given out more than compensates for the decrease in concentration.)

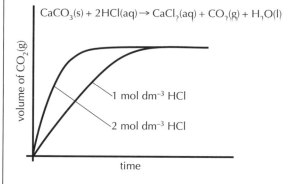

$$CaCO_3(s) + 2HCl(aq) \rightarrow CaCl_2(aq) + CO_2(g) + H_2O(l)$$

Note: this graph assumes that calcium carbonate is the limiting reagent or that equal amounts (mol) of acid have been added.

CATALYST

Catalysts increase the rate of a chemical reaction without themselves being chemically changed at the end of the reaction. They work essentially by bringing the reactive parts of the reactant particles into close contact with each other. This provides an alternative pathway for the reaction with a lower activation energy. More of the reactants will possess this lower activation energy, so the rate increases.

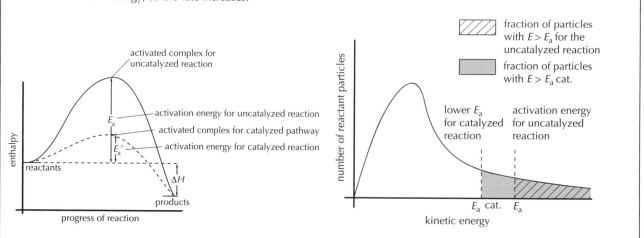

Order of reaction and half-life

RATE EXPRESSIONS

The rate of reaction between two reactants, A and B, can be followed experimentally. The rate will be found to be proportional to the concentration of A raised to some power and also to the concentration of B raised to a power. If square brackets are used to denote concentration this can be written as rate $\propto [A]^x$ and rate $\propto [B]^y$. They can be combined to give the rate expression:

$$rate = k[A]^x[B]^y$$

where k is the constant of proportionality and is known as the **rate constant**.

x is known as the **order of the reaction** with respect to A.

y is known as the order of the reaction with respect to B.

The overall order of the reaction $= x + y$.

Note: the order of the reaction and the rate expression can only be determined experimentally. They cannot be deduced from the balanced equation for the reaction.

UNITS OF RATE CONSTANT

The units of the rate constant depend on the overall order of the reaction.

First order: rate $= k[A]$

$$k = \frac{rate}{[A]} = \frac{mol\ dm^{-3}\ s^{-1}}{mol\ dm^{-3}} = s^{-1}$$

Second order: rate $= k[A]^2$ or $k = [A][B]$

$$k = \frac{rate}{[A]^2} = \frac{mol\ dm^{-3}\ s^{-1}}{(mol\ dm^{-3})^2} = dm^3\ mol^{-1}\ s^{-1}$$

Third order: rate $= k[A]^2[B]$ or rate $= k[A][B]^2$

$$k = \frac{rate}{[A]^2[B]} = \frac{mol\ dm^{-3}\ s^{-1}}{(mol\ dm^{-3})^3} = dm^6\ mol^{-2}\ s^{-1}$$

GRAPHICAL REPRESENTATIONS OF REACTIONS

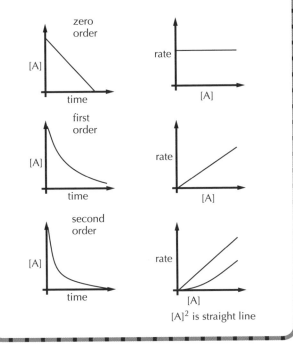

$[A]^2$ is straight line

DERIVING A RATE EXPRESSION BY INSPECTION OF DATA

Experimental data obtained from the reaction between hydrogen and nitrogen monoxide at 1073 K:

$$2H_2(g) + 2NO(g) \rightarrow 2H_2O(g) + N_2(g)$$

Experiment	Initial concentration of $H_2(g)$ / mol dm^{-3}	Initial concentration of NO(g) / mol dm^{-3}	Initial rate of formation of $N_2(g)$ / mol dm^{-3} s^{-1}
1	1×10^{-3}	6×10^{-3}	3×10^{-3}
2	2×10^{-3}	6×10^{-3}	6×10^{-3}
3	6×10^{-3}	1×10^{-3}	0.5×10^{-3}
4	6×10^{-3}	2×10^{-3}	2.0×10^{-3}

From experiments 1 and 2 doubling $[H_2]$ doubles the rate so rate $\propto [H_2]$.

From experiments 3 and 4 doubling [NO] quadruples the rate so rate $\propto [NO]^2$.

Rate expression given by rate $= k[H_2][NO]^2$.

The rate is first order with respect to hydrogen, second order with respect to nitrogen monoxide, and third order overall. The value of k can be found by substituting the values from any one of the four experiments:

$$k = \frac{rate}{[H_2][NO]^2} = 8.33 \times 10^4\ dm^6\ mol^{-2}\ s^{-1}$$

HALF-LIFE $t_{\frac{1}{2}}$

For a first order reaction the rate of change of concentration of A is equal to $k[A]$. This can be expressed as $\frac{d[A]}{dt} = k[A]$.

If this expression is integrated then $kt = \ln [A]_o - \ln [A]$ where $[A]_o$ is the initial concentration and $[A]$ is the concentration at time t. This expression is known as the integrated form of the rate equation.

The half-life is defined as the time taken for the concentration of a reactant to fall to half of its initial value.

At $t_{\frac{1}{2}}$ $[A] = \frac{1}{2}[A]_o$ the integrated rate expression then becomes

$kt_{\frac{1}{2}} = \ln [A]_o - \ln \frac{1}{2}[A]_o = \ln 2$ since $\ln 2 = 0.693$ this simplifies to $t_{\frac{1}{2}} = \frac{0.693}{k}$

From this expression it can be seen that the half-life of a first order reaction is independent of the original concentration of A, i.e. first order reactions have a constant half-life.

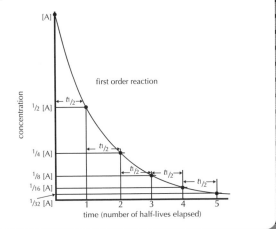

REACTION MECHANISMS

Many reactions do not go in one step. This is particularly true when there are more than two reactant molecules as the chances of a successful collision between three or more particles is extremely small. When there is more than one step then each step will proceed at its own rate. No matter how fast the other steps are the overall rate of the reaction will depend only upon the rate of the slowest step. This slowest step is known as the **rate determining step**.

e.g. consider the reaction between A and B to form A_2B: $2A + B \rightarrow A_2B$. A possible mechanism might be:

$$\text{Step 1.} \quad A + A \xrightarrow{\text{slow}} A\text{–}A \qquad \text{rate determining step}$$

$$\text{Step 2.} \quad A\text{–}A + B \xrightarrow{\text{fast}} A_2B$$

However fast A–A reacts with B the rate of production of A_2B will only depend on how fast A–A is formed.

When the separate steps in a chemical reaction are analysed there are essentially only two types of processes. Either a single species can break down into two or more products by what is known as a **unimolecular process**, or two species can collide and interact by a **bimolecular process**.

In a bimolecular process the species collide with the necessary activation energy to give initially an **activated complex**. An activated complex is not a chemical substance which can be isolated, but consists of an association of the reacting particles in which bonds are in the process of being broken and formed. An activated complex breaks down to form either the products or reverts back to the original reactants.

The number of species taking part in any specified step in the reaction is known as the **molecularity**. In most cases the molecularity refers to the slowest step, that is the rate determining step.

In the reaction on the previous page, between nitrogen monoxide and hydrogen, the stoichiometry of the reaction involves two molecules of hydrogen and two molecules of nitrogen monoxide. Any proposed mechanism must be consistent with the rate expression. For third order reactions, such as this, the rate determining step will never be the first step. The proposed mechanism is:

$$NO(g) + NO(g) \xrightarrow{\text{fast}} N_2O_2(g)$$
$$N_2O_2(g) + H_2(g) \xrightarrow{\text{slow}} N_2O(g) + H_2O(g) \quad \text{rate determining step}$$
$$\underline{N_2O(g) + H_2(g) \xrightarrow{\text{fast}} N_2(g) + H_2O(g)}$$

Overall $\qquad 2NO(g) + 2H_2(g) \rightarrow N_2(g) + 2H_2O(g)$

If the first step was the slowest step the the rate expression would be rate = $k[NO]^2$ and the rate would be zero order with respect to hydrogen. The rate for the second step depends on $[H_2]$ and $[N_2O_2]$. However, the concentration of N_2O_2 depends on the first step. So the rate expression for the second step becomes rate = $k[H_2][NO]^2$, which is consistent with the experimentally determined rate expression. The molecularity of the reaction is two, as two reacting species are involved in the rate determining step.

ARRHENIUS EQUATION

The rate constant for a reaction is only constant if the temperature remains constant. As the temperature increases the reactants possess more energy and the rate constant increases. The relationship between rate constant and absolute temperature is given by the Arrhenius equation:

$$k = Ae^{(-E_a/RT)}$$

where E_a is the activation energy and R is the gas constant. A is known as the Arrhenius constant and is related to the orientation of the reactants at the point of collision. This equation is often expressed in its logarithmic form:

$$\ln k = \frac{-E_a}{RT} + \ln A$$

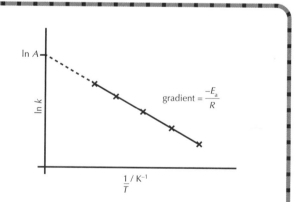

The equation can be used to determine both the Arrhenius constant and the activation energy for the reaction. This can be done either by substitution using simultaneous equations or by plotting $\ln k$ against $\frac{1}{T}$ to give a straight line graph. The gradient of the graph will be equal to $\frac{-E_a}{R}$ from which the activation energy can be calculated. Extrapolating the graph back to the $\ln k$ axis will give an intercept with a value equal to $\ln A$.

IB QUESTIONS – KINETICS

1. Which graph best represents the change in concentration of products with time for a reaction as it goes to completion?

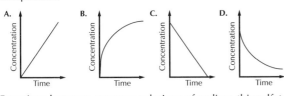

2. Reactions between aqueous solutions of sodium thiosulfate and acid can be followed by timing the appearance of the solid sulfur that is produced. The time required for the appearance of sulfur would be increased by which of the following changes?

   A. Raising the temperature

   B. Diluting the solution

   C. Adding a catalyst

   D. Increasing the concentration of the sodium thiosulfate

3. The reaction between excess calcium carbonate and hydrochloric acid can be followed by measuring the volume of carbon dioxide produced with time. The results or one such reaction are shown below. How does the rate of this reaction change with time and what is the main reason for this change?

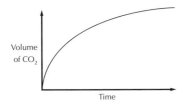

   A. The rate increases with time because the calcium carbonate particles get smaller.

   B. The rate increases with time because the acid becomes more dilute.

   C. The rate decreases with time because the calcium carbonate particles get smaller.

   D. The rate decreases with time because the acid becomes more dilute.

4. The rate of reaction of a strip of magnesium and 50 cm^3 of 1 mol dm^{-3} HCl is determined at 25 °C. In which case would **both** new conditions contribute to an increase in the rate of reaction?

   A. Mg powder and 100 cm^3 of 1 mol dm^{-3} HCl

   B. Mg powder and 50 cm^3 of 0.8 mol dm^{-3} HCl

   C. 100 cm^3 of 1 mol dm^{-3} HCl at 30 °C

   D. 50 cm^3 of 1.2 mol dm^{-3} HCl at 30 °C

5. Calcium carbonate and hydrochloric acid react according to the equation below:

   $$CaCO_3(s) + 2HCl(aq) \rightarrow CO_2(g) + CaCl_2(aq) + H_2O(l)$$

   Which conditions will produce the fastest rate of reaction?

   A. 1 mol dm^{-3} HCl and CaCO$_3$ pieces

   B. 2 mol dm^{-3} HCl and CaCO$_3$ pieces

   C. 1 mol dm^{-3} HCl and CaCO$_3$ powder

   D. 2 mol dm^{-3} HCl and CaCO$_3$ powder

6. When ammonia is manufactured commercially a catalyst is used. What is the effect of this catalyst?

   $$N_2(g) + 3H_2(g) \rightleftharpoons 2NH_3(g) \quad \Delta H = -92 \text{ kJ}$$

   I. To increase the rate of only the forward reaction

   II. To increase the rates of the forward and reverse reactions

   III. To shift the position of equilibrium and increase the yield of ammonia

   A. I only   C. III only

   B. II only   D. I and III only

7. The energy diagram for a particular reaction under catalysed and uncatalysed conditions is shown below:

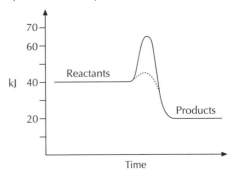

   What are the value and sign of the enthalpy change for the catalysed forward reaction?

   A. –55 kJ   B. –30 kJ   C. –25kJ   D. –20 kJ

8. Some collisions between reactant molecules do not form products. This is most likely because

   A. the molecules do not collide in the proper ratio.

   B. the molecules do not have enough energy.

   C. the concentration is too low.

   D. the reaction is at equilibrium.

9. Which graph shows the energy distribution for the particles of a gas when the temperature is **increased** from T$_1$ to T$_2$?

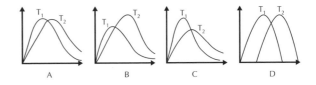

10. Which of the following is (are) important in determining whether a reaction occurs?

    I. Energy of molecules

    II. Orientation of the molecules

    A. I only

    B. II only

    C. Both I and II

    D. Neither I nor II

11. Doubling which one of the following will double the rate of a first order reaction?

    A. Concentrations of the reactant

    B. Size of solid particles

    C. Volume of solution in which the reaction is carried out

    D. Activation energy

12. The kinetic data in the table below are for the reaction:

    $A + B \rightarrow C$

    From these data what are the orders of the reaction with respect to A and B?

[A]	[B]	Initial Rate (mol dm^{-3} sec^{-1})
0.1	0.1	1×10^{-5}
0.1	0.2	4×10^{-5}
0.2	0.1	1×10^{-5}

    A. order of A = 1 order of B = 0

    B. order of A = 0 order of B = 2

    C. order of A = 0 order of B = 4

    D. order of A = 1 order of B = 2

13. The rate information below was obtained for the following reaction:

    $2NO_2(g) + F_2(g) \rightarrow 2NO_2F(g)$

[NO$_2$]/mol dm^{-3}	[F$_2$]/mol dm^{-3}	Rate/mol dm^{-3}s^{-1}
0.001	0.005	2×10^{-4}
0.002	0.005	4×10^{-4}
0.002	0.010	8×10^{-4}

    What are the orders for NO_2 and F_2?

    A. NO_2 is first order, F_2 is first order

    B. NO_2 is first order, F_2 is second order

    C. NO_2 is second order, F_2 is first order

    D. NO_2 is second order, F_2 is second order

14. Which factor(s) will influence the rate of the reaction shown below?

    $NO_2(g) + CO(g) \rightarrow NO(g) + CO_2(g)$

    I.   The number of collisions per second

    II. The energy of the collisions

    III. The geometry with which the molecules collide

    A. I only   **C.** I and II only

    B. II only   **D.** I, II, and III

15. The rate constant for a certain reaction has the units **concentration time** $^{-1}$. What is the order of reaction?

    A. 0   **B.** 1   **C.** 2   **D.** 3

16. Which reaction is an example of homogenous catalysis?

    A. $3H_2(g) + N_2(g) \xrightleftharpoons{\text{Fe(s)}} 2NH_3(g)$

    B. $2H_2O_2(aq) \xrightarrow{\text{MnO}_2\text{(s)}} 2H_2O(l) + O_2(g)$

    C. $2SO_2(g) + O_2(g) \xrightleftharpoons{\text{NO(g)}} 2SO_3(g)$

    D. $C_2H_4(g) + H_2(g) \xrightarrow{\text{Ni(s)}} C_2H_6(g)$

17. The rate-determining step of a reaction is the:

    A. fastest step in the reaction

    B. first step in the reaction

    C. final step in the reaction

    D. step with the highest activation energy.

The equilibrium law

DYNAMIC EQUILIBRIUM

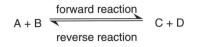

$$A + B \; \underset{\text{reverse reaction}}{\overset{\text{forward reaction}}{\rightleftharpoons}} \; C + D$$

Most chemical reactions do not go to completion. Once some products are formed the reverse reaction can take place to reform the reactants. In a closed system the concentrations of all the reactants and products will eventually become constant. Such a system is said to be in a state of **dynamic equilibrium**. The forward and reverse reactions continue to occur, but at equilibrium the rate of the forward reaction is equal to the rate of the reverse reaction.

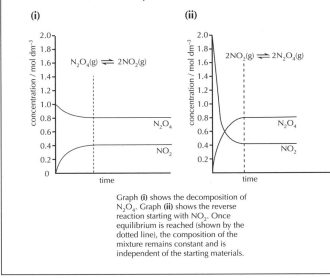

Graph **(i)** shows the decomposition of N_2O_4. Graph **(ii)** shows the reverse reaction starting with NO_2. Once equilibrium is reached (shown by the dotted line), the composition of the mixture remains constant and is independent of the starting materials.

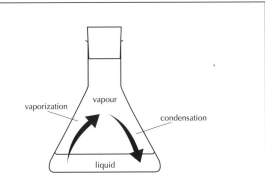

Dynamic equilibrium also occurs when physical changes take place. In a closed flask, containing some water, equilibrium will be reached between the liquid water and the water vapour. The faster moving molecules in the liquid will escape from the surface to become vapour and the slower moving molecules in the vapour will condense back into liquid. Equilibrium will be established when the rate of vaporization equals the rate of condensation.

$$H_2O(l) \rightleftharpoons H_2O(g)$$

CLOSED SYSTEM

A closed system is one in which neither matter nor energy can be lost or gained from the system, that is, the macroscopic properties remain constant. If the system is open some of the products from the reaction could escape and equilibrium would never be reached.

THE EQUILIBRIUM CONSTANT

Consider the following general reversible reaction in which w moles of A react with x moles of B to form y moles of C and z moles of D.

$$wA + xB \rightleftharpoons yC + zD$$

At equilibrium the concentrations of A, B, C, and D can be written as $[A]_{eqm}$, $[B]_{eqm}$, $[C]_{eqm}$, and $[D]_{eqm}$ respectively. The equilibrium law states that for this reaction at a particular temperature

$$K_c = \frac{[C]^y_{eqm} \times [D]^z_{eqm}}{[A]^w_{eqm} \times [B]^x_{eqm}}$$

where K_c is known as the equilibrium constant.

Examples Formation of sulfur trioxide in the Contact process

$$2SO_2(g) + O_2(g) \rightleftharpoons 2SO_3(g)$$

$$K_c = \frac{[SO_3]^2_{eqm}}{[SO_2]^2_{eqm} \times [O_2]_{eqm}}$$

Formation of an ester from ethanol and ethanoic acid

$$C_2H_5OH(l) + CH_3COOH(l) \rightleftharpoons CH_3COOC_2H_5(l) + H_2O(l)$$

$$K_c = \frac{[CH_3COOC_2H_5]_{eqm} \times [H_2O]_{eqm}}{[C_2H_5OH]_{eqm} \times [CH_3COOH]_{eqm}}$$

In both of these examples all the reactants and products are in the same phase. In the first example they are all in the gaseous phase and in the second example they are all in the liquid phase. Such reactions are known as **homogeneous reactions**. Another example of a homogeneous system would be where all the reactants and products are in the aqueous phase.

MAGNITUDE OF THE EQUILIBRIUM CONSTANT

Since the equilibrium expression has the concentration of products on the top and the concentration of reactants on the bottom it follows that the magnitude of the equilibrium constant is related to the position of equilibrium. When the reaction goes nearly to completion $K_c \gg 1$. If the reaction hardly proceeds then $K_c \ll 1$. If the value for K_c lies between about 10^{-2} and 10^2 then both reactants and products will be present in the system in noticeable amounts. The value for K_c in the esterification reaction above is 4 at 100 °C. From this it can be inferred that the concentration of the products present in the equilibrium mixture is roughly twice that of the reactants.

LE CHATELIER'S PRINCIPLE

Provided the temperature remains constant the value for K_c must remain constant. If the concentration of the reactants is increased, or one of the products is removed from the equilibrium mixture then more of the reactants must react in order to keep K_c constant, i.e the position of equilibrium will shift to the right (towards more products). This is the explanation for Le Chatelier's principle, which states that if a system at equilibrium is subjected to a small change the equilibrium tends to shift so as to minimize the effect of the change.

2. equilibrium shifts to minimize effect

$$A + B \rightleftharpoons C + D$$

1. remove product

Applications of the equilibrium law

FACTORS AFFECTING THE POSITION OF EQUILIBRIUM

Change in concentration

$$C_2H_5OH(l) + CH_3COOH(l) \rightleftharpoons CH_3COOC_2H_5(l) + H_2O(l)$$

If more ethanoic acid is added the concentration of ethanoic acid increases so that at the point of addition:

$$K_c \neq \frac{[ester] \times [water]}{[acid] \times [alcohol]}$$

To restore the system so that the equilibrium law is obeyed the equilibrium will move to the right, so that the concentration of ester and water increases and the concentration of the acid and alcohol decreases.

Change in pressure

If there is an overall volume change in a gaseous reaction then increasing the pressure will move the equilibrium towards the side with less volume. This shift reduces the total number of molecules in the equilibrium system and so tends to minimize the pressure.

$$2NO_2(g) \rightleftharpoons N_2O_4(g)$$
brown   colourless
(2 vols)   (1 vol)

If the pressure is increased the mixture will initially go darker as the concentration of NO_2 increases then become lighter as the position of equilibrium is re-established with a greater proportion of N_2O_4.

Change in temperature

In exothermic reactions heat is also a product. Taking the heat away will move the equilibrium towards the right, so more products are formed. The forward reaction in exothermic reactions is therefore increased by lowering the temperature

$$2NO_2(g) \rightleftharpoons N_2O_4(g) \qquad \Delta H^\ominus = -24 \text{ kJ mol}^{-1}$$
brown   colourless

Lowering the temperature will cause the mixture to become lighter as the equilibrium shifts to the right.

For an endothermic reaction the opposite will be true.

Unlike changing the concentration or pressure, a change in temperature will also change the value of K_c. For an exothermic reaction the concentration of the products in the equilibrium mixture decreases as the temperature increases, so the value of K_c will decrease. The opposite will be true for endothermic reactions.

e.g. $H_2(g) + CO_2(g) \rightleftharpoons H_2O(g) + CO(g) \; \Delta H^\ominus = +41 \text{ kJ mol}^{-1}$

T/K	K_c	
298	1.00×10^{-5}	
500	7.76×10^{-3}	increase
700	1.23×10^{-1}	
900	6.01×10^{-1}	

Adding a catalyst

A catalyst will increase the rate at which equilibrium is reached, as it will speed up both the forward and reverse reactions equally, but it will have no effect on the position of equilibrium and hence on the value of K_c.

APPLICATION OF EQUILIBRIUM AND KINETICS TO INDUSTRIAL PROCESSES

The aim in industry is to produce the highest possible yield of the required product in the shortest time for the least cost (both financial and to the environment) in order to maximize profits.

Haber process

Ammonia is used in the manufacture of fertilizers and in the production of nitric acid.

$$N_2(g) + 3H_2(g) \rightleftharpoons 2NH_3(g) \; \Delta H^\ominus = -92 \text{ kJ mol}^{-1}$$

The hydrogen is obtained from natural gas and the nitrogen from the fractional distillation of liquid air.

Conditions Four volumes of reactants produce two volumes of product, so a high pressure will be required. Increasing the pressure will also increase the number of particles per unit volume. This increases the rate at which equilibrium is reached. In practice a pressure of about 250 atm is used. Since it is an exothermic reaction a low temperature is required to give a high yield of ammonia. However, lowering the temperature will decrease the rate of reaction and it will take longer to reach equilibrium. What is required is the **optimum temperature** where the best compromise between yield and rate is reached. A temperature of about 450 °C is usually used. In order to increase the rate at which equilibrium is reached (but not the yield) an iron catalyst is used. It is used in a finely divided form (small pieces) so that the surface area is maximized to increase its efficiency. Even when all these conditions are in place the yield is only about 15%.

Contact process

Sulfuric acid is manufactured by the Contact process. It is the most industrially produced chemical, with over 150 million tonnes being produced world-wide every year. It is used for fertilizers, paints, detergents, and fibres, and as a feedstock for other chemicals.

$$2SO_2(g) + O_2(g) \rightleftharpoons 2SO_3(g) \; \Delta H^\ominus = -197 \text{ kJ mol}^{-1}$$

The sulfur dioxide is obtained from burning sulfur or sulfide ores, and the oxygen is obtained from the fractional distillation of liquid air.

Conditions Three volumes are converted into two volumes, so a high pressure will favour the production of sulfur trioxide. The reaction is exothermic so an optimum temperature, which is a compromise between yield and rate, is required. In practice a yield of more than 99% is obtained when the pressure is 2 atm at a temperature of 450 °C. The catalyst used is vanadium(V) oxide. Since the yield is so high at 2 atm pressure it is uneconomical and unnecessary to build the converter to withstand higher pressures.

HL Equilibrium calculations and phase equilibrium

UNITS OF THE EQUILIBRIUM CONSTANT

The units of K_c depend on the powers of the concentrations in the equilibrium expression.
Haber process: units of K_c: $dm^6\ mol^{-2}$

$$K_c = \frac{[NH_3]^2}{[H_2]^3 \times [N_2]} = \frac{concentration^2}{concentration^4} = concentration^{-2} = dm^6\ mol^{-2}$$

If they are the same on the top and bottom then K_c has no units.
Esterification: K_c: no units

$$K_c = \frac{[acid] \times [alcohol]}{[ester] \times [water]} = \frac{concentration^2}{concentration^2}$$

EQUILIBRIUM CALCULATIONS

The equilibrium law can be used either to find the value for the equilibrium constant, or to find the value of an unknown equilibrium concentration.

(a) 23.0 g (0.50 mol) of ethanol was reacted with 60.0 g (1.0 mol) of ethanoic acid and the reaction allowed to reach equilibrium at 373 K. 37.0 g (0.42 mol) of ethyl ethanoate was found to be present in the equilibrium mixture. Calculate K_c to the nearest integer at 373 K.

	$C_2H_5OH(l)$ +	$CH_3COOH(l)$ $\rightleftharpoons$	$CH_3COOC_2H_5(l)$ +	$H_2O(l)$
Initial amount / mol	0.50	1.00	–	–
Equilibrium amount / mol	(0.50 – 0.42)	(1.00 – 0.42)	0.42	0.42
Equilibrium concentration / mol dm^{-3}	$\dfrac{(0.50 - 0.42)}{V}$	$\dfrac{(1.00 - 0.42)}{V}$	$\dfrac{0.42}{V}$	$\dfrac{0.42}{V}$
(where V = total volume)				

$$K_c = \frac{[ester] \times [water]}{[alcohol] \times [acid]} = \frac{(0.42/V) \times (0.42/V)}{(0.08/V) \times (0.58/V)} = 4 \text{ (to the nearest integer)}$$

(b) What mass of ester will be formed at equilibrium if 2.0 moles of ethanoic acid and 1.0 moles of ethanol are reacted under the same conditions?

Let x moles of ester be formed and let the total volume be $V\ dm^3$.

$$K_c = 4 = \frac{[ester] \times [water]}{[alcohol] \times [acid]} = \frac{x^2/V^2}{(1.0 - x)/V \times (2.0 - x)/V} = \frac{x^2}{(x^2 - 3x + 2)}$$

$$\Rightarrow 3x^2 - 12x + 8 = 0$$

solve by substituting into the quadratic expression $\quad x = \dfrac{-b \pm \sqrt{b^2 - 4ac}}{2a} \Rightarrow x = \dfrac{12 \pm \sqrt{144 - 96}}{6}$

$x = 0.845$ or ~~3.15~~ (it cannot be 3.15 as only 1.0 mol of ethanol was taken)
Mass of ester = $0.845 \times 88.08 = 74.4$ g (Note: IB Diploma Programme chemistry does not examine the use of the quadratic expression.)

PHASE EQUILIBRIUM

Dynamic equilibrium between a liquid and its vapour occurs when the rate of vaporization is equal to the rate of condensation. The vapour pressure of a liquid is the pressure exerted by the particles in the vapour phase. It is independent of the surface area of the liquid or of the size of the container, although in a bigger container it may take longer for the equilibrium to become established. The vapour pressure of any liquid does depend both on the strength of the molecular forces holding the liquid particles together and on the temperature.

The stronger the intermolecular forces the lower the vapour pressure at a particular temperature. Vaporization is an endothermic process, as energy is absorbed to break these intermolecular forces. The enthalpy change required to overcome these forces is known as the enthalpy of vaporization. Water is a covalent substance with a low molar mass, but it has strong hydrogen bonding between its molecules. This explains why water has a relatively low vapour pressure and a relatively high enthalpy of vaporization.

As the temperature increases so does the number of particles with sufficient energy to overcome the attractive forces, and the vapour pressure also increases. A liquid boils when its vapour pressure is equal to the external pressure, as this allows bubbles of vapour to form in the body of the liquid. The boiling point of a liquid can be lowered simply by lowering the external pressure. This principle is useful to purify substances which decompose at or near their normal boiling point, by distilling them under reduced pressure. In mountainous regions where the external pressure is low the boiling point of water can be increased by using a pressure cooker.

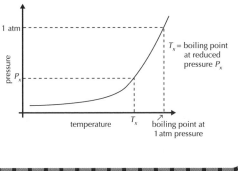

IB QUESTIONS – EQUILIBRIUM

1. Which statement is true about chemical reactions at equilibrium?

   A. The forward and backward reactions proceed at equal rates

   B. The forward and backward reactions have stopped

   C. The concentrations of the reactants and products are equal

   D. The forward reaction is exothermic

2. Chemical equilibrium is referred to as **dynamic** because, at equilibrium, the

   A. equilibrium constant changes.

   B. reactants and products keep reacting.

   C. rates of the forward and backward reactions change.

   D. concentrations of the reactants and products continue to change.

3. What is the equilibrium expression for the reaction: $N_2(g) + 3H_2(g) \rightleftharpoons 2NH_3(g)$?

   A. $K_c = \dfrac{[NH_3]}{[N_2][H_2]}$

   C. $K_c = \dfrac{2[NH_3]}{3[N_2][H_2]}$

   B. $K_c = \dfrac{2[NH_3]}{[N_2][H_2]}$

   D. $K_c = \dfrac{[NH_3]^2}{[N_2][H_2]^3}$

4. For a reaction which goes to completion, the equilibrium constant, K_c, is:

   A. $>>1$ B. $<<1$ C. $=1$ D. $=0$

5. The equilibrium constant for the reaction below is 1.0×10^{-14} at 25 °C and 2.1×10^{-14} at 35 °C. What can be concluded from this information?

   $2H_2O(l) \rightleftharpoons H_3O^+(aq) + OH^-(aq)$

   A. $[H_3O^+]$ decreases as the temperature is raised.

   B. $[H_3O^+]$ is greater than $[OH^-]$ at 35 °C.

   C. Water is a stronger electrolyte at 25 °C.

   D. The ionization of water is endothermic.

6. Ethanol is manufactured from ethene using the reaction below:

   $C_2H_4(g) + H_2O(g) \rightleftharpoons C_2H_5OH(g)$ $\Delta H = -46$ kJ

   Which conditions favour the highest yield of ethanol?

   A. High pressure and low temperature

   B. High pressure and high temperature

   C. Low pressure and low temperature

   D. Low pressure and high temperature

7. N_2O_4 and NO_2 produce an equilibrium mixture according to the equation below:

   $N_2O_4(g) \rightleftharpoons 2NO_2(g)$ $\Delta H = 57.2$ kJ mol^{-1}

   An increase in the equilibrium concentration of NO_2 can be produced by increasing which of the factors below?

   I. Pressure   II. Temperature

   A. I only   C. Both I and II

   B. II only   D. Neither I nor II

8. Which change(s) will increase the amount of $SO_3(g)$ at equilibrium?

   $2SO_2(g) + O_2(g) \rightleftharpoons 2SO_3(g)$ $\Delta H^\ominus = -197$ kJ

   I. Increasing the temperature

   II. Decreasing the volume

   III. Adding a catalyst

   A. I only   C. I and III only

   B. II only   D. I, II and III

9. For the reaction $N_2(g) + 3H_2(g) \rightleftharpoons 2NH_3(g)$ $\Delta H = -92$ kJ

   What conditions will produce the highest percentage of NH_3 at equilibrium?

   A. High pressure and high temperature

   B. High pressure and low temperature

   C. Low pressure and high temperature

   D. Low pressure and low temperature

---

10. The reaction between sulfur dioxide and oxygen occurs according to the equation below:

    $2SO_2(g) + O_2(g) \rightleftharpoons 2SO_3(g)$ $\Delta H^\ominus = -197$ kJ

    A higher equilibrium concentration of SO_3 will be produced by all of the following changes in reaction conditions EXCEPT

    A. increasing the pressure.   C. adding a catalyst.

    B. adding more O_2.   D. decreasing the temperature.

11. The reaction between methane and hydrogen sulfide is represented by the equation below:

    $CH_4(g) + 2H_2S(g) \rightleftharpoons CS_2(g) + 4H_2(g)$

    What is the equilibrium expression for this reaction?

    A. $[CS_2][H_2]/[CH_4][H_2S]$

    B. $4[CS_2][H_2]/2[CH_4][H_2S]$

    C. $[CS_2] + 4[H_2]/[CH_4] + 2[H_2S]$

    D. $[CS_2][H_2]^4/[CH_4][H_2S]^2$

12. At 35 °C $K_c = 1.6 \times 10^{-5}$ mol dm^{-3} for the reaction

    $2NOCl(g) \rightleftharpoons 2NO(g) + Cl_2(g)$

    Which relationship must be correct at equilibrium?

    A. $[NO] = [NOCl]$ C. $[NOCl]<[Cl_2]$

    B. $2[NO]=[Cl_2]$ D. $[NO]<[NOCl]$

13. Methanol can be produced from carbon monoxide and hydrogen according to the equation:

    $CO(g) + 2H_2(g) \rightleftharpoons CH_3OH(g)$

    In a certain equilibrium mixture, $[CO] = 0.2$ mol dm^{-3}, $[H_2] = 0.1$ mol dm^{-3}, $[CH_3OH] = 2$ mol dm^{-3}. What is the value of K_c (dm^6 mol^{-2}) for this reaction?

    A. 1×10^{-3} C. 1×10^{2}

    B. 1×10^{-2} D. 1×10^{3}

8 ACIDS AND BASES

Theories of acids and bases and salt hydrolysis

THE IONIC THEORY

An acid was originally distinguished by its sour taste. Later it was said to be the oxide of a non-metal combined with water although hydrochloric acid does not fit into this definition. The ionic theory which is still commonly used today states that an acid is a substance which produces hydrogen ions, $H^+(aq)$, in aqueous solution, e.g.

$$HCl(aq) \rightarrow H^+(aq) + Cl^-(aq)$$

In aqueous solution hydrogen ions are hydrated to form hydroxonium ions, $H_3O^+(aq)$. In the International Baccalaureate it is correct to write either $H^+(aq)$ or $H_3O^+(aq)$ to represent the hydrogen ions in an aqueous solution. Strictly speaking an acid gives a hydrogen ion concentration in aqueous solution greater than 1.0×10^{-7} mol dm^{-3}. A base is a substance that can neutralize an acid. An alkali is a base that is soluble in water.

BRØNSTED–LOWRY ACIDS AND BASES

A Brønsted–Lowry acid is a substance that can *donate* a proton. A Brønsted–Lowry base is a substance that can *accept* a proton.

Consider the reaction between hydrogen chloride gas and water.

$$HCl(g) + H_2O(l) \rightleftharpoons H_3O^+(aq) + Cl^-(aq)$$
$$\text{acid} \qquad \text{base} \qquad \text{acid} \qquad \text{base}$$

Under this definition both HCl and H_3O^+ are acids as both can donate a proton. Similarly both H_2O and Cl^- are bases as both can accept a proton. Cl^- is said to be the **conjugate base** of HCl and H_2O is the conjugate base of H_3O^+. The conjugate base of an acid is the species remaining after the acid has lost a proton. Every base also has a conjugate acid, which is the species formed after the base has accepted a proton. In the reaction with hydrogen chloride water is behaving as a base. Water can also behave as an acid.

$$NH_3(g) + H_2O(l) \rightleftharpoons NH_4^+(aq) + OH^-(aq)$$
$$\text{base} \qquad \text{acid} \qquad \text{acid} \qquad \text{base}$$

Substances such as water, which can act both as an acid and as a base, are described as **amphiprotic**.

Strong acids (see next page) have weak conjugate bases, whereas weak acids have strong conjugate bases.

Acid	Equilibrium constant	Conjugate base
ethanoic acid, CH_3COOH	1.8×10^{-5}	ethanoate ion, CH_3COO^-
phenol, C_6H_5OH	1.0×10^{-10}	phenoxide ion, $C_6H_5O^-$
water, H_2O	1.0×10^{-14}	hydroxide ion, OH^-
ethanol, C_2H_5OH	approx. 1×10^{-16}	ethoxide ion, $C_2H_5O^-$

acid strength ↑   base strength ↓

LEWIS ACIDS AND BASES

Brønsted–Lowry bases must contain a non-bonding pair of electrons to accept the proton. The Lewis definition takes this further and describes bases as substances which can *donate* a pair of electrons, and acids as substances which can *accept* a pair of electrons. In the process a co-ordinate (both electrons provided by one species) covalent bond is formed between the base and the acid.

The Lewis theory is all-embracing, so the term Lewis acid is usually reserved for substances which are not also Brønsted–Lowry acids. Many Lewis acids do not even contain hydrogen.

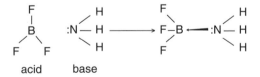

$$\text{acid} \qquad \text{base}$$

BF_3 is a good Lewis acid as there are only six electrons around the central boron atom which leaves room for two more. Other common Lewis acids are aluminium chloride, $AlCl_3$, and also transition metal ions in aqueous solution which can accept a pair of electrons from each of six surrounding water molecules, e.g. $[Fe(H_2O)_6]^{3+}$.

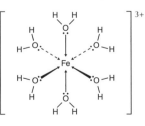

Properties of acids and bases

TYPICAL PROPERTIES OF ACIDS AND BASES

A simple definition of an acid is that it is a substance that produces H^+ ions in aqueous solution. A base is a substance that can neutralize an acid. An alkali is a base that is soluble in water.

The typical reactions of acids are:

1. With indicators.

Acid–base indicators can be used to determine whether or not a solution is acidic. Common indicators include:

Indicator	Colour in acidic solution	Colour in alkaline solution
litmus	red	blue
phenolphthalein	colourless	pink
methyl orange	red	yellow

2. Neutralization reactions with bases.

  (a) With hydroxides to form a salt and water,

    e.g. $CH_3COOH(aq) + NaOH(aq) \rightarrow NaCH_3COO(aq) + H_2O(l)$

  (b) With metal oxides to form a salt and water,

    e.g. $H_2SO_4(aq) + CuO(s) \rightarrow CuSO_4(aq) + H_2O(l)$

  (c) With ammonia to form a salt.

  e.g. $HCl(aq) + NH_3(aq) \rightarrow NH_4Cl(aq)$

3. With reactive metals (those above copper in the reactivity series) to form a salt and hydrogen, e.g.

  $2HCl(aq) + Mg(s) \rightarrow$
  $MgCl_2(aq) + H_2(g)$

4. With carbonates (soluble or insoluble) to form a salt, carbon dioxide, and water, e.g.

  $2HNO_3(aq) + Na_2CO_3(aq) \rightarrow$
  $2NaNO_3(aq) + CO_2(g) + H_2O(l)$
  $2HCl(aq) + CaCO_3(s) \rightarrow$
  $CaCl_2(aq) + CO_2(g) + H_2O(l)$

5. With hydrogencarbonates to form a salt, carbon dioxide, and water, e.g.

  $HCl(aq) + NaHCO_3(aq) \rightarrow$
  $NaCl(aq) + CO_2(g) + H_2O(aq)$

STRONG AND WEAK ACIDS AND BASES

A strong acid is completely dissociated (ionized) into its ions in aqueous solution. Similarly a strong base is completely dissociated into its ions in aqueous solution. Examples of strong acids and bases include:

Strong acids	Strong bases
hydrochloric acid, HCl	sodium hydroxide, NaOH
nitric acid, HNO_3	potassium hydroxide, KOH
sulfuric acid, H_2SO_4	barium hydroxide, $Ba(OH)_2$

Note: because one mole of HCl produces one mole of hydrogen ions it is known as a **monoprotic** acid. Sulfuric acid is known as a **diprotic** acid as one mole of sulfuric acid produces two moles of hydrogen ions.

Weak acids and bases are only slightly dissociated (ionized) into their ions in aqueous solution.

Weak acids	Weak bases
ethanoic acid, CH_3COOH	ammonia, NH_3
'carbonic acid' (CO_2 in water), H_2CO_3	aminoethane, $C_2H_5NH_2$

The difference can be seen in their reactions with water:

Strong acid: $HCl(g) + H_2O(l) \rightarrow H_3O^+(aq) + Cl^-(aq)$
      reaction goes to completion

Weak acid: $CH_3COOH(aq) + H_2O(l) \rightleftharpoons CH_3COO^-(aq) + H_3O^+(aq)$
      equilibrium lies on the left

i.e. a solution of hydrochloric acid consists only of hydrogen ions and chloride ions in water, whereas a solution of ethanoic acid contains mainly undissociated ethanoic acid with only very few hydrogen and ethanoate ions.

Strong base: $KOH(s) \xrightarrow{H_2O(l)} K^+(aq) + OH^-(aq)$

Weak base: $NH_3(g) + H_2O(l) \rightleftharpoons NH_4^+(aq) + OH^-(aq)$
      equilibrium lies on the left

EXPERIMENTS TO DISTINGUISH BETWEEN STRONG AND WEAK ACIDS AND BASES

1. pH measurement
Because a strong acid produces a higher concentration of hydrogen ions in solution than a weak acid, with the same concentration, the pH of a strong acid will be lower than a weak acid. Similarly a strong base will have a higher pH in solution than a weak base, with the same concentration. The most accurate way to determine the pH of a solution is to use a pH meter.

  0.10 mol dm^{-3} HCl(aq)   pH = 1.0
  0.10 mol dm^{-3} CH_3COOH pH = 2.9

2. Conductivity measurement
Strong acids and strong bases in solution will give much higher readings on a conductivity meter than **equimolar** (equal concentration) solutions of weak acids or bases, because they contain more ions in solution.

The pH scale

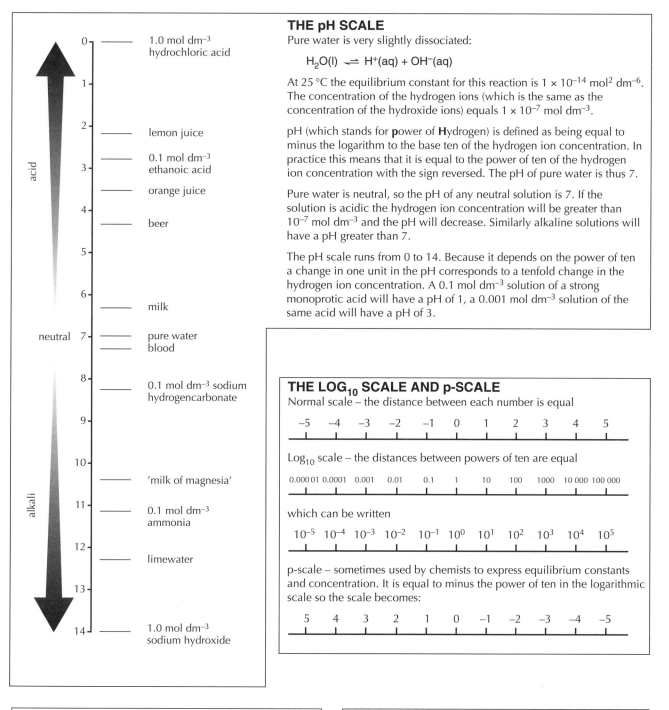

THE pH SCALE

Pure water is very slightly dissociated:

$$H_2O(l) \rightleftharpoons H^+(aq) + OH^-(aq)$$

At 25 °C the equilibrium constant for this reaction is 1×10^{-14} mol^2 dm^{-6}. The concentration of the hydrogen ions (which is the same as the concentration of the hydroxide ions) equals 1×10^{-7} mol dm^{-3}.

pH (which stands for **p**ower of **H**ydrogen) is defined as being equal to minus the logarithm to the base ten of the hydrogen ion concentration. In practice this means that it is equal to the power of ten of the hydrogen ion concentration with the sign reversed. The pH of pure water is thus 7.

Pure water is neutral, so the pH of any neutral solution is 7. If the solution is acidic the hydrogen ion concentration will be greater than 10^{-7} mol dm^{-3} and the pH will decrease. Similarly alkaline solutions will have a pH greater than 7.

The pH scale runs from 0 to 14. Because it depends on the power of ten a change in one unit in the pH corresponds to a tenfold change in the hydrogen ion concentration. A 0.1 mol dm^{-3} solution of a strong monoprotic acid will have a pH of 1, a 0.001 mol dm^{-3} solution of the same acid will have a pH of 3.

THE LOG$_{10}$ SCALE AND p-SCALE

Normal scale – the distance between each number is equal

−5 −4 −3 −2 −1 0 1 2 3 4 5

Log$_{10}$ scale – the distances between powers of ten are equal

0.000 01 0.0001 0.001 0.01 0.1 1 10 100 1000 10 000 100 000

which can be written

10^{-5} 10^{-4} 10^{-3} 10^{-2} 10^{-1} 10^0 10^1 10^2 10^3 10^4 10^5

p-scale – sometimes used by chemists to express equilibrium constants and concentration. It is equal to minus the power of ten in the logarithmic scale so the scale becomes:

5 4 3 2 1 0 −1 −2 −3 −4 −5

DETERMINATION OF pH

The pH of a solution can be determined by using a pH meter or by using 'universal' indicator, which contains a mixture of indicators which give a range of colours at different pH values.

pH	[H$^+$]/ mol dm^{-3}	[OH$^-$]/ mol dm^{-3}	Description	Colour of universal indicator
0	1	1×10^{-14}	very acidic	red
4	1×10^{-4}	1×10^{-10}	acidic	orange
7	1×10^{-7}	1×10^{-7}	neutral	green
10	1×10^{-10}	1×10^{-4}	basic	blue
14	1×10^{-14}	1	very basic	purple

STRONG, CONCENTRATED, AND CORROSIVE

In English the words strong and concentrated are often used interchangeably. In chemistry they have very precise meanings:

- **strong**: completely dissociated into ions
- **concentrated**: a high number of moles of solute per litre (dm^3) of solution
- **corrosive**: chemically reactive.

Similarly weak and dilute also have very different chemical meanings:

- **weak**: only slightly dissociated into ions
- **dilute**: a low number of moles of solute per litre of solution.

THE IONIC PRODUCT OF WATER

Pure water is very slightly ionized:

$$H_2O(l) \rightleftharpoons H^+(aq) + OH^-(aq) \qquad \Delta H^{\ominus} = +57.3 \text{ kJ mol}^{-1}$$

$$K_c = \frac{[H^+(aq)] \times [OH^-(aq)]}{[H_2O(l)]}$$

Since the equilibrium lies far to the left the concentration of water can be regarded as constant so

$$K_w = [H^+(aq)] \times [OH^-(aq)] = 1.00 \times 10^{-14} \text{ mol}^2 \text{ dm}^{-6} \text{ at 298 K, where } K_w \text{ is known as the ionic product of water.}$$

The dissociation of water into its ions is an endothermic process, so the value of K_w will increase as the temperature is increased.

Variation of K_w with temperature

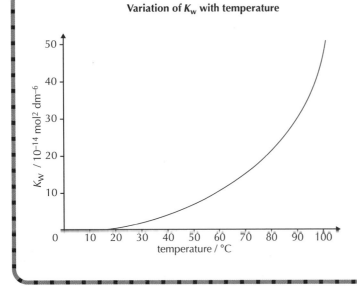

For pure water $[H^+(aq)] = [OH^-(aq)]$
$= 1.00 \times 10^{-7}$ mol dm^{-3} at 298 K

From the graph the value for $K_w = 1.00 \times 10^{-13}$ at 334 K (61 °C)

At this temperature $[H^+(aq)] = \sqrt{1.00 \times 10^{-13}}$
$= 3.16 \times 10^{-7}$ mol dm^{-3}

pH, pOH, AND pK_w

The pH of a solution depends only on the hydrogen ion concentration and is independent of the volume of solution.

$$\text{pH} = -\log_{10} [H^+]$$

Strong acids

For strong monoprotic acids $[H^+]$ will be equal to the concentration of the acid,

e.g. for 0.100 mol dm^{-3} HCl

$[H^+] = 0.100$ mol dm^{-3}
pH $= -\log_{10} 0.100 = 1.0$

For a strong diprotic acid the hydrogen ion concentration will be equal to twice the acid concentration,

e.g. for 0.025 mol dm^{-3} H$_2$SO$_4$

$[H^+] = 2 \times 0.025 = 0.050$ mol dm^{-3}
pH $= 1.3$

Note: it has been assumed that sulfuric acid is a strong acid. In reality the HSO$_4^-$ ion is only partially dissociated in aqueous solution.

pOH for strong bases

For a strong base the hydrogen ion concentration can be calculated using the ionic product of water,

e.g. for 1.00×10^{-3} mol dm^{-3} NaOH

$[OH^-] = 1.00 \times 10^{-3}$ mol dm^{-3}
$[H^+] \times [OH^-] = 1.00 \times 10^{-14}$

$$\Rightarrow [H^+] = \frac{1.00 \times 10^{-14}}{1.00 \times 10^{-3}} = 1.00 \times 10^{-11} \text{ mol dm}^{-3}$$

pH $= -\log_{10} 1.00 \times 10^{-11} = 11.0$

The pH of alkaline solutions can also be calculated by using pOH and the relationship between pOH, pH, and pK_w.

$$[H^+] \times [OH^-] = K_w \quad \text{pOH} = -\log_{10} [OH^-] \quad \text{p}K_w = -\log_{10} K_w = 14$$
$$\text{pH} + \text{pOH} = 14$$

e.g for 4.00×10^{-2} mol dm^{-3} Ba(OH)$_2$

$[OH^-] = 2 \times (4.00 \times 10^{-2}) = 8.00 \times 10^{-2}$ mol dm^{-3}
pOH $= 1.10$
$\Rightarrow$ pH $= 14 - 1.10 = 12.9$

Calculations with weak acids and bases

WEAK ACIDS

The dissociation of a weak acid HA in water can be written:

$$HA(aq) \rightleftharpoons H^+(aq) + A^-(aq)$$

The equilibrium expression for this reaction is:

$$K_a = \frac{[H^+] \times [A^-]}{[HA]}$$ where K_a is known as the acid dissociation constant

For example, to calculate the pH of 0.10 mol dm^{-3} CH$_3$COOH given that $K_a = 1.8 \times 10^{-5}$ mol dm^{-3} at 298 K:

$$CH_3COOH(aq) \rightleftharpoons CH_3COO^-(aq) + H^+(aq)$$

Initial concentration / mol dm^{-3}

0.10 — —

Equilibrium concentration / mol dm^{-3}

(0.10 − x) x x

$$K_a = \frac{[CH_3COO^-] \times [H^+]}{[CH_3COOH]} = \frac{x^2}{(0.10 - x)}$$
$$= 1.8 \times 10^{-5} \text{ mol dm}^{-3}$$
$$\Rightarrow x^2 + (1.8 \times 10^{-5}x) - 1.8 \times 10^{-6} = 0$$

by solving the quadratic equation

$$x = 1.33 \times 10^{-3} \text{ mol dm}^{-3}$$
$$pH = -\log_{10} 1.33 \times 10^{-3} = 2.88$$

If the acids are quite weak the equilibrium concentration of the acid can be assumed to be the same as its initial concentration. Provided the assumption is stated it is usual to simplify the expression in calculations to avoid a quadratic equation. In the above example:

$$K_a = \frac{[CH_3COO^-] \times [H^+]}{[CH_3COOH]} \approx \frac{[H^+]^2}{0.10}$$

$$= 1.8 \times 10^{-5} \text{ mol dm}^{-3}$$

$$\Rightarrow [H^+] = \sqrt{1.8 \times 10^{-6}} = 1.34 \times 10^{-3} \text{ mol dm}^{-3}$$
$$pH = 2.87$$

Examples of other weak acid calculations

1. The pH of a 0.020 mol dm^{-3} solution of a weak acid is 3.9. Find the K_a of the acid.

$$K_a = \frac{[H^+]^2}{(0.020 - [H^+])} \approx \frac{10^{-3.9} \times 10^{-3.9}}{0.020}$$
$$= 7.92 \times 10^{-7} \text{ mol dm}^{-3}$$

2. An acid whose K_a is 4.1×10^{-6} mol dm^{-3} has a pH of 4.5. Find the concentration of the acid.

$$[HA] = \frac{[H^+]^2}{K_a} = \frac{10^{-4.5} \times 10^{-4.5}}{4.1 \times 10^{-6}}$$
$$= 2.44 \times 10^{-4} \text{ mol dm}^{-3}$$

WEAK BASES

The reaction of a weak base can be written:

$$B(aq) + H_2O(l) \rightleftharpoons BH^+(aq) + OH^-(aq)$$

Since the concentration of water is constant:

$$K_b = \frac{[BH^+] \times [OH^-]}{[B]}$$

where K_b is the base dissociation constant

If one considers the reverse reaction of BH$^+$ acting as an acid to give B and H$^+$ then:

$$K_a = \frac{[B] \times [H^+]}{[BH^+]}$$

then

$$K_a \times K_b = \frac{[B] \times [H^+]}{[BH^+]} \times \frac{[BH^+] \times [OH^-]}{[B]}$$

$$= [H^+] \times [OH^-] = K_w$$

since $pK_a = -\log_{10} K_a$; $pK_b = -\log_{10} K_b$ and $pK_w = -\log_{10} K_w = 14$ this can also be expressed as:

$$pK_a + pK_b = 14$$

Examples of calculations

1. The K_b value for ammonia is 1.8×10^{-5} mol dm^{-3}. Find the pH of a 1.00×10^{-2} mol dm^{-3} solution.

Since [NH$_4^+$] = [OH$^-$] then

$$K_b = \frac{[OH^-]^2}{[NH_3]} \approx \frac{[OH^-]^2}{1.00 \times 10^{-2}} = 1.8 \times 10^{-5}$$

$$[OH^-] = \sqrt{1.8 \times 10^{-7}} = 4.24 \times 10^{-4} \text{ mol dm}^{-3}$$

$$\Rightarrow pOH = -\log_{10} 4.24 \times 10^{-4} = 3.37$$

$$\Rightarrow pH = 14 - 3.37 = 10.6$$

2. The pH of a 3.00×10^{-2} mol dm^{-3} solution of weak base is 10.0. Calculate the pK_b value of the base.

pH = 10.0 so pOH = 4.0

$$K_b = \frac{10^{-4} \times 10^{-4}}{3.00 \times 10^{-2}} = 3.33 \times 10^{-7} \text{ mol dm}^{-3}$$

$$\Rightarrow pK_b = 6.48$$

3. The IB data booklet value for the pK_a of methylamine (aminomethane) is 10.64. Calculate the concentration of a solution of methylamine with a pH of 10.8.

$$pK_b = 14 - 10.64 = 3.36; \quad pOH = 3.2$$

$$[CH_3NH_2] = \frac{[OH^-]^2}{K_b} = \frac{10^{-3.2} \times 10^{-3.2}}{10^{-3.36}}$$

$$= 9.13 \times 10^{-4} \text{ mol dm}^{-3}$$

(HL) Salt hydrolysis and buffer solutions

SALT HYDROLYSIS

Sodium chloride is neutral in aqueous solution. It is the salt of a strong acid and a strong base. Salts made from a weak acid and a strong base, such as sodium ethanoate, are alkaline in solution. This is because the ethanoate ions will combine with hydrogen ions from water to form mainly undissociated ethanoic acid, leaving excess hydroxide ions in solution.

$$NaCH_3COO(aq) \rightarrow Na^+(aq) \quad + \quad CH_3COO^-(aq)$$
$$+$$
$$H_2O(l) \rightleftharpoons OH^-(aq) \quad + \quad H^+(aq)$$
$$\text{strong base so completely dissociated} \quad CH_3COOH(aq)$$

Similarly salts derived from a strong acid and a weak base will be acidic in solution.

$$NH_4Cl(aq) \rightarrow NH_4^+(aq) + Cl^-(aq) \quad \text{strong acid}$$
$$+ \qquad\qquad\qquad\qquad \text{so completely}$$
$$H_2O(l) \rightleftharpoons OH^-(aq) + H^+(aq) \quad \text{dissociated}$$
$$NH_3(aq) + H_2O(l)$$

The acidity of salts also depends on the size and charge of the cation. Aluminium chloride reacts vigorously with water to give an acidic solution.

$$AlCl_3(s) + 3H_2O(l) \rightarrow Al(OH)_3(s) + 3HCl(aq)$$

The +3 charge is spread over a very small ion, which gives the Al^{3+} ion a very high charge density. The lone pair of one of the six water molecules surrounding the ion will be strongly attracted to the ion and the water molecule will lose a hydrogen ion in the process. This process will continue until aluminium hydroxide is formed. The equilibrium can be moved further to the right by adding $OH^-(aq)$ ions or back to the left by adding $H^+(aq)$ ions, which exemplifies the amphoteric nature of aluminium hydroxide. Similar examples of hydrolysis in aqueous solution occur with other small highly charged ions such as Fe^{3+}.

$$[Fe(H_2O)_6]^{3+} \underset{H^+}{\overset{-H^+}{\rightleftharpoons}} [Fe(H_2O)_5OH]^{2+} \underset{H^+}{\overset{-H^+}{\rightleftharpoons}}$$

$$[Fe(H_2O)_4(OH)_2]^+ \underset{H^+}{\overset{-H^+}{\rightleftharpoons}} Fe(H_2O)_3(OH)_3$$
$$OH^- \updownarrow H^+$$
$$[Fe(H_2O)_2(OH)_4]^-$$

Even $MgCl_2$ is slightly acidic in aqueous solution for the same reason.

	Charge	Ionic radius / nm	Aqueous solution
Na^+	+1	0.098	neutral
Mg^{2+}	+2	0.065	acidic
Al^{3+}	+3	0.045	acidic

BUFFER SOLUTIONS

A buffer solution resists changes in pH when small amounts of acid or alkali are added to it.

An acidic buffer solution can be made by mixing a weak acid together with the salt of that acid and a strong base. An example is a solution of ethanoic acid and sodium ethanoate. The weak acid is only slightly dissociated in solution, but the salt is fully dissociated into its ions, so the concentration of ethanoate ions is high.

$$NaCH_3COO(aq) \rightarrow Na^+(aq) + CH_3COO^-(aq)$$
$$CH_3COOH(aq) \rightleftharpoons CH_3COO^-(aq) + H^+(aq)$$

If an acid is added the extra H^+ ions coming from the acid are removed as they combine with ethanoate ions to form undissociated ethanoic acid, so the concentration of H^+ ions remains unaltered.

$$CH_3COO^-(aq) + H^+(aq) \rightleftharpoons CH_3COOH(aq)$$

If an alkali is added the hydroxide ions from the alkali are removed by their reaction with the undissociated acid to form water, so again the H^+ ion concentration stays constant.

$$CH_3COOH(aq) + OH^-(aq) \rightarrow CH_3COO^-(aq) + H_2O(l)$$

In practice acidic buffers are often made by taking a solution of a strong base and adding excess weak acid to it, so that the solution contains the salt and the unreacted weak acid.

$$NaOH(aq) + CH_3COOH(aq) \rightarrow NaCH_3COO(aq) + H_2O(l) + CH_3COOH(aq)$$
limiting reagent salt   excess weak acid

buffer solution

An alkali buffer with a fixed pH greater than 7 can be made from a weak base together with the salt of that base with a strong acid. An example is ammonia with ammonium chloride.

$$NH_4Cl(aq) \rightarrow NH_4^+(aq) + Cl^-(aq)$$
$$NH_3(aq) + H_2O(l) \rightleftharpoons NH_4^+(aq) + OH^-(aq)$$

If H^+ ions are added they will combine with OH^- ions to form water and more of the ammonia will dissociate to replace them. If more OH^- ions are added they will combine with ammonium ions to form undissociated ammonia. In both cases the hydroxide ion concentration and the hydrogen ion concentration remain constant.

Buffer calculations

BUFFER CALCULATIONS

The equilibrium expression for weak acids also applies to acidic buffer solutions.

In the buffer made from an aqueous solution of sodium ethanoate and ethanoic acid the salt is completely dissociated but the acid is only slightly dissociated.

$$NaCH_3COO(aq) \rightarrow Na^+(aq) + CH_3COO^-(aq)$$

$$CH_3COOH(aq) \rightleftharpoons CH_3COO^-(aq) + H^+(aq)$$

$$K_a = \frac{[H^+] \times [CH_3COO^-]}{[CH_3COOH]}$$

The essential difference is that now the concentrations of the two ions from the acid will not be equal. Since the sodium ethanoate is completely dissociated the concentration of the ethanoate ions in solution will be almost the same as the concentration of the sodium ethanoate as very little will come from the acid.

If logarithms to the base ten are taken then:

$$\log_{10}K_a = \log_{10}[H^+] + \log_{10}\frac{[CH_3COO^-]}{[CH_3COOH]}$$

or:

$$-\log_{10}[H^+] = -\log_{10}K_a + \log_{10}\frac{[CH_3COO^-]}{[CH_3COOH]}$$

Which gives:

$$pH = pK_a + \log_{10}\frac{[CH_3COO^-]}{[CH_3COOH]}$$

Two facts can be deduced from this expression. Firstly the pH of the buffer does not change on dilution as the concentration of the ethanoate ions and the acid will be affected equally. Secondly the buffer will be most efficient when $[CH_3COO^-] = [CH_3COOH]$. At this point, which equates to the half equivalence point when ethanoic acid is titrated with sodium hydroxide, the pH of the solution will equal the pK_a value of the acid.

WORKED EXAMPLES

(a) Calculate the pH of a buffer solution containing 0.2 mol of sodium ethanoate in 500 cm³ of 0.10 mol dm⁻³ ethanoic acid (given that K_a for ethanoic acid = 1.8 x 10-5 mol dm⁻³).

Step 1. Find the concentration of the ethanoate ion and the ethanoic acid.

$[CH_3COO^-]$ = 0.4 mol dm⁻³;
$[CH_3COOH]$ = 0.1 mol dm⁻³

Step 2. Make the assumptions that the equilibrium concentration of the acid is the same as the undissociated acid and that all the ethanoate ions come from the salt.

Step 3. Substitute the values into the expression for the acid dissociation constant.

$$K_a = \frac{[H^+] \times 0.4}{[CH_3COOH]}$$

$$= 1.8 \times 10^{-5} \text{ mol dm}^{-3}$$

Step 4. Calculate the value for [H⁺] and hence the pH

$[H^+] = 4.5 \times 10^{-6}$ mol dm⁻³

pH = $-\log_{10}(4.5 \times 10^{-6})$ = 5.4

(b) Calculate what mass of sodium propanoate must be dissolved in 1.0 dm³ of 1.0 mol dm³ propanoic acid (pK_a = 4.87) to give a buffer solution with a pH of 4.5.

Step 1. Determine the hydrogen ion concentration from the pH

$[H^+] = 10^{-4.5} = 3.16 \times 10^{-5}$ mol dm⁻³ (although this can be kept as $10^{-4.5}$)

Step 2. Assume the acid concentration in the buffer is the same as the undissociated acid.

Step 3. Use the expression for K_a to find the propanoate ion concentration.

$$[C_2H_5COO^-] = \frac{K_a \times [C_2H_5COOH]}{[H^+]}$$

$$= \frac{10^{-4.87} \times 1.0}{10^{-4.5}} = 0.427 \text{ mol dm}^{-3}$$

Step 4. Calculate the M_r for the salt and assume that all the propanoate ions come from the salt to determine the mass of salt required.

M_r (NaC₂H₅COO) = [23 + (3 x 12.01) + (5 x 1.01) + (2 x 16.00)] = 96.08

Mass of NaC₂H₅COO required = 0.427 x 96.08 = 41.0g

(c) Calculate the pH of the buffer formed when 18.0 cm³ of 0.100 mol dm⁻³ hydrochloric acid is added to 32.0 cm³ of 0.100 mol dm⁻³ ammonia solution. (pK_b for NH_3 = 4.75).

Step 1. Calculate the amount of the excess ammonia and the amount of salt formed.

$$HCl(aq) + NH_3(aq) \rightarrow NH_4^+(aq) + Cl^-(aq)$$

Initial amount of
HCl = $\frac{18.0 \times 0.100}{1000}$ = 1.80 x 10⁻³ mol

Initial amount of
NH₃ = $\frac{32 \times 0.100}{1000}$ = 3.20 x 10⁻³ mol

Amount of ammonium salt = 1.80 x 10⁻³ mol

Amount of excess ammonia
= 3.20 x 10⁻³ – 1.80 x 10⁻³ = 1.6 x 10⁻³ mol

Step 2. Calculate the equilibrium concentrations making the usual assumptions.

Since the combined volume = 50.0 cm³

$[NH_4^+]$ = 3.6 x 10⁻² mol dm⁻³ and $[NH_3]$ = 3.2 x 10⁻² mol dm⁻³

Step 3. Substitute into the expression for K_b

$$K_b = \frac{[NH_4^+][OH^-]}{[NH_3]}$$

$$= \frac{3.6 \times 10^{-2} \times [OH^-]}{3.2 \times 10^{-2}} = 10^{-4.75}$$

Step 4. Calculate [OH⁻] then pOH then pH

$[OH^-] = 0.889 \times 10^{-4.75} = 1.58 \times 10^{-5}$ mol dm⁻³

pOH = $-\log_{10} 1.58 \times 10^{-5}$ = 4.80

pH = 14 – pOH = 9.2

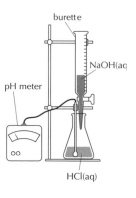

HL Titration curves and indicators

STRONG ACID – STRONG BASE TITRATION

The change in pH during an acid–base titration can be followed using a pH meter. Consider starting with 50 cm³ of 1.0 mol dm⁻³ hydrochloric acid. Since $[H^+(aq)] = 1.0$ mol dm⁻³ the initial pH will be 0. After 49 cm³ of 1.0 mol dm⁻³ NaOH have been added there will be 1.0 cm³ of the original 1.0 mol dm⁻³ hydrochloric acid left in 99 cm³ of solution. At this point $[H^+(aq)] \approx 1.0 \times 10^{-2}$ mol dm⁻³ so the pH = 2.

When 50 cm³ of the NaOH solution has been added the solution will be neutral and the pH will be 7. This is indicated by the point of inflexion, which is known as the equivalence point. It can be seen that there is a very large change in pH around the equivalence point. Almost all of the common acid–base indicators change colour (reach their end point) within this pH region. This means that it does not matter which indicator is used.

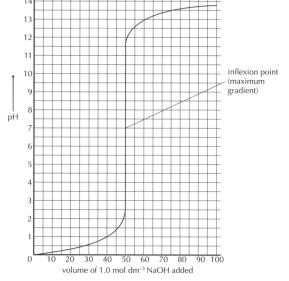

This curve shows what happens when 1.0 mol dm⁻³ sodium hydroxide is added to 50 cm³ of 1.0 mol dm⁻³ hydrochloric acid

$$NaOH(aq) + HCl(aq) \rightarrow NaCl(aq) + H_2O(l)$$

inflexion point (maximum gradient)

volume of 1.0 mol dm⁻³ NaOH added

WEAK ACID – STRONG BASE TITRATION

Consider titrating 50.0 cm³ of 1.0 mol dm⁻³ CH₃COOH with 1.0 mol dm⁻³ NaOH.

$K_a = 1.8 \times 10^{-5}$. Making the usual assumptions the initial $[H^+] = \sqrt{K_a \times [CH_3COOH]}$ and pH = 2.37.

When 49.0 cm³ of the 1.0 mol dm⁻³ NaOH has been added $[CH_3COO^-] \approx 0.05$ mol dm⁻³ and $[CH_3COOH] \approx 1.0 \times 10^{-2}$ mol dm⁻³.

$$[H^+] = \frac{K_a \times [CH_3COOH]}{[CH_3COO^-]} \approx \frac{1.8 \times 10^{-5} \times 1 \times 10^{-3}}{0.05}$$

$$= 3.6 \times 10^{-7} \text{ mol dm}^{-3} \text{ and pH} = 6.44$$

After the equivalence point the graph will follow the same pattern as the strong acid – strong base curve as more sodium hydroxide is simply being added to the solution.

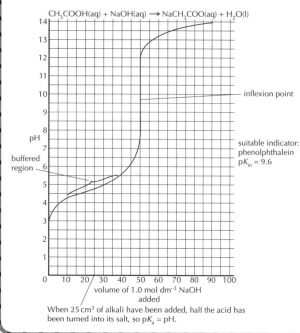

$$CH_3COOH(aq) + NaOH(aq) \rightarrow NaCH_3COO(aq) + H_2O(l)$$

inflexion point

suitable indicator: phenolphthalein $pK_{in} = 9.6$

buffered region

volume of 1.0 mol dm⁻³ NaOH added

When 25 cm³ of alkali have been added, half the acid has been turned into its salt, so pK_a = pH.

INDICATORS

An indicator is a weak acid (or base) in which the dissociated form is a different colour to the undissociated form.

$$\underset{\substack{\text{colour A} \\ \text{(colour in acid solution)}}}{HIn(aq)} \quad \rightleftharpoons \quad \underset{\substack{\text{colour B} \\ \text{(colour in alkali solution)}}}{H^+(aq) + In^-(aq)}$$

$K_{in} = [H^+] \times \dfrac{[In^-]}{[HIn]}$ Assuming the colour changes when $[In^-] \approx [HIn]$ then the end point of the indicator will be when $[H^+] \approx K_{in}$, i.e. when pH $\approx pK_{in}$. Different indicators have different K_{in} values and so change colour within different pH ranges.

Indicator	pK_{in}	pH range	Use
methyl orange	3.7	3.1–4.4	titrations with strong acids
phenolphthalein	9.6	8.3–10.0	titrations with strong bases

Similar arguments can be used to explain the shapes of pH curves for strong acid – weak base, and weak acid – weak base titrations. Since there is no sharp inflexion point titrations involving weak acids with weak bases should not be used in analytical chemistry.

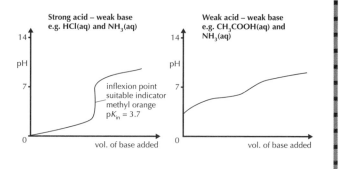

Strong acid – weak base e.g. HCl(aq) and NH₃(aq)

inflexion point suitable indicator methyl orange $pK_{in} = 3.7$

Weak acid – weak base e.g. CH₃COOH(aq) and NH₃(aq)

vol. of base added

IB QUESTIONS – ACIDS AND BASES

1. Which statement about hydrochloric acid is false?

   A. It can react with copper to give hydrogen

   B. It can react with sodium carbonate to give carbon dioxide

   C. It can react with ammonia to give ammonium chloride

   D. It can react with copper oxide to give water

2. $1.00 \ cm^3$ of a solution has a pH of 3. $100 \ cm^3$ of the same solution will have pH of:

   A. 1   C. 5

   B. 3   D. Impossible to calculate from the data given.

3. Which statement(s) is/are true about separate solutions of a strong acid and a weak acid both with the same concentration?

   I. They both have the same pH.

   II. They both have the same electrical conductivity.

   A. I and II   B. I only   C. II only   D. Neither I nor II

4. Identify the correct statement about $25 \ cm^3$ of a solution of $0.1 \ mol \ dm^{-3}$ ethanoic acid CH_3COOH.

   A. It will contain more hydrogen ions than $25 \ cm^3$ of $0.1 \ mol \ dm^{-3}$ hydochloric acid.

   B. It will have a pH greater than 7.

   C. It will react exactly with $25 \ cm^3$ of $0.1 \ mol \ dm^{-3}$ sodium hydroxide.

   D. It is completely dissociated into ethanoate and hydrogen ions in solution.

5. Which species cannot act as a Lewis acid?

   A. NH_3 B. BF_3 C. Fe^{2+} D. $AlCl_3$

6. $NH_3(aq)$, $HCl(aq)$, $NaOH(aq)$, $CH_3COOH(aq)$

   When $1.0 \ mol \ dm^{-3}$ solutions of the substances above are arranged in order of **decreasing** pH the order is:

   A. $NaOH(aq)$, $NH_3(aq)$, $CH_3COOH(aq)$, $HCl(aq)$

   B. $NH_3(aq)$, $NaOH(aq)$, $HCl(aq)$, $CH_3COOH(aq)$

   C. $CH_3COOH(aq)$, $HCl(aq)$, $NaOH(aq)$, $NH_3(aq)$

   D. $HCl(aq)$, $CH_3COOH(aq)$, $NH_3(aq)$, $NaOH(aq)$

7. A solution with a pH of 8.5 would be described as:

   A. very basic.   C. slightly acidic.

   B. slightly basic.   D. very acidic.

8. Which statement is true about two solutions one with a pH of 3 and the other with a pH of 6?

   A. The solution with a pH of 3 is twice as acidic as the solution with a pH of 6

   B. The solution with a pH of 6 is twice as acidic as the solution with a pH of 3

   C. The hydrogen ion concentration in the solution with a pH of 6 is one thousand times greater than that in the solution with a pH of 3

   D. The hydrogen ion concentration in the solution with a pH of 3 is one thousand times greater than that in the solution with a pH of 6

9. Which of the following is not a conjugate acid–base pair?

   A. HNO_3/NO_3^- C. NH_3/NH_2^-

   B. H_2SO_4/HSO_4^- D. H_3O^+/OH^-

10. During the titration of a known volume of a strong acid with a strong base:

    A. there is a steady increase in pH

    B. there is a sharp increase in pH around the end point

    C. there is a steady decrease in pH

    D. there is a sharp decrease in pH around the end point.

11. Three acids, HA, HB, and HC have the following K_a values

    $K_a(HA) = 1 \times 10^{-5}$ $K_a(HB) = 2 \times 10^{-5}$ $K_a(HC) = 1 \times 10^{-6}$

    What is the correct order of increasing acid strength (weakest first)?

    A. HA, HB, HC   C. HC, HA, HB

    B. HC, HB, HA   D. HB, HA, HC

12. Which of the following reagents could not be added together to make a buffer solution?

    A. $NaOH(aq)$ and $CH_3COOH(aq)$

    B. $NaCH_3COO(aq)$ and $CH_3COOH(aq)$

    C. $NaOH(aq)$ and $NaCH_3COO(aq)$

    D. $NH_4Cl(aq)$ and $NH_3(aq)$

13. When $1.0 \ cm^3$ of a weak acid solution is added to $100 \ cm^3$ of a buffer solution:

    A. the volume of the resulting mixture will be $100 \ cm^3$

    B. there will be almost no change in the pH of the solution

    C. the pH of the solution will increase noticeably

    D. the pH of the solution will decrease noticeably.

14. What is the pH of a buffer solution in which the concentration of the acid HX and the salt NaX are both $0.1 \ mol \ dm^{-3}$ ($K_a = 1 \times 10^{-5}$)?

    A. 3   B. 4   C. 5   D. 6

15. Which salt does not form an acidic solution in water?

    A. $MgCl_2$ B. Na_2CO_3 C. $FeCl_3$ D. NH_4NO_3

16. An indicator changes colour in the pH range 8.3–10.0. This indicator should be used when titrating a known volume of:

    A. a strong acid with a weak base

    B. a weak acid with a weak base

    C. a weak base with a strong acid

    D. a weak acid with a strong base.

Redox reactions (1)

DEFINITIONS OF OXIDATION AND REDUCTION

Oxidation used to be narrowly defined as the addition of oxygen to a substance. For example, when magnesium is burned in air the magnesium is oxidized to magnesium oxide.

$$2Mg(s) + O_2(g) \rightarrow 2MgO(s)$$

The electronic configuration of magnesium is 2.8.2. During the oxidation process it loses two electrons to form the Mg^{2+} ion with the electronic configuration of 2.8. **Oxidation** is now defined as the *loss of one or more electrons from a substance*. This is a much broader definition, as it does not necessarily involve oxygen. Bromide ions, for example, are oxidized by chlorine to form bromine.

$$2Br^-(aq) + Cl_2(aq) \rightarrow Br_2(aq) + 2Cl^-(aq)$$

If a substance loses electrons then something else must be gaining electrons. *The gain of one or more electrons* is called **reduction**. In the first example oxygen is reduced as it is gaining two electrons from magnesium to form the oxide ion O^{2-}. Similarly, in the second example chlorine is reduced as each chlorine atom gains one electron from a bromide ion to form a chloride ion.

Since the processes involve the transfer of electrons oxidation and reduction must occur simultaneously. Such reactions are known as **redox reactions**. In order to distinguish between the two processes half-equations are often used:

$2Mg(s) \rightarrow 2Mg^{2+}(s) + 4e^-$	— OXIDATION —	$2Br^-(aq) \rightarrow Br_2(aq) + 2e^-$
$O_2(g) + 4e^- \rightarrow 2O^{2-}(s)$	— REDUCTION —	$Cl_2(aq) + 2e^- \rightarrow 2Cl^-(aq)$
$2Mg(s) + O_2(g) \rightarrow 2MgO(s)$	OVERALL REDOX EQUATION	$2Br^-(aq) + Cl_2(aq) \rightarrow Br_2(aq) + 2Cl^-(aq)$

Understanding that magnesium must lose electrons and oxygen must gain electrons when magnesium oxide MgO is formed from its elements is a good way to remember the definitions of oxidation and reduction. Some students prefer to use the mnemonic OILRIG: **O**xidation **I**s the **L**oss of electrons, **R**eduction **I**s the **G**ain of electrons.

RULES FOR DETERMINING OXIDATION NUMBERS

It is not always easy to see how electrons have been transferred in redox processes. Oxidation numbers can be a useful tool to identify which species have been oxidized and which reduced. Oxidation numbers are assigned according to a set of rules:

1. In an ionic compound between two elements the oxidation number of each element is equal to the charge carried by the ion, e.g.

   $$Na^+Cl^- \qquad Ca^{2+}Cl^-_2$$
   $$(Na = +1; Cl = -1) \quad (Ca = +2; Cl = -1)$$

2. For covalent compounds assume the compound is ionic with the more electronegative element forming the negative ion, e.g.

   $$CCl_4 \qquad NH_3$$
   $$(C = +4; Cl = -1) \quad (N = -3; H = +1)$$

3. The algebraic sum of all the oxidation numbers in a compound = zero, e.g.

   $$CCl_4 [(+4) + 4 \times (-1) = 0];$$
   $$H_2SO_4 [2 \times (+1) + (+6) + 4 \times (-2) = 0]$$

4. The algebraic sum of all the oxidation numbers in an ion = the charge on the ion, e.g.

   $$SO_4^{2-} [(+6) + 4 \times (-2) = -2];$$
   $$MnO_4^- [(+7) + 4 \times (-2) = -1]; NH_4^+ [(-3) + 4 \times (+1) = +1]$$

5. Elements not combined with other elements have an oxidation number of zero, e.g. O_2; P_4; S_8.

6. Oxygen when combined always has an oxidation number of –2 except in peroxides (e.g. H_2O_2) when it is –1.

7. Hydrogen when combined always has an oxidation number of +1 except in certain metal hydrides (e.g. NaH) when it is –1.

Many elements can show different oxidation numbers in different compounds, e.g. nitrogen in:

NH_3	N_2H_4	N_2	N_2O	NO	NO_2	NO_3^-
(–3)	(–2)	(0)	(+1)	(+2)	(+4)	(+5)

When elements show more than one oxidation state the oxidation number is represented by using Roman numerals when naming the compound,

e.g. $FeCl_2$ iron(II) chloride; $FeCl_3$ iron(III) chloride
$K_2Cr_2O_7$ potassium dichromate(VI); $KMnO_4$ potassium manganate(VII)
Cu_2O copper(I) oxide; CuO copper(II) oxide.

Redox reactions (2)

OXIDATION AND REDUCTION IN TERMS OF OXIDATION NUMBERS

When an element is oxidized its oxidation number *increases*,

e.g. $Mg(s) \rightarrow Mg^{2+}(aq) + 2e^-$
  (0)   (+2)

When an element is reduced its oxidation number *decreases*,

e.g. $SO_4^{2-}(aq) + 2H^+(aq) + 2e^- \rightarrow SO_3^{2-}(aq) + H_2O(l)$
  (+6)   (+4)

The change in the oxidation number will be equal to the number of electrons involved in the half-equation.

Using oxidation numbers makes it easy to identify whether or not a reaction is a redox reaction.

Redox reactions (change in oxidation numbers)

$CuO(s) + H_2(g) \rightarrow Cu(s) + H_2O(l)$
(+2)   (0)   (0)   (+1)

$5Fe^{2+}(aq) + MnO_4^-(aq) + 8H^+(aq) \rightarrow Fe^{3+}(aq) + Mn^{2+}(aq) + 4H_2O(l)$
(+2)   (+7)   (+3)   (+2)

Not redox reactions (no change in oxidation numbers)

precipitation $Ag^+(aq) + Cl^-(aq) \rightarrow AgCl(s)$
        (+1)   (−1)   (+1) (−1)

neutralization $HCl(aq) + NaOH(aq) \rightarrow NaCl(aq) + H_2O(l)$
        (+1)(−1) (+1)(−2)(+1) (+1)(−1)   (+1)(−2)

Note: reactions where an element is uncombined on one side of the equation and combined on the other side *must* be redox reactions since there must be a change in oxidation number,

e.g. $Mg(s) + 2HCl(aq) \rightarrow MgCl_2(aq) + H_2(g)$

OXIDIZING AGENTS AND REDUCING AGENTS

A substance that readily oxidizes other substances is known as an **oxidizing agent**. Oxidizing agents are thus substances that readily accept electrons. Usually they contain elements that are in their highest oxidation state,

e.g. O_2, Cl_2, F_2, SO_3 (SO_4^{2-} in solution), MnO_4^-, and $Cr_2O_7^{2-}$.

Reducing agents readily donate electrons and include H_2, Na, C, CO, and SO_2 (SO_3^{2-} in solution),

e.g. $Cr_2O_7^{2-}(aq)$ + $3SO_3^{2-}(aq)$ $+ 8H^+(aq) \rightarrow 2Cr^{3+}(aq) + 3SO_4^{2-}(aq) + 4H_2O(l)$
    (+6)   (+4)   (+3)   (+6)
   (orange)   (green)
  (oxidizing agent)   (reducing agent)

        $Cl_2(aq)$ + $2Br^-(aq)$ $\rightarrow$ $2Cl^-(aq)$ + $Br_2(aq)$
         (0)   (−1)   (−1)   (0)
    (oxidizing agent)   (reducing agent)

Reactivity series

REACTIVITY

Lithium, sodium, and potassium all react with cold water to give similar products but the reactivity increases down the group.

$$2M(s) + 2H_2O(l) \rightarrow 2M^+(aq) + 2OH^-(aq) + H_2(g) \quad (M = Li, Na, or K)$$

Slightly less reactive metals react with steam and will give hydrogen with dilute acids, e.g.

$$Mg(s) + 2H_2O(g) \rightarrow Mg(OH)_2 + H_2(g)$$

$$Mg(s) + 2HCl(aq) \rightarrow Mg^{2+}(aq) + 2Cl^-(aq) + H_2(g)$$

In all of these reactions the metal is losing electrons – that is, it is being oxidized and in the process it is acting as a reducing agent. A reactivity series of reducing agents can be deduced by considering the reactivity of metals with water and acids, and the reactions of metals with the ions of other metals.

Reactivity series of reducing agents

Increasing reactivity

$$K(s) \rightleftharpoons e^- + K^+(aq)$$
$$Na(s) \rightleftharpoons e^- + Na^+(aq)$$
$$Li(s) \rightleftharpoons e^- + Li^+(aq)$$
$$Ca(s) \rightleftharpoons 2e^- + Ca^{2+}(aq)$$
$$Mg(s) \rightleftharpoons 2e^- + Mg^{2+}(aq)$$
$$Al(s) \rightleftharpoons 3e^- + Al^{3+}(aq)$$
$$Zn(s) \rightleftharpoons 2e^- + Zn^{2+}(aq)$$
$$Fe(s) \rightleftharpoons 2e^- + Fe^{2+}(aq)$$
$$Pb(s) \rightleftharpoons 2e^- + Pb^{2+}(aq)$$
$$\tfrac{1}{2}H_2(g) \rightleftharpoons e^- + H^+(aq)$$
$$Cu(s) \rightleftharpoons 2e^- + Cu^{2+}(aq)$$
$$Ag(s) \rightleftharpoons e^- + Ag^+(aq)$$

The more readily the metal loses its outer electrons the more reactive it is. Metals higher in the series can displace metal ions lower in the series from solution, e.g. zinc can react with copper ions to form zinc ions and precipitate copper metal.

$$Zn(s) + Cu^{2+}(aq) \rightarrow Zn^{2+}(aq) + Cu(s)$$

$$Zn(s) \rightarrow Zn^{2+}(aq) + 2e^- \qquad \text{Zn loses electrons in preference to Cu}$$
$$Cu^{2+}(aq) + 2e^- \rightarrow Cu(s) \qquad \text{Cu}^{2+} \text{ gains electrons in preference to Zn}^{2+}$$

This also explains why only metals above hydrogen can react with acids (displace hydrogen ions) to produce hydrogen gas, e.g.

$$Zn(s) + 2H^+(aq) \rightarrow Zn^{2+}(aq) + H_2(g)$$

The series can be extended for oxidizing agents. The most reactive oxidizing agent will be the species that gains electrons the most readily. For example, in group 7

$$I^-(aq) = e^- + \tfrac{1}{2}I_2(aq)$$
$$Br^-(aq) = e^- + \tfrac{1}{2}Br_2(aq)$$
$$Cl^-(aq) = e^- + \tfrac{1}{2}Cl_2(aq)$$
$$F^-(aq) = e^- + \tfrac{1}{2}F_2(aq)$$

increasing oxidizing ability

Oxidizing agents lower in the series gain electrons from species higher in the series, e.g.

$$Cl_2(aq) + 2Br^-(aq) \rightarrow 2Cl^-(aq) + Br_2(aq)$$

SIMPLE VOLTAIC CELLS

A half-cell is simply a metal in contact with an aqueous solution of its own ions. A voltaic cell consists of two different half-cells, connected together to enable the electrons transferred during the redox reaction to produce energy in the form of electricity. The cells are connected by an external wire and by a salt bridge, which allows the free movement of ions.

A good example of a voltaic cell is a zinc half-cell connected to a copper half-cell. Because zinc is higher in the reactivity series the electrons will flow from the zinc half-cell towards the copper half-cell. To complete the circuit and to keep the half-cells electrically neutral, ions will flow through the salt bridge. The voltage produced by a voltaic cell depends on the relative difference between the two metals in the reactivity series. Thus the voltage from a $Mg(s)/Mg^{2+}(aq)$ half-cell connected to a $Cu(s)/Cu^{2+}(aq)$ half-cell will be greater than that obtained from a $Zn(s)/Zn^{2+}(aq)$ half-cell connected to a $Fe(s)/Fe^{2+}(aq)$ half-cell.

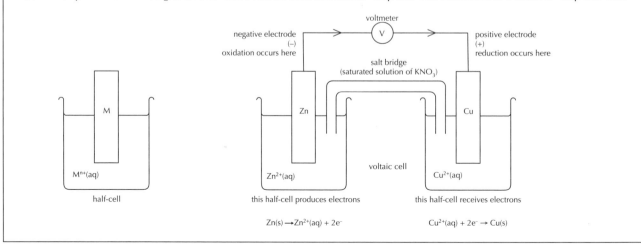

Electrolysis (1)

ELECTROLYTIC CELL

In a voltaic cell electricity is produced by the spontaneous redox reaction taking place. Electrolytic cells are used to make non-spontaneous redox reactions occur by providing energy in the form of electricity from an external source. In an electrolytic cell electricity is passed through an **electrolyte** and electrical energy is converted into chemical energy. An electrolyte is a substance which does not conduct electricity when solid, but does conduct electricity when molten or in aqueous solution and is chemically decomposed in the process. A simple example of an electrolytic cell is the electrolysis of molten sodium chloride.

During the electrolysis:
- sodium is formed at the negative electrode (cathode)
  $Na^+(l) + e^- \rightarrow Na(l)$ reduction

- chlorine is formed at the positive electrode (anode)
  $2Cl^-(l) \rightarrow Cl_2(g) + 2e^-$ oxidation

- The current is due to the movement of electrons in the external circuit and the movement of ions in the electrolyte.

Electrolysis is an important industrial process used to obtain reactive metals, such as sodium, from their common ores.

FACTORS AFFECTING THE DISCHARGE OF IONS DURING ELECTROLYSIS

During the electrolysis of molten salts there are only usually two ions present, so the cation will be discharged at the negative electrode (cathode) and the anion at the positive electrode (anode). However, for aqueous electrolytes there will also be hydrogen ions and hydroxide ions from the water present. There are three main factors which influence which ions will be discharged at their respective electrodes.

1. Position in the electrochemical series

The lower the metal ion is in the electrochemical series the more readily it will gain electrons (be reduced) to form the metal at the cathode. Thus in the electrolysis of a solution of sodium hydroxide, hydrogen will be evolved at the negative electrode in preference to sodium, whereas in a solution of copper(II) sulfate, copper will be deposited at the negative electrode in preference to hydrogen.

For negative ions the order of discharge follows $OH^- > Cl^- > SO_4^{2-}$.

2. Concentration

If one of the ions is much more concentrated than another ion then it may be preferentially discharged. For example, when electricity is passed through an aqueous solution of sodium chloride both oxygen and chlorine are evolved at the positive electrode. For dilute solutions mainly oxygen is evolved, but for concentrated solutions of sodium chloride more chlorine than oxygen is evolved.

3. The nature of the electrode

It is normally safe to assume that the electrode is inert, i.e. does not play any part in the reaction. However, if copper electrodes are used during the electrolysis of a solution of copper sulfate then the positive electrode is itself oxidized to release electrons and form copper(II) ions. Since copper is simultaneously deposited at the negative electrode the concentration of the solution will remain constant throughout the electrolysis.

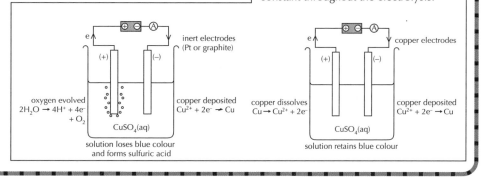

ELECTROPLATING

Electrolysis can also be used in industry to coat one metal with a thin layer of another metal. This process is known as **electroplating**.

For example, in copper plating the negative electrode (cathode) is made from the metal to be copper plated. It is placed into a solution of copper(II) sulfate together with a positive electrode (anode) made from a piece of copper. As electricity is passed through the solution the copper anode dissolves in the solution to form $Cu^{2+}(aq)$ ions and the $Cu^{2+}(aq)$ ions in solution are deposited onto the cathode. By making the anode of impure copper and the cathode from a small piece of pure copper this process can also be used to purify impure copper. This is an important industrial process as one of the main uses of copper is for electrical wiring, where purity is important since impure copper has a much higher electrical resistance.

ⓗ Electrolysis (2) and standard electrode potentials

FACTORS AFFECTING THE QUANTITY OF PRODUCTS DISCHARGED DURING ELECTROLYSIS

The amount of substance deposited will depend on:

1. The number of electrons flowing through the system, i.e. the charge passed. This in turn depends on the current and the time for which it flows. If the current is doubled then twice as many electrons pass through the system and twice as much product will be formed. Similarly if the time is doubled twice as many electrons will pass through the system and twice as much product will be formed.

   charge = current × time
   (1 coulomb = 1 ampere × 1 second)

2. The charge on the ion. To form one mole of sodium in the electrolysis of molten sodium chloride requires one mole of electrons to flow through the cell. However, the formation of one mole of lead during the electrolysis of molten lead(II) bromide requires two moles of electrons.

   $$Na^+(l) + e^- \rightarrow Na(l)$$
   $$Pb^{2+}(l) + 2e^- \rightarrow Pb(l)$$

   If cells are connected in series…

   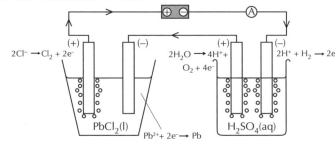

   molten lead(II) chloride   dilute sulphuric acid
   molar ratios of products evolved $2Cl_2 : 2Pb : O_2 : 2H_2$

   If cells are connected in series then the same amount of electricity will pass through both cells and the relative amounts of products obtained can be determined.

STANDARD ELECTRODE POTENTIALS

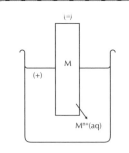

There are two opposing tendencies in a half-cell. The metal may dissolve in the solution of its own ions to leave the metal with a negative potential compared with the solution, or the metal ions may deposit on the metal, which will give the metal a positive potential compared with the solution. It is impossible to measure this potential, as any attempt to do so interferes with the system being investigated. However, the electrode potential of one half-cell can be compared against another half-cell. The hydrogen half-cell is normally used as the standard. Under standard conditions of 1 atm pressure, 298 K, and 1.0 mol dm⁻³ hydrogen ion concentration the standard electrode potential of the hydrogen electrode is assigned a value of zero volts.

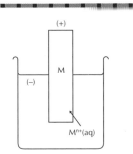

When the half-cell contains a metal above hydrogen in the reactivity series electrons flow from the half-cell to the hydrogen electrode, and the electrode potential is given a negative value. If the half-cell contains a metal below hydrogen in the reactivity series electrons flow from the hydrogen electrode to the half-cell, and the electrode potential has a positive value. The standard electrode potentials are arranged in increasing order to form the electrochemical series.

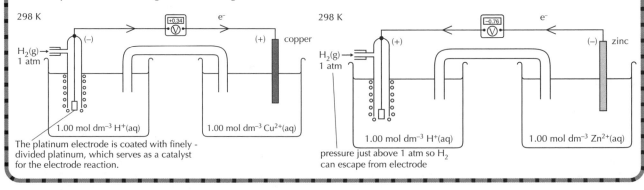

ELECTROCHEMICAL SERIES

(more complete series can be found in the IB data booklet)

Couple	$E^{\ominus}$ / V
$K(s)/K^+(aq)$	−2.92
$Ca(s)/Ca^{2+}(aq)$	−2.87
$Na(s)/Na^+(aq)$	−2.71
$Mg(s)/Mg^{2+}(aq)$	−2.36
$Al(s)/Al^{3+}(aq)$	−1.66
$Zn(s)/Zn^{2+}(aq)$	−0.76
$Fe(s)/Fe^{2+}(aq)$	−0.44
$\frac{1}{2}H_2(g)/H^+(aq)$	0.00
$Cu(s)/Cu^{2+}(aq)$	+0.34
$I^-(aq)/\frac{1}{2}I_2(aq)$	+0.62
$Ag(s)/Ag^+(aq)$	+0.80
$Br^-(aq)/\frac{1}{2}Br_2(aq)$	+1.09
$Cl^-(aq)/\frac{1}{2}Cl_2(aq)$	+1.36
$F^-(aq)/\frac{1}{2}F_2(aq)$	+2.87

SHORTHAND NOTATION FOR A CELL

To save drawing out the whole cell a shorthand notation has been adopted. A half-cell is denoted by a / between the metal and its ions, and two vertical lines are used to denote the salt bridge between the two half cells,

e.g. $Cu(s)/Cu^{2+}(aq) \,||\, H^+(aq)/H_2(g)$ and $Zn(s)/Zn^{2+}(aq) \,||\, H^+(aq)/H_2(g)$

The standard electromotive force (emf) of any cell $E^{\ominus}_{cell}$ is simply the difference between the standard electrode potentials of the two half-cells,

e.g. $Cu(s)/Cu^{2+}(aq) \,||\, Zn(s)/Zn^{2+}(aq)$

$E^{\ominus}$ + 0.34 V   −0.76 V

$E^{\ominus}_{cell}$ = 1.10 V

(Diagram: energy level scale showing Zn/Zn^{2+} at −0.76, $\frac{1}{2}H_2/H^+$ at 0, Cu/Cu^{2+} at +0.34, $E^{\ominus}$ / V, spanning 1.10 V)

ELECTRON FLOW AND SPONTANEOUS REACTIONS

By using standard electrode potentials it is easy to determine what will happen when two half-cells are connected together. The electrons will always flow from the more negative half-cell to the more positive half-cell, e.g. consider an iron half-cell connected to a magnesium half-cell:

$$\overset{e^-}{Fe(s)/Fe^{2+}(aq) \,||\, Mg^{2+}(aq)/Mg(s)}$$

$E^{\ominus}$ −0.44 V   −2.36 V

more positive   more negative

$Fe^{2+} + 2e^- \rightarrow Fe$ $Mg \rightarrow Mg^{2+} + 2e^-$

Spontaneous reaction: $Mg(s) + Fe^{2+}(aq) \rightarrow$
$Mg^{2+}(aq) + Fe(s)$ $E^{\ominus}_{cell}$ = 1.92 V

Positive $E^{\ominus}_{cell}$ values give negative $\Delta G^{\ominus}$ values as the reaction can provide electrical energy, i.e. do work, and the reaction is spontaneous. The reverse reaction ($Fe(s) + Mg^{2+}(aq) \rightarrow$ $Fe^{2+}(aq) + Mg(s)$) has a negative $E^{\ominus}_{cell}$ value which gives a positive value for $\Delta G^{\ominus}$ and the reaction will not be spontaneous, so can only proceed if an external voltage greater than 1.92 V is applied to force the reaction in the opposite direction. Note: just because a reaction has a positive $E^{\ominus}_{cell}$ value does not mean that it will necessarily proceed, as there may be a large activation energy which needs to be overcome.

REDOX EQUATIONS

Standard electrode potentials can be extended to cover any half-equation. The values can then be used to determine whether a particular redox reaction is spontaneous, e.g. $Cr_2O_7^{2-}(aq)/Cr^{3+}(aq)$, $E^{\ominus}$ = +1.33 V, but this only takes place in acid solution, so hydrogen ions and water are required to balance the half-equation. Consider the reaction between dichromate(VI) ions and sulfite ions SO_3^{2-}. Using standard electrode potentials the cell becomes:

$$\overset{e^-}{Cr_2O_7^{2-}(aq)/Cr^{3+}(aq) \,||\, SO_3^{2-}(aq)/SO_4^{2-}(aq)}$$

$E^{\ominus}$ +1.33 V   +0.17 V

more positive   more negative

$Cr_2O_7^{2-}$ will gain electrons SO_3^{2-} will lose electrons

$Cr_2O_7^{2-} + 14H^+ + 6e^- \rightarrow$ $SO_3^{2-} + H_2O \rightarrow$
$2Cr^{3+} + 7H_2O$ $SO_4^{2-} + 2H^+ + 2e^-$

Since the number of electrons transferred must be the same for both equations the overall equation is obtained by multiplying the sulfite equation by three, adding the two equations together, and simplifying the water and hydrogen ion components which appear on both sides.

$Cr_2O_7^{2-}(aq) + 8H^+(aq) + 3SO_3^{2-}(aq) \rightarrow$
$2Cr^{3+}(aq) + 3SO_4^{2-}(aq) + 4H_2O(l)$ $E^{\ominus}_{cell}$ = 1.16 V

CONVENTION FOR WRITING CELLS

By convention in a cell diagram the half-cell undergoing oxidation is placed on the left of the diagram and the half-cell undergoing reduction on the right of the diagram. The two aqueous solutions are then placed either side of the salt bridge e.g.

$Zn(s)/Zn^{2+}(aq) \,||\, Cu^{2+}(aq)/Cu(s)$

Some text books then state that:

$E^{\ominus}_{cell} = E^{\ominus}_{\text{right hand side}} - E^{\ominus}_{\text{left hand side}}$

This can cause much confusion. Essentially a cell consists of two half-cells connected by a salt bridge and it makes no difference which is placed on the left or on the right. Provided you remember that the electron flow is always *from* the half-cell with the more negative electrode potential *to* the half-cell with the more positive electrode potential the convention can be safely ignored.

IB QUESTIONS – OXIDATION AND REDUCTION

1. $5Fe^{2+}(aq) + MnO_4^-(aq) + 8H^+(aq) \rightarrow 5Fe^{3+}(aq) + Mn^{2+}(aq) + 4H_2O(l)$

   In the equation above:

   A. $Fe^{2+}(aq)$ is the oxidizing agent

   B. H^+ (aq) ions are reduced

   C. $Fe^{2+}(aq)$ ions are oxidized

   D. $MnO_4^-(aq)$ is the reducing agent

2. The oxidation numbers of nitrogen in NH_3, HNO_3, and NO_2 are, respectively

   A. $-3, -5, +4$ C. $-3, +5, -4$

   B. $+3, +5, +4$ D. $-3, +5, +4$

3. Which one of the following reactions is **not** a redox reaction?

   A. $Ag^+(aq) + Cl^-(aq) \rightarrow AgCl(s)$

   B. $2Na(s) + Cl_2(g) \rightarrow 2NaCl(s)$

   C. $Mg(s) + 2HCl(aq) \rightarrow MgCl_2(aq) + H_2(g)$

   D. $Cu^{2+}(aq) + Zn(s) \rightarrow Cu(s) + Zn^{2+}(aq)$

4. Which substance does not have the correct formula?

   A. iron(III) sulfate $Fe_2(SO_4)_3$

   B. iron(II) oxide Fe_2O

   C. copper(I) sulfate Cu_2SO_4

   D. copper(II) nitrate $Cu(NO_3)_2$

5. For which conversion is an oxidising agent required?

   A. $2H^+(aq) \rightarrow H_2(g)$ C. $SO_3(g) \rightarrow SO_4^{2-}(aq)$

   B. $2Br^-(aq) \rightarrow Br_2(aq)$ D. $MnO_2(s) \rightarrow Mn^{2+}(aq)$

6. $Mn(s) + Hg^{2+}(aq) \rightarrow Mn^{2+}(aq) + Hg(l)$

   $Ni^{2+}(aq) + Mn(s) \rightarrow Ni(s) + Mn^{2+}(aq)$

   $Hg^{2+}(aq) + Ni(s) \rightarrow Hg(l) + Ni^{2+}(aq)$

   From the above reactions the **increasing** reactivity of the three metals is:

   A. Hg, Ni, Mn   C. Ni, Mn, Hg

   B. Mn, Hg, Ni   D. Mn, Ni, Hg

7. When an $Fe(s)/Fe^{2+}(aq)$ half-cell is connected to a $Cu(s)/Cu^{2+}(aq)$ half-cell by a salt bridge and a current allowed to flow between them

   A. the electrons will flow from the copper to the iron.

   B. the salt bridge allows the flow of ions to complete the circuit.

   C. the salt bridge allows the flow of electrons to complete the circuit.

   D. the salt bridge can be made of copper or iron.

8. During the electrolysis of molten sodium chloride using platinum electrodes

   A. sodium is formed at the negative electrode

   B. chlorine is formed at the negative electrode

   C. sodium is formed at the positive electrode

   D. oxygen is formed at the postitive electrode

9. Which statement is true?

   A. Lead chloride is ionic so solid lead chloride will conduct electricity.

   B. When a molten ionic compound conducts electricity free electrons pass through the liquid.

   C. When liquid mercury conducts electricity mercury ions move towards the negative electrode.

   D. During the electrolysis of a molten salt reduction will always occur at the negative electrode.

10. The following information is given about reactions involving the metals X, Y and Z and solutions of their sulfates.

    $X(s) / YSO_4(aq) \rightarrow$ no reaction

    $Z(s) + YSO_4(aq) \rightarrow Y(s) + ZSO_4(aq)$

    When the metals are listed in decreasing order of reactivity (most reactive first), what is the correct order?

    A. $Z > Y > X$

    B. $X > Y > Z$

    C. $Y > X > Z$

    D. $Y > Z > X$

HL

---

11. Ethanol can be oxidised to ethanal by an acidic solution of dichromate(VI) ions.

    $\_C_2H_5OH(aq) + \_H^+(aq) + \_Cr_2O_7^{2-}(aq) \rightarrow$
    $\_CH_3CHO(aq) + \_Cr^{3+}(aq) + \_H_2O(l)$

    The sum of all the co-efficients in the balanced equation is:

    A. 24   B. 26   C. 28   D. 30

12. Which statement(s) is/are true about the standard hydrogen electrode?

    I. The electrode potential is assigned a value of 0.00 volts.

    II. The hydrogen ion concentration is 0.100 mol dm^{-3}.

    III. The temperature is 273 K.

    A. I, II and III   C. I only

    B. I and II only   D. II and III only

Use the following information to answer questions 13 and 14.

$Sn^{2+}(aq) + 2e^- \rightleftharpoons Sn(s)$ $E^{\ominus} = -0.14\,V$

$Sn^{4+}(aq) + 2e^- \rightleftharpoons Sn^{2+}(aq)$ $E^{\ominus} = +0.15\,V$

$Fe^{2+}(aq) + 2e^- \rightleftharpoons Fe(s)$ $E^{\ominus} = -0.44\,V$

$Fe^{3+}(aq) + e^- \rightleftharpoons Fe^{2+}(aq)$ $E^{\ominus} = +0.77\,V$

13. Under standard conditions which statement is correct?

    A. $Sn^{2+}(aq)$ can reduce $Fe^{3+}(aq)$.

    B. $Fe(s)$ can oxidise $Sn^{2+}(aq)$.

    C. $Sn(s)$ can reduce $Fe(s)$.

    D. $Fe^{3+}(aq)$ can reduce $Sn^{4+}(aq)$.

14. When a half-cell of $Fe^{2+}(aq)/Fe^{3+}(aq)$ is connected by a salt bridge to a half-cell of $Sn^{2+}(aq)/Sn^{4+}(aq)$ under standard conditions and a current allowed to flow in an external circuit the total e.m.f. of the spontaneous reaction will be:

    A. $+0.92\,V$ B. $-0.92\,V$ C. $+0.62\,V$ D. $-0.62\,V$

Functional groups and homologous series

NAMING ORGANIC COMPOUNDS

Organic chemistry is concerned with the compounds of carbon. Since there are more compounds of carbon known than all the other elements put together, it is helpful to have a systematic way of naming them.

1. Identify the longest carbon chain.
   - 1 carbon = **meth-**
   - 2 carbons = **eth-**
   - 3 carbons = **prop-**
   - 4 carbons = **but-**
   - 5 carbons = **pent-**
   - 6 carbons = **hex-**
   - 7 carbons = **hept-**
   - 8 carbons = **oct-**

2. Identify the type of bonding in the chain or ring
   All single bonds in the carbon chain = **-an-**
   One double bond in the carbon chain = **-en-**
   One triple bond in the carbon chain = **-yn-**

3. Identify the functional group joined to the chain or ring.

This may come at the beginning or at the end of the name, e.g.

alkane: only hydrogen (-H) joined to chain = **-e**
alcohol: –OH = **-ol**
amine: –NH₂ = **amino-**

halogenoalkane: -X: **chloro-, bromo, or iodo-**

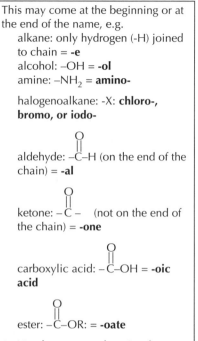

aldehyde: –C–H (on the end of the chain) = **-al**

ketone: –C – (not on the end of the chain) = **-one**

carboxylic acid: –C–OH = **-oic acid**

ester: –C–OR: = **-oate**

4. Numbers are used to give the positions of groups or bonds along the chain.

HOMOLOGOUS SERIES

The alkanes form a series of compounds all with the general formula C_nH_{2n+2}, e.g.

methane CH_4

ethane C_2H_6

propane C_3H_8

butane C_4H_{10}

If one of the hydrogen atoms is removed what is left is known as an alkyl radical R – (e.g methyl CH_3–; ethyl C_2H_5–). When other atoms or groups are attached to an alkyl radical they can form a different series of compounds. These atoms or groups attached are known as functional groups and the series formed are all homologous series.

Homologous series have the same general formula with the neighbouring members of the series differing by –CH_2; for example the general formula of alcohols is $C_nH_{2n+1}OH$. The chemical properties of the individual members of an homologous series are similar and they show a gradual change in physical properties.

SOME COMMON FUNCTIONAL GROUPS

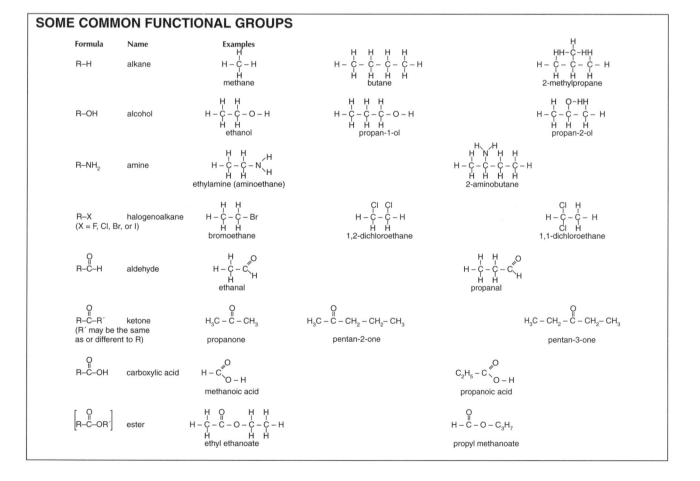

Properties of different functional groups

BOILING POINTS

As the carbon chain gets longer the mass of the molecules increases and the van der Waals' forces of attraction increase. A plot of boiling point against number of carbon atoms shows a sharp increase at first, as the percentage increase in mass is high, but as successive $-CH_2-$ groups are added the rate of increase in boiling point decreases.

When branching occurs the molecules become more spherical in shape, which reduces the contact surface area between them and lowers the boiling point.

Other homologous series show similar trends but the actual temperatures at which the compounds boil will depend on the types of attractive forces between the molecules. The volatility of the compounds also follows the same pattern. The lower members of the alkanes are all gases as the attractive forces are weak and the next few members are volatile liquids. Methanol, the first member of the alcohols is a liquid at room temperature, due to the presence of hydrogen bonding. Methanol is classed as volatile as its boiling point is 64.5 °C but when there are four or more carbon atoms in the chain the boiling points exceed 100 °C and the higher alcohols have low volatility.

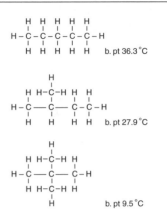

Compound	Formula	M_r	Functional group	Strongest type of attraction	B. pt / °C
butane	C_4H_{10}	58	alkane	van der Waals'	−0.5
butene	C_4H_8	56	alkene	van der Waals'	−6.2
butyne	C_4H_6	54	alkyne	van der Waals'	8.1
methyl methanoate	$HCOOCH_3$	60	ester	dipole:dipole	31.5
propanal	CH_3CH_2CHO	58	aldehyde	dipole:dipole	48.8
propanone	CH_3COCH_3	58	ketone	dipole:dipole	56.2
aminopropane	$CH_3CH_2CH_2NH_2$	59	amine	hydrogen bonding	48.6
propan-1-ol	$CH_3CH_2CH_2OH$	60	alcohol	hydrogen bonding	97.2
ethanoic acid	CH_3COOH	60	carboxylic acid	hydrogen bonding	118

SOLUBILITY IN WATER

Whether or not an organic compound will be soluble in water depends on the polarity of the functional group and on the chain length. The lower members of alcohols, amines, aldehydes, ketones, and carboxylic acids are all water soluble. However, as the length of the non-polar hydrocarbon chain increases the solubility in water decreases. For example, ethanol and water mix in all proportions, but hexan-1-ol is only slightly soluble in water. Compounds with non-polar functional groups, such as alkanes, and alkenes, do not dissolve in water but are soluble in other non-polar solvents. Propan-1-ol is a good solvent because it contains both polar and non-polar groups and can to some extent dissolve both polar and non-polar substances.

STRUCTURAL FORMULAS

The difference between the empirical, molecular and structural formulas of a compound has been covered in Topic 1 - quantitative chemistry. Because the physical and chemical properties of organic compounds are determined by the functional group and the arrangement of carbon atoms within the molecule, the structural formulas for organic compounds are often used.

The structural formula unambiguously shows how the atoms are bonded together. All the hydrogen atoms must be shown when drawing organic structures. The skeletal formula showing just the carbon atoms without the hydrogen atoms is not acceptable except for benzene (see below). However, unless specifically asked, Lewis structures showing all the valence electrons are not necessary. The bonding must be clearly indicated. Structures may be shown using lines as bonds or in their shortened form e.g. $CH_3CH_2CH_2CH_2CH_3$ or $CH_3-(CH_2)_3-CH_3$ for pentane but the molecular formula C_5H_{12} will not suffice.

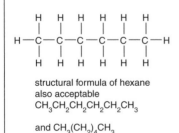

structural formula of hexane
also acceptable
$CH_3CH_2CH_2CH_2CH_2CH_3$

and $CH_3(CH_2)_4CH_3$

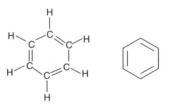

skeletal formula of hexane
not acceptable as structural formula

three different ways of showing the structural formula of benzene, all are acceptable

Structural isomers

STRUCTURES OF HYDROCARBONS

Isomers of alkanes
Each carbon atom contains four single bonds. There is only one possible structure for each of methane, ethane, and propane however two structures of butane are possible.

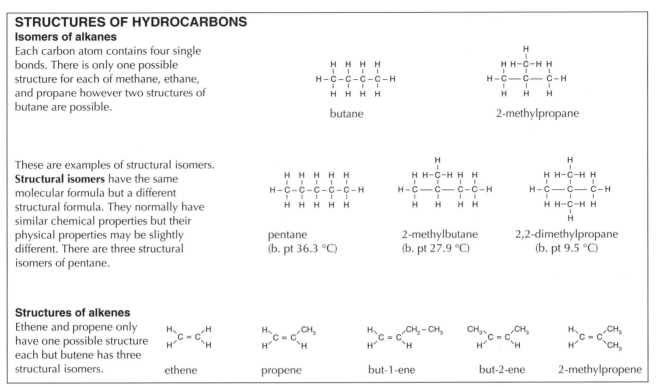

butane

2-methylpropane

These are examples of structural isomers. **Structural isomers** have the same molecular formula but a different structural formula. They normally have similar chemical properties but their physical properties may be slightly different. There are three structural isomers of pentane.

pentane
(b. pt 36.3 °C)

2-methylbutane
(b. pt 27.9 °C)

2,2-dimethylpropane
(b. pt 9.5 °C)

Structures of alkenes
Ethene and propene only have one possible structure each but butene has three structural isomers.

ethene

propene

but-1-ene

but-2-ene

2-methylpropene

CLASSIFICATION OF ALCOHOLS AND HALOGENOALKANES

Alcohols and halogenoalkanes may be classified according to how many R- groups and how many hydrogen atoms are bonded to the carbon atom containing the functional group.

primary (on R-group bonded to C atom)

secondary (two R-group bonded to C atom)
R may be the same as R' or different

Tertiary (three R-group bonded to C atom)

NAMING STRUCTURAL ISOMERS

The naming system explained on page 61 is known as the IUPAC (International Union of Pure and Applied Chemistry) system. The IUPAC names to distinguish between structural isomers of alcohols, aldehydes, ketones, carboxylic acids and halogenoalkanes containing up to six carbon atoms are required.

For example, four different structural isomers with the molecular formula $C_6H_{12}O$ are shown.

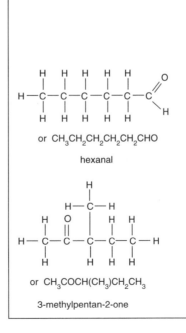

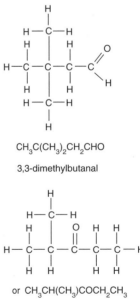

or $CH_3CH_2CH_2CH_2CH_2CHO$

hexanal

$CH_3C(CH_3)_2CH_2CHO$

3,3-dimethylbutanal

or $CH_3COCH(CH_3)CH_2CH_3$

3-methylpentan-2-one

or $CH_3CH(CH_3)COCH_2CH_3$

2-methylpentan-3-one

Alkanes

LOW REACTIVITY OF ALKANES

Because of the relatively strong C–C and C–H bonds and because they have low polarity, alkanes tend to be quite unreactive. They only readily undergo combustion reactions with oxygen and substitution reactions with halogens in ultraviolet light.

COMBUSTION

Alkanes are hydrocarbons - compounds that contain carbon and hydrogen only. All hydrocarbons burn in a plentiful supply of oxygen to give carbon dioxide and water. The general equation for the combustion of any hydrocarbon is:

$$C_xH_y + (x + \frac{y}{4})O_2 \rightarrow xCO_2 + \frac{y}{2}H_2O$$

Although the C–C and C–H bonds are strong the C=O and O–H bonds in the products are even stronger so the reaction is very exothermic and much use is made of the alkanes as fuels.

e.g natural gas (methane)

$$CH_4(g) + 2O_2(g) \rightarrow CO_2(g) + 2H_2O(l) \quad \Delta H^\ominus = -890.4 \text{ kJ mol}^{-1}$$

gasoline (petrol)

$$C_8H_{18}(l) + 12\frac{1}{2}O_2(g) \rightarrow 8CO_2(g) + 9H_2O(l) \quad \Delta H^\ominus = -5512 \text{ kJ mol}^{-1}$$

If there is an insufficient supply of oxygen then incomplete combustion occurs and carbon monoxide and carbon are also produced as products.

SUBSTITUTION REACTIONS

Alkanes can react with chlorine (or other halogens) in the presence of ultraviolet light to form hydrogen chloride and a substituted alkane, e.g. methane can react with chlorine to form chloromethane and ethane can react with bromine to form bromoethane.

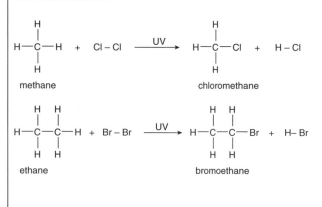

MECHANISM OF CHLORINATION OF METHANE

The mechanism of an organic reaction describes the individual steps. When chemical bonds break they may break **heterolytically** or **homolytically**. In heterolytic fission both of the shared electrons go to one of the atoms resulting in a negative and a positive ion. In homolytic fission each of the two atoms forming the bond retains one of the shared electrons resulting in the formation of two **free radicals**. The bond between two halogen atoms is weaker than the C–H or C–C bond in methane and can break homolytically in the presence of ultraviolet light.

$$Cl_2 \rightarrow Cl^\bullet + Cl^\bullet$$

This stage of the mechanism is called **initiation**.

Free radicals contain an unpaired electron and are highly reactive. When the chlorine free radicals come into contact with a methane molecule they combine with a hydrogen atom to produce hydrogen chloride and a methyl radical.

$$H_3C–H + Cl^\bullet \rightarrow H_3C^\bullet + Cl^\bullet$$

Since a new radical is produced this stage of the mechanism is called **propagation**. The methyl free radical is also extremely reactive and reacts with a chlorine molecule to form the product and regenerate another chlorine radical. This is a further propagation step and enables a chain reaction to occur as the process can repeat itself.

$$CH_3^\bullet + Cl_2 \rightarrow CH_3–Cl + Cl^\bullet$$

In theory a single chlorine radical may cause up to 10 000 molecules of chloromethane to be formed. **Termination** occurs when two radicals react together.

$$\left.\begin{array}{l} Cl^\bullet + Cl^\bullet \rightarrow Cl_2 \\ CH_3^\bullet + Cl^\bullet \rightarrow CH_3Cl \\ CH_3^\bullet + CH_3^\bullet \rightarrow C_2H_6 \end{array}\right\} \quad \text{termination}$$

Further substitution can occur when chlorine radicals react with the substituted products. For example:

The substitution can continue even further to produce trichloromethane and then tetrachloromethane.

The overall mechanism is called **free radical substitution**. [Note that in this mechanism hydrogen radicals H$^\bullet$ are not formed.]

Alkenes

ADDITION REACTIONS

The bond enthalpy of the C=C double bond in alkenes has a value of 612 kJ mol⁻¹. This is less than twice the average value of 348 kJ mol⁻¹ for the C–C single bond and accounts for the relative reactivity of alkenes compared to alkanes. The most important reactions of alkenes are addition reactions. Reactive molecules are able to add across the double bond. The double bond is said to be **unsaturated** and the product, in which each carbon atom is bonded by four single bonds, is said to be **saturated**.

Addition reactions include the addition of hydrogen, bromine, hydrogen halides, and water.

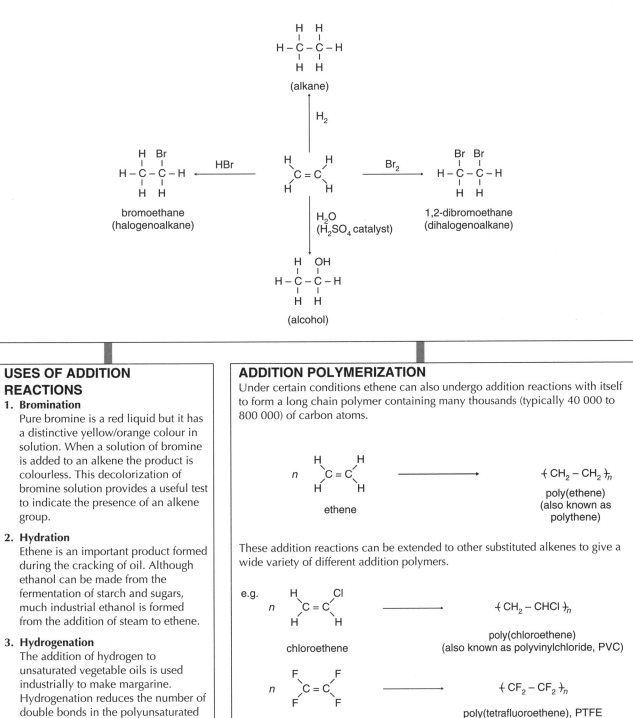

USES OF ADDITION REACTIONS

1. Bromination

Pure bromine is a red liquid but it has a distinctive yellow/orange colour in solution. When a solution of bromine is added to an alkene the product is colourless. This decolorization of bromine solution provides a useful test to indicate the presence of an alkene group.

2. Hydration

Ethene is an important product formed during the cracking of oil. Although ethanol can be made from the fermentation of starch and sugars, much industrial ethanol is formed from the addition of steam to ethene.

3. Hydrogenation

The addition of hydrogen to unsaturated vegetable oils is used industrially to make margarine. Hydrogenation reduces the number of double bonds in the polyunsaturated vegetable oils present in the margarine, which causes it to become a solid at room temperature.

ADDITION POLYMERIZATION

Under certain conditions ethene can also undergo addition reactions with itself to form a long chain polymer containing many thousands (typically 40 000 to 800 000) of carbon atoms.

These addition reactions can be extended to other substituted alkenes to give a wide variety of different addition polymers.

Alcohols

COMBUSTION

Ethanol is used both as a solvent and as a fuel. It combusts completely in a plentiful supply of oxygen to give carbon dioxide and water.

$$C_2H_5OH(l) + 3O_2(g) \rightarrow 2CO_2(g) + 3H_2O(l) \quad \Delta H^\ominus = -1371 \text{ kJ mol}^{-1}$$

Ethanol is already partially oxidized so it releases less energy than burning an alkane of comparable mass. However, it can be obtained by the fermentation of biomass so in some countries it is mixed with petrol to produce 'gasohol' which decreases the dependence on crude oil.

The general equation for an alcohol combusting completely in oxygen is:

$$C_nH_{(2n+1)}OH + (2n-1)O_2 \rightarrow nCO_2 + (n+1)H_2O$$

OXIDATION OF ETHANOL

Ethanol can be readily oxidized by warming with an acidified solution of potassium dichromate(VI). During the process the orange dichromate(VI) ion $Cr_2O_7^{2-}$ is reduced from an oxidation state of +6 to the green Cr^{3+} ion. Use is made of this in simple breathalyser tests, where a motorist who is suspected of having exceeded the alcohol limit blows into a bag containing crystals of potassium dichromate(VI).

Ethanol is initially oxidized to ethanal. The ethanal is then oxidized further to ethanoic acid.

Unlike ethanol (b. pt 78.5 °C) and ethanoic acid (b. pt 118 °C) ethanal (b. pt 20.8 °C) does not have hydrogen bonding between its molecules, and so has a lower boiling point. To stop the reaction at the aldehyde stage the ethanal can be distilled from the reaction mixture as soon as it is formed. If the complete oxidation to ethanoic acid is required, then the mixture can be heated under reflux so that none of the ethanal can escape.

OXIDATION OF ALCOHOLS

Ethanol is a primary alcohol, that is the carbon atom bonded to the –OH group is bonded to two hydrogen atoms and one alkyl group. The oxidation reactions of alcohols can be used to distinguish between primary, secondary, and tertiary alcohols.

All **primary alcohols** are oxidized by acidified potassium dichromate(VI), first to aldehydes then to carboxylic acids.

Secondary alcohols are oxidized to ketones, which cannot undergo further oxidation.

Tertiary alcohols cannot be oxidized by acidified dichromate(VI) ions as they have no hydrogen atoms attached directly to the carbon atom containing the –OH group. It is not true to say that tertiary alcohols can never be oxidized, as they burn readily, but when this happens the carbon chain is destroyed.

Substitution reactions and reaction pathways

SUBSTITUTION REACTIONS OF HALOGENOALKANES

Because of the greater electronegativity of the halogen atom compared with the carbon atom halogenoalkanes have a polar bond. Reagents that have a non-bonding pair of electrons are attracted to the carbon atom in halogenoalkanes and a substitution reaction occurs. Such reagents are called **nucleophiles**

A double-headed curly arrow represents the movement of a pair of electrons. It shows where they come from and where they move to.

MECHANISM OF NUCLEOPHILIC SUBSTITUTION

Primary halogenoalkanes (one alkyl group attached to the carbon atom bonded to the halogen)

e.g. the reaction between bromoethane and warm dilute sodium hydroxide solution.

$$C_2H_5Br + OH^- \rightarrow C_2H_5OH + Br^-$$

The experimentally determined rate expression is:
rate = $k[C_2H_5Br][OH^-]$

The proposed mechanism involves the formation of a transition state which involves both of the reactants.

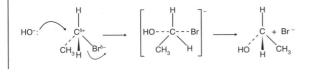

Because the molecularity of this single-step mechanism is two it is known as an S_N2 mechanism (bimolecular nucleophilic substitution).

Tertiary halogenoalkanes (three alkyl groups attached to the carbon atom bonded to the halogen)

e.g. the reaction between 2-bromo-2-methylpropane and warm dilute sodium hydroxide solution.

$$C(CH_3)_3Br + OH^- \longrightarrow C(CH_3)_3OH + Br^-$$

The experimentally determined rate expression for this reaction is: rate = $k[C(CH_3)_3Br]$

A two-step mechanism is proposed that is consistent with this rate expression.

$$C(CH_3)_3Br \xrightarrow{\text{slow}} C(CH_3)_3^+ + Br^-$$

$$C(CH_3)_3^+ + OH^- \xrightarrow{\text{fast}} C(CH_3)_3OH$$

In this reaction it is the first step that is the rate determining step. The molecularity of this step is one and the mechanism is known as S_N1 (unimolecular nucleophilic substitution).

The mechanism for the hydrolysis of secondary halogenoalkanes (e.g 2-bromopropane $CH_3CHBrCH_3$) is more complicated as they can proceed by either S_N1 or S_N2 pathways or a combination of both.

REACTION PATHWAYS

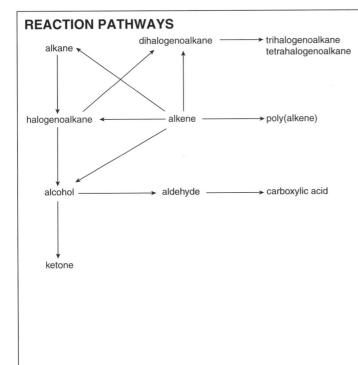

Using the scheme on the left which summarizes the organic reactions in the text, it is possible to devise reaction pathways. These should involve no more than two steps and should include the reagents, conditions and relevant equations.

e.g. to convert but-2-ene to butanone

Step 1. Heat but-2-ene in the presence of H_2SO_4 as a catalyst to form butan-2-ol

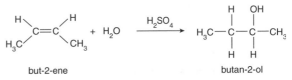

Step 2. Oxidize butan-2-ol to but-2-ene by warming with acidified potassium dichromate(VI) solution

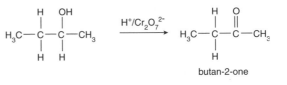

Identifying and naming more functional groups

AMINES (R-NH₂)

IUPAC accepts several different ways of naming amines. The most straightforward system is to prefix the longest chain alkane by the word amino- with the location of the NH_2– group being indicated. For example, 2-aminopentane and 1-aminohexane. It is also correct to call them by the longest alkane with the suffix –amine e.g. pentan-2-amine. If the number of carbon atoms is small (one, two or three) then the old names of methylamine, ethylamine and propylamine tend to be used rather than aminomethane, aminoethane and aminopropane. IUPAC accepts 1-butylamine, 1-butanamine and 1-aminobutane for $CH_3CH_2CH_2CH_2NH_2$.

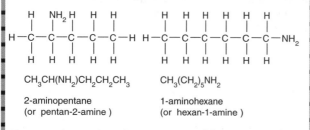

CH₃CH(NH₂)CH₂CH₂CH₃

2-aminopentane
(or pentan-2-amine)

CH₃(CH₂)₅NH₂

1-aminohexane
(or hexan-1-amine)

For secondary amines the main name of the amine is taken from the longest carbon chain attached to the nitrogen atom. The other chain is prefixed as an alkyl group with the location prefix given as an italic N. Examples include N-methylethanamine and N-ethylpropanamine. Tertiary amines conatin two prefixes with an italic N, for example $CH_3CH_2N(CH_3)_2$ is N,N-dimethylethanamine.

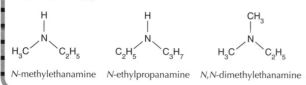

N-methylethanamine   N-ethylpropanamine   N,N-dimethylethanamine

ESTERS (R-COO-R')

Esters take their IUPAC name from the acid and alcohol from which they are derived. The first part of the ester is named after the R- group from the alcohol. There is then a space followed by the name for the carboxylic acid anion. For example, methyl ethanoate, ethyl propanoate and propyl methanoate.

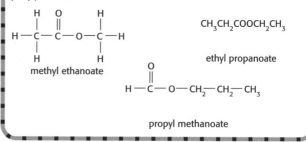

methyl ethanoate

CH₃CH₂COOCH₂CH₃

ethyl propanoate

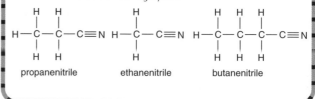

propyl methanoate

NITRILES (R-CN)

Nitriles used to be called cyanides so that C_2H_5CN was known as ethyl cyanide. IUPAC bases the name on the longest carbon chain (which includes the carbon atom of the nitrile group) with the word –nitrile is added to the alkane. For example, the IUPAC name for C_2H_5CN is propanenitrile. Ethanenitrile has the formula CH_3CN, and butanenitrile the formula C_3H_7CN.

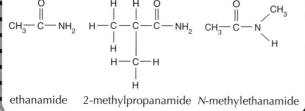

propanenitrile   ethanenitrile   butanenitrile

AMIDES (R-CO-NH₂)

Amides are named after the longest carbon chain (which includes the carbon atom in the functional group) followed by –amide. For example, ethanamide and 2-methylpropanamide. Secondary amides are named rather like amines in that the other alkyl group attached to the nitrogen atom is prefixed by an N, e.g., N-methylethanamide

ethanamide   2-methylpropanamide   N-methylethanamide

HL Nucleophilic substitution

NUCLEOPHILIC SUBSTITUTION

The reaction between halogenoalkanes and a warm dilute aqueous solution of sodium hydroxide is a nucleophilic substitution reaction. Other nucleophiles are CN^-, NH_3 and H_2O. The nucleophiles are attracted to the δ^+ carbon atom and substitute the halogen atom in halogenoalkanes.

Primary halogenoalkanes react by an S_N2 mechanism:

$$C_2H_5Br + OH^- \rightarrow C_2H_5OH + Br^-$$

and tertiary halogenoalkanes react by an S_N1 mechanism.

$$C(CH_3)_3Br + OH^- \longrightarrow C(CH_3)_3OH + Br^-$$

There are several factors which affect the rate of the substitution reactions.

FACTORS AFFECTING THE RATE OF NUCLEOPHILIC SUBSTITUTION

THE NATURE OF THE NUCLEOPHILE

The effectiveness of a nucleophile depends on its electron density. Anions tend to be more reactive than the corresponding neutral species. For example, the rate of substitution with the hydroxide ion is faster than with water. Among species with the same charge a less electronegative atom carrying a non bonded pair of electrons is a better nucleophile than a more electronegative one. Thus ammonia is a better nucleophile than water. This is because the less electronegative atom can more easily donate its pair of electrons as they are held less strongly.

$$CN^- > OH^- > NH_3 > H_2O$$

order of reactivity of common nucleophiles

THE NATURE OF THE HALOGEN

For both S_N1 and S_N2 reactions the iodoalkanes react faster than bromoalkanes, which in turn react faster than chloroalkanes. This is due to the relative bond energies, as the C–I bond is much weaker than the C–Cl bond and therefore breaks more readily.

Bond enthalpy / kJ mol^{-1}

C–I	238
C–Br	276
C–Cl	338

THE NATURE OF THE HALOGENOALKANE

Tertiary halogenoalkanes react faster than secondary halogenoalkanes, which in turn react faster than primary halogenoalkanes. The S_N1 route, which involves the formation of an intermediate carbocation, is faster than the S_N2 route, which involves a transition state with a relatively high activation energy.

SUBSTITUTION WITH AMMONIA AND POTASSIUM CYANIDE

In addition to forming alcohols when water or hydroxide ions are used as the nucleophile, halogenoalkanes can react with ammonia to form amines and with cyanide ions to form nitriles. With primary halogenoalkanes the mechanism is S_N2 in both cases, e.g. with bromoethane and cyanide ions propanenitrile is produced.

The nucleophilic substitution reactions of halogenoalkanes makes them particularly useful in organic synthesis. The reaction with potassium cyanide provides a useful means of increasing the length of the carbon chain by one carbon atom. The nitrile can then be converted either into amines by reduction using hydrogen with a nickel catalyst or into carboxylic acids by acid hydrolysis, e.g.

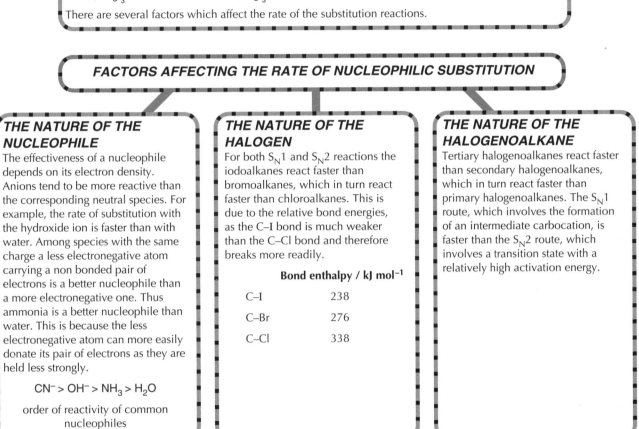

When bromoethane reacts with ammonia, ethylamine is produced. However ethylamine also contains a nitrogen atom with a non- bonding pair of electrons so this too can act as a nucleophile and secondary and tertiary amines can be formed. Even the tertiary amine is still a nucleophile and can react further to form the quaternary salt.

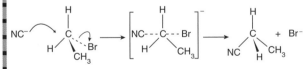

Elimination and condensation reactions

HL

ELIMINATION REACTIONS OF HALOGENOALKANES

The reactions of halogenoalkanes with hydroxide ions provide an example of how altering the reaction conditions can cause the same reactants to produce completely different products. (Note that another good example is the reaction of methylbenzene with chlorine.) With dilute sodium hydroxide solution the OH⁻ ion acts as a nucleophile and substitution occurs to produce an alcohol, e.g.

$$HO^-: \quad R-Br \longrightarrow R-OH + Br^-$$

However with hot alcoholic sodium hydroxide solution (i.e. sodium hydroxide dissolved in ethanol) **elimination** occurs and an alkene is formed, e.g.

$$-\underset{H}{\overset{H}{C}}-\underset{Br}{\overset{Br}{C}}- + OH^- \longrightarrow \quad C=C \quad + H_2O + Br^-$$

In this reaction the hydroxide ion reacts as a base. The elimination of HBr can proceed either by a carbocation or as a concerted process, e.g.

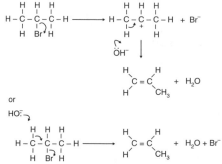

or

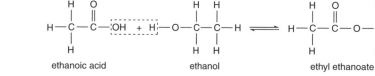

In the presence of ethanol there will also be some ethoxide ions present. Ethoxide is a stronger base than hydroxide so the equilibrium lies to the left but some ethoxide ions will be present and these may be the actual species acting as the base.

$$HO^- + C_2H_5OH \rightleftharpoons H_2O + C_2H_5O^-$$

CONDENSATION REACTIONS

A condensation reaction involves the reaction between two molecules to produce a larger molecule with the elimination of a small molecule such as water or hydrogen chloride. One important condensation reaction is the formation of esters when an alcohol reacts with a carboxylic acid.

e.g.

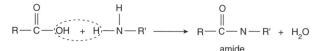

Most esters have a distinctive, pleasant, fruity smell and are used both as natural and artificial flavouring agents in food. For example, ethyl methanoate HCOOCH$_2$CH$_3$ is added to chocolate to give it the characteristic flavour of 'rum truffle'. Esters are also used as solvents in perfumes and as plasticizers (substances used to modify the properties of polymers by making them more flexible).

Another example of a condensation reaction is the formation of secondary amides by reacting a carboxylic acid with an amine.

$$R-\overset{O}{\overset{\|}{C}}-OH + H-\overset{H}{\overset{|}{N}}-R' \longrightarrow R-\overset{O}{\overset{\|}{C}}-N-R' + H_2O$$
amide

This reaction is important in biological reactions as amino acids contain an amine group and a carboxylic acid group so that amino acids can condense together in the presence of enzymes to form poly(amides).

HL Condensation polymerization and reaction pathways

CONDENSATION POLYMERIZATION

Condensation involves the reaction between two molecules to eliminate a smaller molecule, such as water or hydrogen chloride. If each of the reacting molecules contain *two* functional groups that can undergo condensation, then the condensation can continue to form a polymer.

An example of a polyester is polyethene terephthalate (known as Terylene in the UK and as Dacron in the USA) used for textiles, which is made from benzene-1,4-dicarboxylic acid and ethane-1,2-diol.

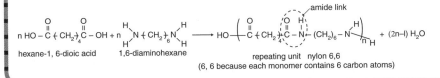

Amines can also condense with carboxylic acids to form an amide link (also known as a peptide bond). One of the best known examples of a polyamide is nylon.

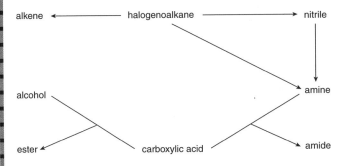

REACTION PATHWAYS

The compounds and reaction types covered in the AHL can be summarized in the following scheme:

```
alkene  ←————————  halogenoalkane  ————————→  nitrile

                                                   │
                                                   ↓
alcohol                                          amine

ester  ←————       carboxylic acid       ————→  amide
```

Given the starting materials, two step syntheses for new products can be devised. For example, the conversion of 1-bromopropane to 1-aminobutane (1-butylamine) can be performed in the following two stages.

Step 1. 1-bromopropane can undergo nucleophilic substitution with potassium cyanide solution to form propanenitrile.

$$CH_3CH_2CH_2Br + CN^- \longrightarrow CH_3CH_2CH_2CN + Br^-$$

Step 2. Propanenitrile can be reduced by heating with hydrogen over a nickel catalyst.

$$CH_3CH_2CH_2CN + 2H_2 \xrightarrow{Ni} CH_3CH_2CH_2CH_2NH_2$$

Another example would be the formation of ethylamine starting with ethane. Now reactions covered in the core can also be included.

Step 1. React ethane with chlorine in ultraviolet light so that chloroethane is formed by free radical substitution.

$$C_2H_6 + Cl_2 \xrightarrow{uv} C_2H_5Cl + HCl$$

Step 2. React chloroethane with ammonia.

$$C_2H_5Cl + NH_3 \longrightarrow C_2H_5NH_2 + HCl$$

ⓗⓛ Stereoisomerism (1)

Structural isomers share the same molecular formula but have different structural formulas. That is, their atoms are bonded in different ways. Stereoisomers have the same structural formula but differ in their spatial arrangement. There are two types of stereoisomerism: geometrical isomerism and optical isomerism.

GEOMETRICAL ISOMERISM

Geometrical isomerism occurs when rotation about a bond is restricted or prevented. The classic examples of geometric isomers occur with asymmetric non-cyclic alkenes. A *cis-* isomer is one in which the substituents are on the same side of the double bond. In a *trans-* isomer the substituents are on opposite sides of the double bond. For example consider *cis*-but-2-ene and *trans*-but-2-ene.

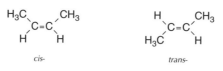

When there is a single bond between two carbon atoms free rotation about the bond is possible. However, the double bond in an alkene is made up of a σ and a π bond. The π bond is formed from the combination of two p orbitals, one from each of the carbon atoms. These two p orbitals must be in the same plane to combine. Rotating the bond would cause the π bond to break so no rotation is possible.

Cis- and *trans-* isomerism will always occur in alkenes when the two groups attached to each of the two carbon atoms are different.

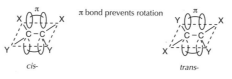

Geometrical isomerism can also occur in disubstituted cycloalkanes. The rotation is restricted because the C–C single bond is part of a ring system. Examples include 1,2-dichlorocyclopropane and 1,3-dichlorocyclobutane.

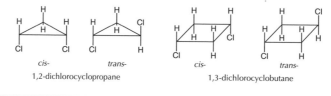

PHYSICAL AND CHEMICAL PROPERTIES OF GEOMETRICAL ISOMERS

The chemical properties of geometric isomers tend to be similar but their physical properties are different. For example, the boiling point of *cis*-1,2-dichloroethene is 60.3 °C whereas *trans*-1,2-dichloroethene boils at the lower temperature of 47.5 °C. Sometimes there can be a marked difference in both chemical and physical properties. This tends to occur when there is some sort of chemical interaction between the substituents. *cis*-but-2-ene-1,4-dioic acid melts with decomposition at 130–131 °C. However, *trans*-but-2-ene,1,4-dioic acid does not melt until 286 °C. In the *cis-* isomer the two carboxylic acid groups are closer together so that intramolecular hydrogen bonding is possible between them. In the *trans-* isomer they are too far apart to attract each other so there are stronger intermolecular forces of attraction between different molecules, resulting in a higher melting point. The *cis-* isomer reacts when heated to lose water and form a cyclic acid anhydride. The *trans-* isomer cannot undergo this reaction.

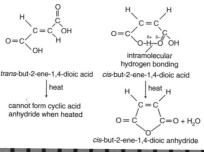

OPTICAL ISOMERISM

Optical isomerism is shown by all compounds that contain at least one asymmetric or chiral carbon atom within the molecule, that is, one that contains four different atoms or groups bonded to it. The two isomers are known as enantiomers and are mirror images of each other. Example include butan-2-ol, $CH_3CH(OH)C_2H_5$, 2-hydroxypropanoic acid (lactic acid), $CH_3CH(OH)COOH$ and all amino acids (except glycine, NH_2CH_2COOH).

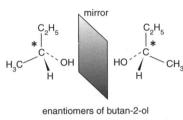

enantiomers of butan-2-ol
(* asymmetric carbon / chiral carbon)

The two different isomers are optically active with plane-polarized light. Normal light consists of electromagnetic radiation which vibrates in all planes. When it is passed through a polarizing filter the waves only vibrate in one plane and the light is said to be plane-polarized.

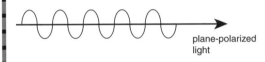

plane-polarized light

The two enantiomers both rotate the plane of plane-polarised light. One of the enantiomers rotates it to the left and the other rotates it by the same amount to the right. Apart from their behaviour towards plane-polarised light enantiomers have identical physical properties. Their chemical properties are identical too except when they interact with other optically active substances. This is often the case in the body where the different enantiomers can have completely different physiological effects. For example one of the enantiomers of the amino acid asparagine $H_2NCH(CH_2CONH_2)COOH$ tastes bitter whereas the other enantiomer tastes sweet.

ENANTIOMERS

The optical activity of enantiomers can be detected and measured by an instrument called a polarimeter. It consists of a light source, two polarizing lenses, and between the lenses a tube to hold the sample of the enantiomer.

When light passes through the first polarizing lens (polarizer) it becomes plane-polarized. That is, it is vibrating in a single plane. With no sample present the observer will see the maximum intensity of light when the second polarizing lens (analyser) is in the same plane. Rotating the analyser by 90° will cut out all the light. When the sample is placed between the lenses the analyser must be rotated by θ degrees, either clockwise (dextrorotatory) or anticlockwise (laevorotatory) to give light of maximum intensity. The two enantiomers rotate the plane of plane-polarized light by the same amount but in opposite directions. If both enantiomers are present in equal amounts the two rotations cancel each other out and the mixture appears to be optically inactive. Such a mixture is known as a **racemic mixture** or **racemate**.

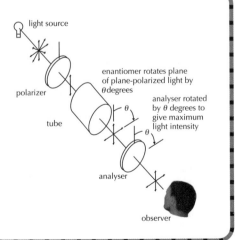

IB QUESTIONS – ORGANIC CHEMISTRY

1. Which of the following is/are true about alkanes?

   I. They form an homologous series with the general formula C_nH_{2n-2}

   II. They all have identical physical properties

   III. They all have similar chemical properties

   A. I, II and III   C. I and III only

   B. II and III only   D. III only

2. Which of the following two compounds both belong to the same homologous series?

   A. CH_3COOH and $HCOOCH_3$

   B. CH_3OH and C_2H_5OH

   C. C_2H_4 and C_2H_6

   D. C_2H_5Cl and $C_2H_4Cl_2$

3. How many different isomers of C_5H_{12} exist?

   A. 1   B. 2   C. 3   D. 4

4. Give the correct name for:

   $$H_3C-\underset{\underset{CH_3}{\overset{|}{CH_2}}}{\overset{\overset{CH_3}{|}}{C}}-CH_3$$

   A. 2-methyl-2-ethylpropane   C. 2,2-dimethylbutane

   B. hexane   D. 2-methylpentane

5. Which statement is correct about the reaction between methane and chlorine?

   A. It involves heterolytic fission and Cl^- ions.

   B. It involves heterolytic fission and $Cl\bullet$ radicals.

   C. It involves homolytic fission and Cl^- ions.

   D. It involves homolytic fission and $Cl\bullet$ radicals.

6. Which compound is an ester?

   A. CH_3COOH C. C_2H_5CHO

   B. $CH_3OC_2H_5$ D. $HCOOCH_3$

7. When ethanol is partially oxidized by an acidified solution of potassium dichromate(VI), the product that can be obtained by distillation as soon as it is formed is:

   A. ethanal   C. ethanoic acid

   B. ethene   D. ethane-1,2-diol

8. Which formula is that of a secondary halogenoalkane?

   A. $CH_3CH_2CH_2CH_2Br$ C. $(CH_3)_2CHCH_2Br$

   B. $CH_3CHBrCH_2CH_3$ D. $(CH_3)_2CBr$

9. Which cannot undergo an addition reaction with ethene?

   A. oxygen   C. water

   B. hydrogen   D. hydrogen bromide

10. Which compound is converted to butanal by acidified potassium dichromate(VI) solution?

    A. butan-1-ol   C. butanone

    B. butan-2-ol   D. butanoic acid

11. Which reaction(s) involve(s) the formation of a positive ion?

    I. $CH_3CH_2CH_2Br + OH^-$

    II. $(CH_3)_3CBr + OH^-$

    A. I only   C. Both I and II

    B. II only   D. Neither I nor II

12. Identify which two compounds can react together to form a condensation polymer.

    A. C_2H_4 and Br_2

    B. C_2H_5OH and CH_3COOH

    C. $H_2N(CH_2)_6NH_2$ and CH_3COOH

    D. $HOOC(CH_2)_4COOH$ and $HOCH_2CH_2OH$

13. Consider the following reaction:

    $CH_3COOH + NH_3 \rightarrow CH_3COONH_4 \overset{heat}{\rightarrow} CH_3CONH_2$

    What will be the final product if aminoethane (ethylamine) is used instead of NH_3?

    A. $CH_3CONHCH_2CH_3$ C. CH_3CONH_2

    B. $CH_3CONHCH_3$ D. $CH_3CONH_2CH_2CH_3$

14. Which compound can exhibit optical isomerism?

    A. $HOOC-\underset{\underset{H}{|}}{\overset{\overset{H}{|}}{C}}-NH_2$ C. $HOOC-\underset{\underset{CH_3}{|}}{\overset{\overset{H}{|}}{C}}-OH$

    B. $HOOC-\underset{\underset{CH_3}{|}}{\overset{\overset{H}{|}}{C}}-COOH$ D. $H_2N-\underset{\underset{H}{|}}{\overset{\overset{H}{|}}{C}}-NH_2$

15. A ketone containing four carbon atoms will be formed from the reaction of an acidified solution of potassium dichromate(IV) solution with:

    A. $CH_3CH_2CH_2CH_2OH$ C. $CH_3CH_2CH_2CHO$

    B. $(CH_3)_3COH$ D. $CH_3CH_2CH(OH)CH_3$

16. Which of the following will give an amine as the product?

    I. $(CH_3)_3CBr + NH_3$ II. $CH_3CN + H_2$ III. $CH_3COOH + NH_3$

    A. I, II and III   C. I only

    B. I and II only   D. I and III only

17. Which statement(s) is/are true about the reactions of halogenoalkanes with warm dilute sodium hydroxide solution

    I. CH_3I reacts faster than CH_3F

    II. $(CH_3)_3CBr$ reacts faster than CH_3Br

    III. $(CH_3)_3CBr$ and $(CH_3)_3CCl$ both react by S_N1 mechanisms.

    A. I and II only   C. II and III only

    B. I, II and III   D. I only

Uncertainty and error in measurement

RANDOM UNCERTAINTIES AND SYSTEMATIC ERRORS

Quantitative chemistry involves measurement. A measurement is a method by which some quantity or property of a substance is compared with a known standard. If the instrument used to take the measurements has been calibrated wrongly or if the person using it consistently misreads it then the measurements will always differ by the same amount. Such an error is known as a **systematic error**. An example might be always reading a pipette from the sides of the meniscus rather than from the middle of the meniscus.

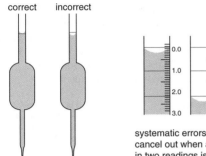

correct   incorrect

a systematic error will be introduced if a pipette is read incorrectly

systematic errors may cancel out when a difference in two readings is taken $\Delta V = 2.0$ cm^3 whether the burette is read correctly or incorrectly.

Random uncertainties occur if there is an equal probability of the reading being too high or too low from one measurement to the next. These might include variations in the volume of glassware due to temperature fluctuations or the decision on exactly when an indicator changes colour during an acid base titration.

PRECISION AND ACCURACY

Precision refers to how close several experimental measurements of the same quantity are to each other. **Accuracy** refers to how close the reading are to the true value. This may be the standard value, or the literature or accepted value. A measuring cylinder used to measure exactly 25 cm^3 is likely to be much less accurate than a pipette that has been carefully calibrated to deliver exactly that volume. It is possible to have very precise readings which are inaccurate due to a systematic error. For example all the students in the class may obtain the same or very close results in a titration but if the standard solution used in all the titrations had been prepared wrongly beforehand the results would be inaccurate due to the systematic error. Because they are always either too high or too low systematic errors cannot be reduced by repeated readings. However random errors can be reduced by repeated readings because there is an equal probability of them being high or low each time the reading is taken. When taking a measurement it is usual practice to report the reading from a scale as the smallest division or the last digit capable of precise measurement even though it is understood that the last digit has been rounded up or down so that there is a random error or uncertainty of ± 0.5 of the last unit.

SIGNIFICANT FIGURES

Whenever a measurement of a physical quantity is taken there will be a random uncertainty in the reading. The measurement quoted should include the first figure that is uncertain. This should include zero if necessary. Thus a reading of 25.30 °C indicates that the temperature was measured with a thermometer that is accurate to + 0.01 °C. If a thermometer accurate to only $\pm$ 0.1 °C was used the temperature should be recorded as 25.3 °C.

Zeros can cause problems when determining the number of significant figures. Essentially zero only becomes significant when it comes *after* a non-zero digit (1,2,3,4,5,6,7,8,9).

 000123.4   0.0001234   1.0234   1.2340
zero not a significant figure   zero is a significant figure
values quoted to 4 sig. figs.   values quoted to 5 sig. figs.

Zeros after a non-zero digit but before the decimal point may or may not be significant depending on how the measurement was made. For example 123 000 might mean exactly one hundred and twenty three thousand or one hundred and twenty three thousand to the nearest thousand. This problem can be neatly overcome by using scientific notation.

  1.23000×10^6 quoted to six significant figures
  1.23×10^6 quoted to three significant figures.

Calculations

1. When adding or subtracting it is the number of decimal places that is important. Thus when using a balance which measures to $\pm$ 0.01 g the answer can also be quoted to two decimal places which may increase or decrease the number of significant figures.

e.g.   7.10 g   +   3.10 g   =   10.20 g
       3 sig. figs.   3 sig. figs.   4 sig. figs.

       22.36 g   –   15.16 g   =   7.20 g
       4 sig. figs.   4 sig. figs.   3 sig. figs.

2. When multiplying or dividing it is the number of significant figures that is important. The number with the least number of significant figures used in the calculation determines how many significant figures should be used when quoting the answer.

e.g.   When the temperature of 0.125 kg of water is increased by 7.2 °C

        the heat required
        = 0.125 kg x 7.2 °C x 4.18 kJ kg^{-1} °C^{-1}
        = 3.762 kJ.

Since the temperature was only recorded to two significant figures the answer should strictly be given as 3.8 kJ.

Uncertainties in calculated results and graphical techniques

ABSOLUTE AND PERCENTAGE UNCERTAINTIES

When making a single measurement with a piece of apparatus the absolute uncertainty and the percentage uncertainty can both be stated relatively easily. For example consider measuring 25.0 cm^3 with a 25 cm^3 pipette which measures to $\pm 0.1 \text{ cm}^3$. The absolute uncertainty is 0.1 cm^3 and the percentage uncertainty is equal to:

$$\frac{0.1}{25.0} \times 100 = 0.4\%$$

If two volumes or two masses are simply added or subtracted then the absolute uncertainties are added. For example suppose two volumes of $25.0 \text{ cm}^3 \pm 0.1 \text{ cm}^3$ are added. In one extreme case the first volume could be 24.9 cm^3 and the second volume 24.9 cm^3 which would give a total volume of 48.8 cm^3. Alternatively the first volume might have been 25.1 cm^3 which when added to a second volume of 25.1 cm^3 gives a total volume of 50.2 cm^3. The final answer therefore can be quoted between 48.8 cm^3 and 50.2 cm^3, that is, $50.0 \text{ cm}^3 \pm 0.2 \text{ cm}^3$.

When using multiplication, division or powers then percentage uncertainties should be used during the calculation and then converted back into an absolute uncertainty when the final result is presented. For example, during a titration there are generally four separate pieces of apparatus, each of which contributes to the uncertainty.

e.g. when using a balance that weighs to $\pm 0.001 \text{ g}$ the uncertainty in weighing 2.500 g will equal

$$\frac{0.001}{2.500} \times 100 = 0.04\%$$

Similarly a pipette measures $25.00 \text{ cm}^3 \pm 0.04 \text{ cm}^3$.

The uncertainty due to the pipette is thus
$$\frac{0.04}{25.00} \times 100 = 0.16\%$$

Assuming the uncertainty due to the burette and the volumetric flask is 0.50% and 0.10% respectively the overall uncertainty is obtained by summing all the individual uncertainties:

Overall uncertainty = 0.04 + 0.16 + 0.50 + 0.10
$$= 0.80\% \simeq 1.0\%$$

Hence if the answer is 1.87 mol dm^{-3} the uncertainty is 1.0% or $0.0187 \text{ mol dm}^{-3}$

The answer should be given as $1.87 \pm 0.02 \text{ mol dm}^{-3}$.

If the generally accepted 'correct' value (obtained from the data book or other literature) is known then the total error in the result is the difference between the literature value and the experimental value divided by the literature value expressed as a percentage. For example, if the 'correct' concentration for the concentration determined above is 1.90 mol dm^{-3} then:

the total error $= \dfrac{(1.90 - 1.87)}{1.90} \times 100 = 1.6\%$.

GRAPHICAL TECHNIQUES

By plotting a suitable graph to give a straight line or some other definite relationship between the variables graphs can be used to predict unknown values. There are various methods to achieve this. They include measuring the intercept, measuring the gradient, extrapolation and interpolation. **Interpolation** involves determining an unknown value within the limits of the values already measured. **Extrapolation** (see example on page 29) requires extending the graph to determine an unknown value which lies outside the range of the values measured. If possible manipulate the data to produce a straight line graph. For example, when investigating the relationship between pressure and volume for a fixed mass of gas a plot of P against V gives a curve whereas a plot of P against 1/V will give a straight line. Once the graph is in the form of $y = mx + c$ then the values for both the gradient (m) and the intercept (c) can be determined.

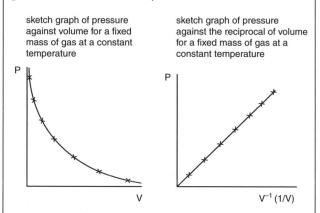

sketch graph of pressure against volume for a fixed mass of gas at a constant temperature

sketch graph of pressure against the reciprocal of volume for a fixed mass of gas at a constant temperature

The following points should be observed when drawing a graph.

- Plot the independent variable on the horizontal axis and the dependent variable on the vertical axis.
- Choose appropriate scales for the axes.
- Use Standard International (SI) units wherever possible.
- Label each axis and include the units.
- Draw the line of best fit.
- Give the graph a title.

Measuring a gradient from a graph

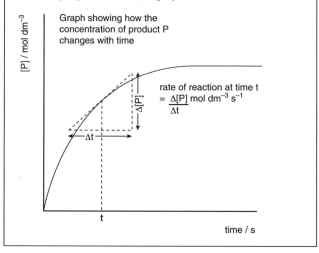

Graph showing how the concentration of product P changes with time

rate of reaction at time t
$= \dfrac{\Delta[P]}{\Delta t} \text{ mol dm}^{-3} \text{ s}^{-1}$

IB QUESTIONS – MEASUREMENT AND DATA PROCESSING

1. A 25.0 cm³ sample of a base solution of unknown concentration is to be titrated with a solution of acid of unknown concentration. Which of the following technique errors would give a value for the concentration of the base that is **too high**?

   I. The pipette that is used to deliver the base solution is rinsed only with distilled water before delivering the sample to be titrated.

   II. The burette that is used to measure the acid solution is rinsed with distilled water but not with the solution of the titration.

   A. I only
   B. II only
   C. Both I and II
   D. Neither I nor II

2. In a school laboratory, which of the items listed below has the greatest relative uncertainty in a measurement?

   A. A 50 cm³ burette when used to measure 25 cm³ of water
   B. A 25 cm³ pipette when used to measure 25 cm³ of water
   C. A 50 cm³ graduated cylinder when used to measure 25 cm³ of water
   D. An analytical balance when used to weigh 25 g of water

3. A piece of metallic indium with a mass of 15.456 g was placed in 49.7 cm³ of ethanol in a graduated cylinder. The ethanol level was observed to rise to 51.8 cm³. From these data, the best value one can report for the density of indium is

   A. 7.360 g cm⁻³
   B. 7.4 g cm⁻³
   C. 1.359 x 10⁻¹ g cm⁻³
   D. 32.4 g cm⁻³

4. A mixture of sodium chloride and potassium chloride is prepared by mixing 7.35 g of sodium chloride with 6.75 g of potassium chloride. The total mass of the salt mixture should be reported to ____ significant figures; the mass ratio of sodium chloride to potassium chloride should be reported to ____ significant figures, and the difference in mass between sodium chloride and potassium chloride should be reported to ____ significant figures. The numbers required to fill the blanks above are, respectively,

   A. 4, 3, 2
   B. 4, 2, 2
   C. 3, 3, 1
   D. 4, 3, 1

5. Repeated measurements of a quantity can reduce the effects of
   A. both random and systematic errors
   B. neither random nor systematic errors
   C. only systematic errors
   D. only random errors

6. A 50.0 cm³ pipette with an uncertainty of 0.1 cm³ is used to measure 50.0 cm³ of 1.00 ± 0.01 mol dm⁻³ sodium hydroxide solution. The amount in moles of sodium hydroxide present in the measured volume is

   A. 0.0500 ± 0.0010
   B. 0.0500 ± 0.0001
   C. 0.0500 ± 0.0012
   D. 0.0500 ± 0.0006

7. Consider the following three sets each of five measurements of the same quantity which has an accurate value of 20.0

I.	II.	III.
19.8	19.2	20.0
17.2	19.1	19.9
18.3	19.3	20.0
20.1	19.2	20.1
18.4	19.2	20.0

   Which results can be described as precise?

   A. I, II and III
   B. II only
   C. II and III
   D. III only

8. A thermometer with an accuracy of ± 0.2 °C was used to record an initial temperature of 20.2 °C and a final temperature of 29.8 °C. The temperature rise was

   A. 9.6 ± 0.4 °C
   B. 9.6 ± 0.2 °C
   C. 10 °C
   D. 9.6 ± 0.1 °C

9. An experiment to determine the molar mass of solid hydrated copper(II) sulfate, CuSO₄.5H₂O gave a result of 240 g. The experimental error was

   A. 0.04%
   B. 4%
   C. 10%
   D. 50%

10. Which sketch graph shows interpolation to find an unknown value?

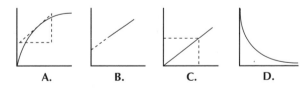

A.   B.   C.   D.

Analytical techniques and the principles of spectroscopy

INFORMATION FROM DIFFERENT ANALYTICAL TECHNIQUES

The classic way to determine the structure of an organic compound was to determine both its empirical formula and relative molar mass experimentally, then deduce the nature of the functional groups from its chemical reactivity. It can still be useful to determine the empirical formula of a substance by burning a known mass of it in excess oxygen, and finding the mass of the oxides formed from the different elements. However modern well-equipped laboratories now employ a variety of instrumental techniques which if used in combination are able to unambiguously determine the exact structural formula. These techniques are becoming ever more refined and some of them (e.g. mass spectrometry) can be used on extremely small samples. This has revolutionized many branches of chemistry, for example combinatorial chemistry in the search for new drugs. This is only possible because very small amounts of thousands of new compounds can be analysed accurately and efficiently.

The main uses for these techniques are structural determination, the analysis of the different composition of compounds, and to determine purity. Before analysis can usually take place it is important to separate any mixture into its individual components – hence the need for chromatography. Often information is not obtained from a single technique but from a combination of several of them. Some examples are:

- Atomic absorption spectroscopy – assaying of metal ions in water, blood, soils and foods.
- Infrared spectroscopy – organic structural determination, information on the strength of bonds, information about the secondary structure of proteins, measuring the degree of unsaturation of oils and fats, and determining the level of alcohol in the breath.
- Mass spectrometry – organic structural determination, isotopic dating (e.g. ^{14}C dating).
- ^{1}H nuclear magnetic resonance – organic structural determination, body scanning.
- Chromatography – drug testing in the blood and urine, food testing, and forensic science.

THE ELECTROMAGNETIC SPECTRUM

The electromagnetic spectrum has already been briefly described in Topic 2 – Atomic structure. You should be familiar with the relationship $c = \lambda f$ and know the different regions of the spectrum.

The electromagnetic spectrum

Wavelength / m	10^{-10}	10^{-9}	10^{-8}	10^{-7}	10^{-6}	10^{-5}	10^{-4}	10^{-3}	10^{-2}	10^{-1}	10^{0}	10^{1}	10^{2}	10^{3}
Frequency / MHz		3×10^{10}		3×10^{8}		3×10^{6}		3×10^{4}		3×10^{2}		3		
Type of radiation	X-rays γ-rays		ultraviolet	visible		infrared		microwaves					radio waves	
Type of transition	inner electron		outer electron				molecular vibrations	molecular rotations					nuclear spin	

←———————————————— Increasing energy ————————————————

ABSORPTION SPECTRA AND EMISSION SPECTRA

Spectroscopy can be divided into two main types. Emission spectroscopy involves the analysis of light emitted by excited atoms or molecules as they return to their ground state. The atomic emission spectrum of hydrogen is a good example of this. Many analytical techniques involve absorption spectroscopy. When radiation is passed through a sample some of the energy is absorbed by the sample to excite an atom or molecule to an excited state. The spectrometer analyses the transmitted energy relative to the incident energy. Since the energy levels are quantized only radiation with a frequency corresponding to the difference in the energy levels will be absorbed. The relationship between energy and frequency is given by:

$E = hf$ where h is Planck's constant, 6.626×10^{-34} J s.

The greater the energy difference between the levels the higher the frequency (or the shorter the wavelength) of the light absorbed. The most energetic absorptions are atomic electronic transitions which involve bond breaking and ionization. Absorptions in the ultraviolet and visible region are due to atomic and molecular transitions in which electrons become excited to higher levels. Molecular vibrations (stretching and bending) occur in the infrared region and molecular rotations in the microwave region. The weakest transitions of all involve nuclear spin. These occur in the radio wave region and form the basis of nuclear magnetic resonance spectroscopy.

Infrared spectroscopy (1)

VIBRATION OF BONDS

When molecules vibrate they absorb energy. This energy lies in the frequency range $1.2 \times 10^{14} - 1.2 \times 10^{13}$ s^{-1} (Hz), i.e. wavelengths of $2.5 \times 10^{-6} - 2.5 \times 10^{-5}$ m which is in the infrared region of the electromagnetic spectrum. For simple diatomic molecules made up of different atoms, such as HCl, there is only one form of vibration. This is stretching where the atoms alternatively move further apart then closer together. Different molecules absorb at different frequencies as the energy needed to excite a vibration depends on the bond enthalpy. Weaker bonds require less energy.

Note that infrared absorptions are usually given in cm^{-1}. Frequency and wavelength are related by the equation $c = \lambda f$ where c is the velocity of light in a vacuum. Since c is a constant the reciprocal of wavelength is a direct measure of frequency. This reciprocal of wavelength ($1/\lambda$) is known as the **wavenumber** and has the units cm^{-1}. Hence an absorption of 2886 cm^{-1} corresponds to a wavelength of 3.465×10^{-6} m. The longer the wavelength the lower the energy and the smaller the value of $1/\lambda$ in cm^{-1}. Conversely the higher the wavenumber the higher the energy.

Molecule	Bond enthalpy / kJ mol^{-1}	Absorption / cm^{-1}
H–Cl	431	2886
H–Br	366	2559
H–I	299	2230

CHANGE IN BOND POLARITY

Not all vibrations absorb infrared radiation. For absorption there must be a change in the dipole moment (bond polarity) as the vibration occurs. Thus diatomic gas molecules containing only one element, such as H_2, Cl_2, and O_2, do not absorb infrared radiation.

Vibrations of H_2O, SO_2, and CO_2

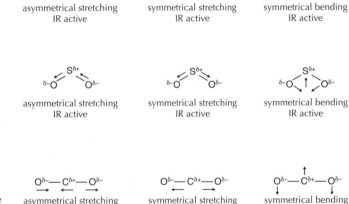

Stretching involves a change in dipole: IR active

Stretching involves no change in dipole: IR inactive

For more complex molecules only those vibrations which result in a dipole change will be infrared active. For example, the symmetrical stretch in carbon dioxide will be infrared inactive whereas the asymmetric stretch and the bending are both infrared active as they result in a dipole change.

Stretching and bending are the main modes of vibration but bending can be sub-divided into rocking, scissoring, twisting and wagging as exemplified by the –CH_2– group.

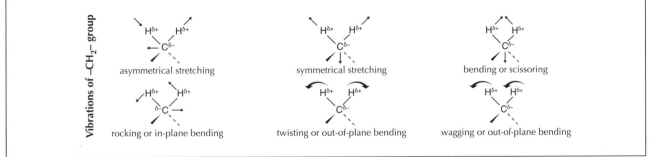

Infrared spectroscopy (2)

THE OPERATING PRINCIPLES OF A DOUBLE BEAM INFRARED SPECTROMETER

Traditional infrared spectrometers work by scanning wavelengths from about 2.5×10^{-6} m (4000 cm^{-1}) to 2.5×10^{-5} m (400 cm^{-1}). By using a rotating mirror the beam of monochromatic radiation is alternately passed through the sample and a reference. A photomultiplier converts photons of radiation into an electrical current. The spectrum is generated by comparison of the currents produced by the sample and the reference beams. Modern spectrometers pass all the wavelengths through the sample at the same time and a mathematical technique known as Fourier transformation is used to automatically analyse the transmission at each wavelength. Modern spectrometers are also linked to computers which store the spectra of known compounds. This enables an unknown compound to be identified by exact matching with the spectrum of a known compound.

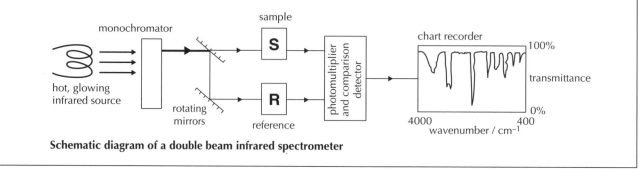

Schematic diagram of a double beam infrared spectrometer

USE OF INFRARED SPECTROSCOPY

Because of all the different vibrations possible most compounds actually have quite complex spectra particularly in the region between about $1400 - 400$ cm^{-1} known as the 'fingerprint' region. This is a characteristic pattern which is specific to a particular compound. Unknown samples can be identified by comparing the fingerprint region with a library of spectra of known samples. However for simple laboratory use there are several main absorptions which can be used to identify particular functional groups. The most common of these are the –OH, C=O, C-H, C=C, and C-O bonds. The precise region at which these absorb is determined by the neighbouring atoms as these will influence the bond enthalpy, hence the reason for the range of absorptions given for each bond in the correlation chart. The main use of infrared spectroscopy is thus to confirm the presence of particular functional groups. The intoximeter for determining the amount of alcohol in the breath (see Option D – Medicine and drugs) provides a good example of how it can also be used to give quantitative data on the amount of a component present in a sample.

A simplified correlation chart

Bond	Wavenumber / cm^{-1}
C–O	1000–1300
C=C	1610–1680
C=O	1680–1750
C≡C	2070–2250
O–H (in carboxylic acids)	2500–3300
C–H	2840–3095
O–H (in alcohols)	3230–3500

Wavenumber is the reciprocal of wavelength.

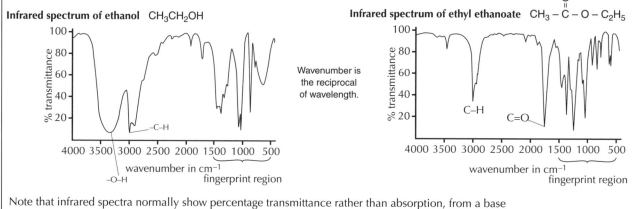

Note that infrared spectra normally show percentage transmittance rather than absorption, from a base of 100% transmittance.

Mass spectrometry

MASS SPECTROMETRY

The principles of mass spectroscopy and its use to determine relative atomic masses have already been explained in Topic 2 – Atomic structure. It can be used in a similar way with organic compounds. However in addition to giving the precise molecular mass of the substance, mass spectroscopy gives considerable information about the actual structure of the compound from the fragmentation patterns.

When a sample is introduced into the machine the vaporized sample becomes ionized to form the molecular ion $M^+(g)$. Inside the mass spectrometer some of the molecular ions break down to give fragments, which are also deflected by the external magnetic field and which then show up as peaks on the detector. By looking at the difference in mass from the parent peak it is often possible to identify particular fragments,

e.g. $(M_r - 15)^+ = $ loss of CH_3

$(M_r - 29)^+ = $ loss of C_2H_5 or CHO

$(M_r - 31)^+ = $ loss of CH_3O

$(M_r - 45)^+ = $ loss of COOH

This can provide a useful way of distinguishing between structural isomers.

The mass spectra of propan-1-ol and propan-2-ol both show a peak at 60 due to the molecular ion $C_3H_8O^+$. However, the mass spectrum of propan-1-ol shows a strong peak at 31 due to the loss of $-C_2H_5$, which is absent in the mass spectrum of propan-2-ol. There is a strong peak at 45 in the spectrum of propan-2-ol as it contains two different methyl groups, which can fragment.

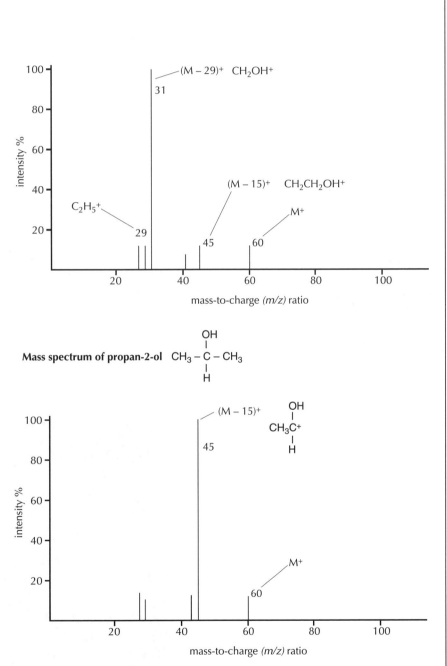

Mass spectrum of propan-1-ol $CH_3 - CH_2 - CH_2{-}OH$

Mass spectrum of propan-2-ol

Nuclear magnetic resonance (NMR) spectroscopy (1)

¹H NMR SPECTROSCOPY

Whereas infrared spectroscopy gives information about the types of bonds in a molecule, ¹H NMR spectroscopy provides information on the chemical environment of all the hydrogen atoms in the molecule. The nuclei of hydrogen atoms possess spin and can exist in two possible states of equal energy. If a strong magnetic field is applied the spin states may align themselves either with the magnetic field, or against it, and there is a small energy difference between them. The nuclei can absorb a photon of energy when transferring from the lower to the higher spin state. The photon's energy is very small and occurs in the radio region of the spectrum. The precise energy difference depends on the chemical environment of the hydrogen atoms.

The position in the ¹H NMR spectrum where the absorption occurs for each hydrogen atom in the molecule is known as the chemical shift, and is measured in parts per million (ppm). The area under each peak corresponds to the number of hydrogen atoms in that particular environment.

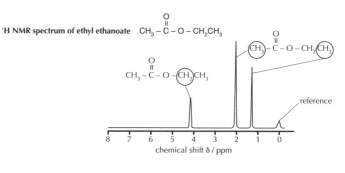

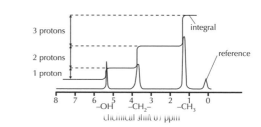

The additional trace integrates the area under each peak. The height of each section is proportional to the number of hydrogen atoms in each chemical environment.

USES OF NMR SPECTROSCOPY

1. Structural determination

¹H NMR is a particularly powerful tool in structural determination as it enables information to be gained on the precise chemical environment of all the protons in the molecule. Similarly ¹³C and other forms of NMR can also provide very detailed structural information including, for example, distinguishing between *cis*- and *trans*-isomers in organometallic compounds.

2. Medicinal uses

NMR is particularly useful in medicine as the energy of the radio waves involved is completely harmless and there are no known side effects. ³¹P is particularly useful in determining the extent of damage following a heart attack and in monitoring the control of diabetes. ¹H NMR is used in body scanning. The whole body of the patient can be placed inside the magnet of a large NMR machine. Protons in water, lipids, carbohydrates, etc. give different signals so that an image of the body can be obtained. This is known as MRI (magnetic resonance imaging). The image can be used to diagnose and monitor conditions, such as cancer, multiple sclerosis, and hydrocephalus.

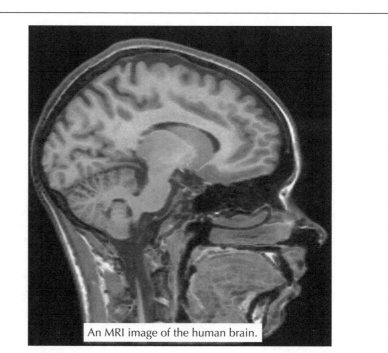

An MRI image of the human brain.

Atomic absorption (AA) spectroscopy and chromatography (1)

PRINCIPLES AND USES OF AA

In Topic 2 – Atomic structure it was explained that each element produces a unique emission spectrum. Some of the lines of the emission spectra lie in the visible region of the spectrum and give rise to the characteristic flame colours of some metal ions, e.g. sodium (yellow) and potassium (lilac). Atomic emission spectra are obtained by giving the atoms of the element sufficient electrical or heat energy and then recording the energy of the light emitted as the electrons fall from an excited state to a state of lower energy. Atomic absorption spectroscopy is the reverse process. In AA it is the energy absorbed by the electrons as they are promoted from a lower to a higher energy level that is measured. By recording the amount of light absorbed the concentration of that element in a particular sample can be precisely quantified. Atomic absorption spectroscopy is extremely sensitive and can measure concentrations as low as 1.0×10^{-6} g dm^{-3}, i.e. one part per billion. For that reason it is used to determine the concentration of metals in water, blood, soils, oils and foodstuffs.

Modern AA machines employ the double beam principle already outlined for an infrared spectrometer. A different light source is used for each element being analyzed. This utilizes the emission spectra of the element under study so that light of a specific frequency is being passed through the sample (analyte). For example, if the concentration of lead in water is to be determined then the light source would consist of excited lead atoms. The sample to be analyzed is first turned into a fine mist or aerosol in a nebulizer. This mixes with a combustion mixture made up from a fuel and an oxidizing agent. The most common combustion mixture is an ethyne (fuel)–air (oxidizing agent) mixture. The monochromatic light from the source is then passed through this vapourized sample contained inside the flame and the amount of light absorbed by the atoms is detected by converting it into an electrical signal using a photomultiplier. By changing the light source the concentration of different elements in the same sample can be determined.

INTRODUCTION TO CHROMATOGRAPHY

There are several ways in which mixtures can be separated into their individual components, for example filtration, crystallization, distillation, and solvent extraction. Chromatography can be used to separate mixtures containing very small amounts of the individual components. Coupled with other techniques, such as mass spectrometry, it can be used to separate and identify complex mixtures both quantitatively and qualitatively. It can also be used to determine how pure a substance is.

There are several different types of chromatography. They include paper, thin layer (TLC), column (LC), gas–liquid (GLC), and high performance liquid chromatography (HPLC). In each case there are two phases: a **stationary phase** that stays fixed and a **mobile phase** that moves. Chromatography relies upon the fact that in a mixture the components have different tendencies to adsorb onto a surface or dissolve in a solvent. This provides a way of separating them.

DETERMINING THE CONCENTRATION OF AN ELEMENT

The concentration of each element can be determined because there is a direct relationship between the amount of light absorbed and the concentration of the element in the path of the light. The amount of light absorbed, known as the absorbance, is measured by comparing the logarithm of initial intensity of the light, I_0, with the logarithm of the intensity of the emerging light, I, after it has passed through the sample. This relationship is known as the Beer-Lambert law which only applies for dilute solutions.

$$\text{Absorbance, } A = \log_{10}\frac{I_0}{I} = \varepsilon cl$$

where ε represents the molar absorptivity coefficient and is a constant for a particular element at a fixed frequency, l represents the path length through the sample cell and c represents the concentration of the element in the sample.

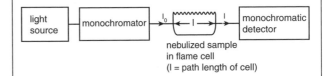

Simplified diagram of atomic absorption spectrometer

Since the path length can be fixed and the molar absorptivity coefficient is constant this means that the absorbance is directly proportional to the concentration.

In practice samples of known concentrations are used to produce a calibration curve. The absorbance of the unknown sample is then measured and the concentration can be read directly by interpolation from the calibration curve.

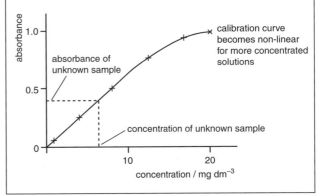

ADSORPTION AND PARTITION

Adsorption involves a solid stationary phase with a moving liquid phase (LC, HPLC, and at times TLC). The rate at which the solute moves through the solid phase depends on the equilibrium between its solubility in the moving liquid phase and its adsorption to the solid phase. The more tightly the component is adsorbed the slower it will elute.

Partition involves a stationary liquid phase and a mobile gaseous or liquid phase (paper, TLC, and GLC). It is dependent on the relative solubility of each component in the two phases if both are liquid. If the mobile phase is gaseous then the rate of movement depends upon the volatility of the components. The more soluble or volatile the component in the mobile phase the faster it will elute.

Chromatography (2)

PAPER CHROMATOGRAPHY

Most people are familiar with a simple example of paper chromatography. When black ink is placed on blotting paper the different colours in the ink move at different rates so it quickly separates into several colours. Paper consists largely of cellulose fibres. These contain a large number of hydroxyl groups making the paper quite polar. Water molecules hydrogen bond to these groups so that a sheet of 'dry' paper actually contains about 10% water. It is this water that acts as the stationary phase. The mobile phase is a solvent, either water itself or a polar organic liquid such as ethanol, propanone, or ethanoic acid. Normally a small amount of the mixture is spotted onto the paper about 1.0 cm from its base. The paper is then suspended in a small quantity of the solvent (known as the eluent) in a closed container. (The container is closed in order to saturate the atmosphere and prevent evaporation of the solvent from the paper to give better and faster separation.) As the solvent rises up the paper the components in the mixture partition between the two phases depending on their relative solubility. As the solvent nears the top of the paper a mark is made to record the level and the paper then removed and dried. Coloured components can be seen with the naked eye. Other components can be made visible by staining (e.g. with iodine) or by irradiating with an ultraviolet lamp.

As with all equilibria the partition of a solute between the two phases is constant at a fixed temperature, thus the solute will always move the same fraction of the distance moved by the solvent. Each solute will have a particular retention factor (R_f) for a given eluent. It is obtained by measuring the distance from the original spot both to the centre of the particular component and to the solvent front.

$$R_f = \frac{\text{distance moved by solute}}{\text{distance moved by solvent (eluent)}} = \frac{x}{y}$$

Substances can be identified by their R_f values. If two substances have similar R_f values in one solvent the paper can be turned 90° and a different solvent used. This is known as two way chromatography and is often used to separate amino acids (see Option B – Human biochemistry).

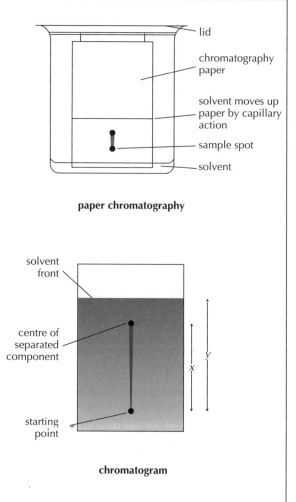

paper chromatography

chromatogram

COLUMN CHROMATOGRAPHY (LC)

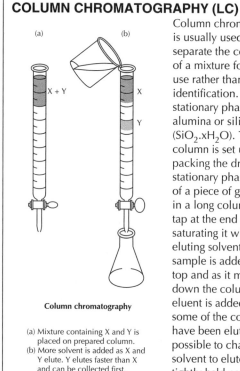

Column chromatography

(a) Mixture containing X and Y is placed on prepared column.
(b) More solvent is added as X and Y elute. Y elutes faster than X and can be collected first.

Column chromatography is usually used to separate the components of a mixture for further use rather than for identification. The stationary phase is alumina or silica gel ($SiO_2 \cdot xH_2O$). The column is set up by packing the dry stationary phase on top of a piece of glass wool in a long column with a tap at the end and then saturating it with the eluting solvent. The sample is added at the top and as it moves down the column more eluent is added. After some of the components have been eluted it is possible to change the solvent to elute the more tightly held components.

THIN LAYER CHROMATOGRAPHY (TLC)

This is very similar to paper chromatography but uses a thin layer of solid, such as alumina (Al_2O_3) or silica (SiO_2), on an inert support, such as glass. When absolutely dry it works by adsorption but, like paper, silica and alumina have a high affinity for water therefore the separation occurs more by partition, with water as the stationary phase. The separated components can be recovered pure by scraping off the section containing the component and dissolving it in a suitable solvent. Pregnancy tests may use TLC to detect pregnanediol in urine.

Visible and ultraviolet spectroscopy (1)

FACTORS AFFECTING THE COLOUR OF TRANSITION METAL COMPLEXES

Transition metals are defined as elements having an incomplete d sub-level in one or more of their oxidation states. Compounds of Sc^{3+} which have no d electrons and of Cu^+ and Zn^{2+} which both have complete d sub-shells are colourless. This strongly suggests that the colour of transition metal complexes is related to an incomplete d level. The actual colour is determined by four different factors.

1. The nature of the transition element. For example $Mn^{2+}(aq)$ and $Fe^{3+}(aq)$ both have the configuration $[Ar]3d^5$. $Mn^{2+}(aq)$ is pink whereas $Fe^{3+}(aq)$ is yellow.
2. The oxidation state. $Fe^{2+}(aq)$ is green whereas $Fe^{3+}(aq)$ is yellow.
3. The identity of the ligand. $[Cu(H_2O)_6]^{2+}$ (sometimes shown as $[Cu(H_2O)_4]^{2+}$, is blue, $[Cu(NH_3)_4(H_2O)_2]^{2+}$ (sometimes shown as $[Cu(NH_3)_4]^{2+}$), is blue/violet whereas $[CuCl_4]^{2-}$ is yellow (green in aqueous solution).
4. The stereochemistry of the complex. The colour is also affected by the shape of the molecule or ion. In the above example $[Cu(H_2O)_6]^{2+}$ is octahedral whereas $[CuCl_4]^{2-}$ is tetrahedral. However for the IB only octahedral complexes in aqueous solution will be considered.

SPLITTING OF THE d ORBITALS

In the free ion the five d orbitals are degenerate. That is they are all of equal energy.

Note that three of the orbitals (d_{xy}, d_{yz}, and d_{zx}) lie *between* the axes whereas the other two ($d_{x^2-y^2}$ and d_{z^2}) lie *along* the axes. Ligands act as Lewis bases and donate a non-bonding pair of electrons to form a co-ordinate bond. As the ligands approach the metal along the axes to form an octahedral complex the non-bonding pairs of electrons on the ligands will repel the $d_{x^2-y^2}$ and d_{z^2} orbitals causing the five

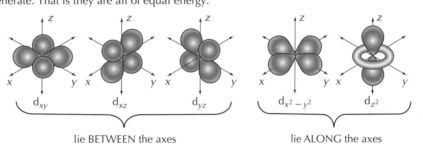

lie BETWEEN the axes   lie ALONG the axes

d orbitals to split, three to lower energy and two to higher energy. The difference in energy between the two levels corresponds to the wavelength of visible light.

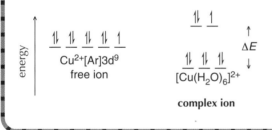

When white light falls on the aqueous solution of the complex the colour corresponding to ΔE is absorbed and the transmitted light will be the complementary colour. For example, $[Cu(H_2O)_6]^{2+}$ absorbs red light so the compound appears blue. The amount that the d orbitals are split will determine the exact colour. Changing the transition metal changes the number of protons in the nucleus which will affect the levels. Similarly changing the oxidation state will affect the splitting as the number of electrons in the level is different. Different ligands will also cause different amounts of splitting depending on their electron density.

COMPLEMENTARY COLOURS

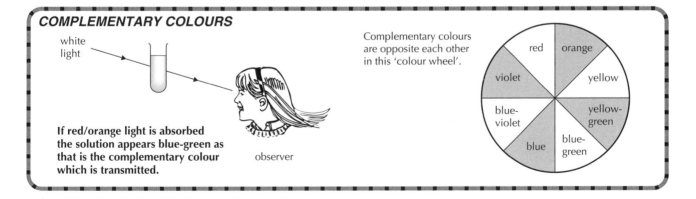

If red/orange light is absorbed the solution appears blue-green as that is the complementary colour which is transmitted.

observer

Complementary colours are opposite each other in this 'colour wheel'.

Modern analytical chemistry 85

THE SPECTROCHEMICAL SERIES

Ligands can be arranged in order of their ability to split the d orbitals.

$$I^- < Br^- < Cl^- < OH^- < H_2O < NH_3 < CN^-$$

This order is known as the spectrochemical series. Iodide ions cause the smallest splitting and cyanide ions the largest splitting. The energy of light absorbed increases when ammonia is substituted for water in Cu^{2+} complexes as the splitting increases, i.e. in going from $[Cu(H_2O)_6]^{2+}$ to $[Cu(NH_3)_4(H_2O)_2]^{2+}$. This means that the wavelength of the light absorbed decreases and this is observed in the colour of the transmitted light which changes from blue to a blue-violet colour.

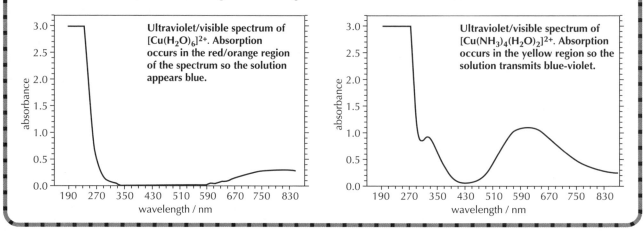

Ultraviolet/visible spectrum of $[Cu(H_2O)_6]^{2+}$. Absorption occurs in the red/orange region of the spectrum so the solution appears blue.

Ultraviolet/visible spectrum of $[Cu(NH_3)_4(H_2O)_2]^{2+}$. Absorption occurs in the yellow region so the solution transmits blue-violet.

ORGANIC MOLECULES

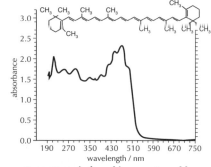

β-carotene is found in carrots and has a characteristic orange colour. It contains eleven conjugated double bonds and absorbs strongly in the violet-blue (400–510 nm) region.

Compounds containing unsaturated groups, such as C=C, C=O, –N=N–, –NO₂, and the benzene ring, can absorb in the ultraviolet or visible part of the spectrum. Such groups are known as chromophores and the precise energy of absorption is affected by the other groups attached to the chromophore. The absorption is due to electrons in the bond being excited to an empty orbital of higher energy, usually an anti-bonding orbital. The energy involved in this process is relatively quite high and most organic compounds absorb in the ultraviolet region and thus appear colourless. For example, ethene absorbs at 185 nm. However if there is extensive conjugation of double bonds in the molecule involving the delocalization of π electrons then less energy is required to excite the electrons and the absorption occurs in the visible region. Good examples include vitamin A (retinol), β-carotene, chlorophyll, and phenolphthalein.

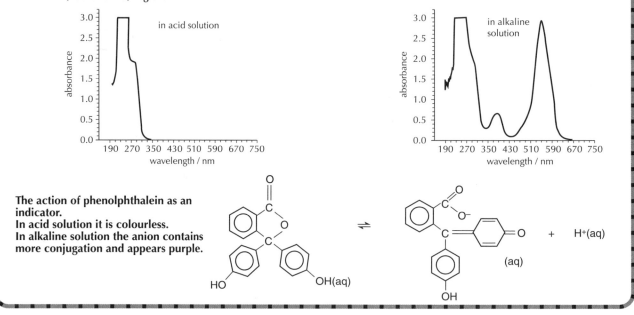

The action of phenolphthalein as an indicator.
In acid solution it is colourless.
In alkaline solution the anion contains more conjugation and appears purple.

Visible and ultraviolet spectroscopy (3)

BEER–LAMBERT LAW

The Beer–Lambert law states that:

$$\log_{10}\frac{I_o}{I} = \varepsilon l c$$

where: I_o is the intensity of the incident radiation and I is the intensity of the transmitted radiation.

ε is the molar absorption co-efficient (a constant for each absorbing substance).

l is the path length of the absorbing solution (usually 1.0 cm) and c is the concentration.

The intensity of the light transmitted will also depend on the wavelength and it is usual to carry out experiments involving the Beer–Lambert law at the wavelength of maximum absorption λ_{max}. Most spectrometers measure $\log_{10} I_o / I$ directly as absorbance. Often knowing the values of ε and λ_{max} is enough to identify a particular substance.

DETERMINING THE CONCENTRATION OF AN UNKNOWN SOLUTION

Most dilute solutions obey the Beer–Lambert law and it can be used to determine the concentration of an unknown solution. Examples include the amount of iron in a sample of blood, the percentage of copper in brass (by first converting it into a copper(II) ion solution), or investigating the reaction kinetics of a reaction involving coloured species. The method is essentially the same in each case and is illustrated by the following example.

A student was asked to plan an experiment to determine the concentration of Cu^{2+} in an unknown solution. She was provided with a separate solution of 1.00 mol dm^{-3} copper(II) sulfate.

Concentration of $CuSO_4$(aq) / mol dm^{-3}	Absorbance at 720 nm
0.100	0.362
0.150	0.498
0.200	0.798
0.250	0.901
0.300	1.002
0.500	1.751

Using a spectrometer the student measured the absorbance of a diluted sample of the copper(II) sulfate solution at different wavelengths to obtain the value of 720 nm for λ_{max}. She then carefully diluted the 1.00 mol dm^{-3} solution to give separate solutions with a range of known concentrations. The absorbance at 720 nm for each of these diluted solutions was then measured.

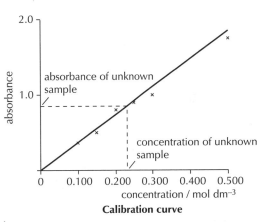

Calibration curve

The student used these results to plot a calibration curve using the line of best fit. The absorbance of the unknown solution at 720 nm was then measured. The value was found to be 0.850. From the graph the unknown concentration was determined to be 0.232 mol dm^{-3}.

Nuclear magnetic resonance (NMR) spectroscopy (2)

NUCLEAR SPIN

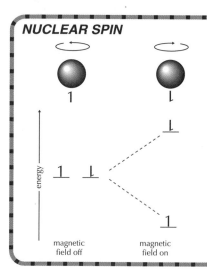

magnetic field off   magnetic field on

Unlike most other branches of spectroscopy NMR involves the interaction of **nuclei**, not electrons, with electromagnetic radiation. NMR concerns nuclei with uneven mass numbers, e.g. ^1H, ^{13}C, ^{19}F, and ^{31}P. All nuclei that possess spin behave as small magnets. In the absence of a magnetic field the two orientations of spin have the same energy. In the presence of a magnetic field the two orientations have a small difference in energy as they line up with or against the magnetic field. Nuclei can absorb energy as they move from the lower to the higher state. This small difference in energy corresponds to radio frequencies in the region of 10^7 to 10^8 s^{-1} (Hz).

To obtain an NMR spectrum the sample, either a pure liquid or in solution, together with a reference is placed in a cylindrical tube. The tube is then placed between the poles of a powerful electromagnet. A probe coil connected to a radio-frequency generator and receiver surrounds the sample. The spectrum is obtained either by varying the radio frequency or more commonly by varying the strength of the magnetic field.

THE MAIN FEATURES OF ^1H NMR SPECTRA

1. **The number of different absorptions (peaks)**
   Each proton in a particular chemical environment absorbs at a particular frequency. The number of peaks thus gives information as to the number of different chemical environments occupied by the protons.
2. **The area under each peak**
   The area under each absorption is proportional to the number of hydrogen atoms in that particular chemical environment. Normally each area is integrated and the heights of the integrated traces can be used to obtain the ratio of the number of hydrogen atoms in each environment.
3. **The chemical shift**
   Because spinning electrons create their own magnetic field the surrounding electrons of neighbouring atoms can exert a shielding effect. The greater the shielding the lower the frequency for the resonance to occur. The 'chemical shift' (δ) of each absorption is measured in parts per million (ppm) relative to a standard. The normal standard is tetramethylsilane (TMS) which is assigned a value of 0 ppm. A table of chemical shifts is to be found in the IB data booklet.
4. **Splitting pattern**
   In ^1H NMR spectroscopy the chemical shift of protons within a molecule is slightly altered by protons bonded to adjacent carbon atoms. This spin–spin coupling shows up in high resolution ^1H NMR as splitting patterns. If the number of adjacent equivalent protons is equal to n then the peak will be split into (n + 1) peaks.

Tetramethylsilane as a reference sample

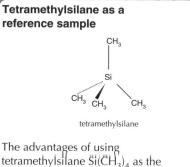

tetramethylsilane

The advantages of using tetramethylsilane $Si(CH_3)_4$ as the standard reference are:

- All the protons are in the same environment so it gives a strong single peak.
- It is not toxic and is very unreactive (so does not interfere with the sample).
- It absorbs upfield well away from most other protons.
- It is volatile (has a low boiling point) so can easily be removed from the sample.

INTERPRETING A ^1H NMR SPECTRUM

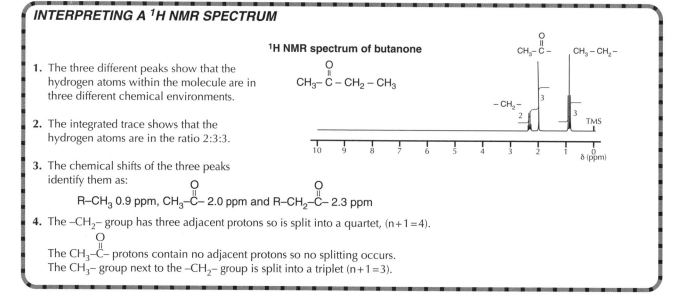

^1H NMR spectrum of butanone

1. The three different peaks show that the hydrogen atoms within the molecule are in three different chemical environments.

2. The integrated trace shows that the hydrogen atoms are in the ratio 2:3:3.

3. The chemical shifts of the three peaks identify them as:
   R–CH$_3$ 0.9 ppm, CH$_3$–C– 2.0 ppm and R–CH$_2$–C– 2.3 ppm

4. The –CH$_2$– group has three adjacent protons so is split into a quartet, (n+1=4).

   The CH$_3$–C– protons contain no adjacent protons so no splitting occurs.
   The CH$_3$– group next to the –CH$_2$– group is split into a triplet (n+1=3).

SPIN–SPIN COUPLING

Splitting patterns are due to spin–spin coupling. For example, if there is one proton (n = 1) adjacent to a methyl group then it will either line up with the magnetic field or against it. The effect will be that the methyl protons will thus experience one slightly stronger and one slightly weaker external magnetic field resulting in an equal splitting of the peak. This is known as a doublet (n + 1 = 2).

This atom splits the methyl absorption into a doublet

If there is a –CH_2– group (n = 2) adjacent to a methyl group then there are three possible energy states available.

1. Both proton spins are aligned with the field
2. One is aligned with the field and one against it (2 possible combinations)
3. Both are aligned against the field.

These two H atoms split the methyl absorption into a triplet

This results in a triplet with peaks in the ratio of 1:2:1.

The pattern of splitting can always be predicted using Pascal's triangle to cover all the possible combinations.

Number of adjacent protons (n)	Splitting pattern								Type of splitting
0					1				singlet
1				1		1			doublet
2			1		2		1		triplet
3		1		3		3		1	quartet
4	1		4		6		4		1	quintet

Thus a methyl group (n = 3) next to a proton will result in the absorption for that proton being split into a quartet with peaks in the ratio 1:3:3:1.

These three H atoms split the absorption due to the single proton into a quartet

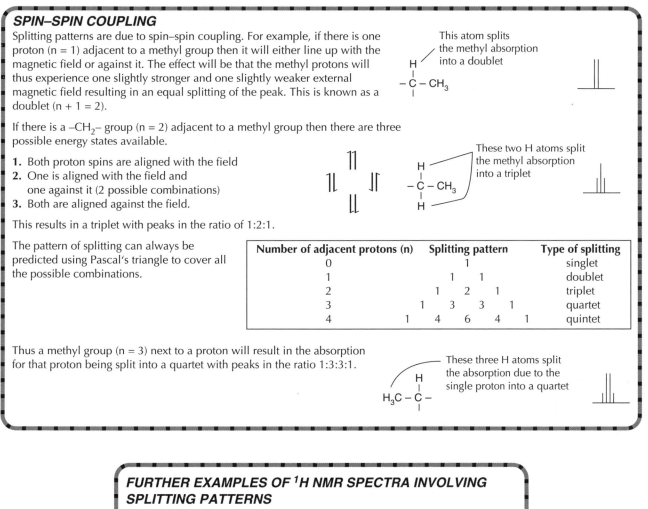

FURTHER EXAMPLES OF ^{1}H NMR SPECTRA INVOLVING SPLITTING PATTERNS

^{1}H NMR spectrum of phenylpropanone $C_6H_5 – \overset{\overset{O}{\|}}{C} – CH_2 – CH_3$

^{1}H NMR spectrum of 1,1-dichloroethane $CHCl_2CH_3$

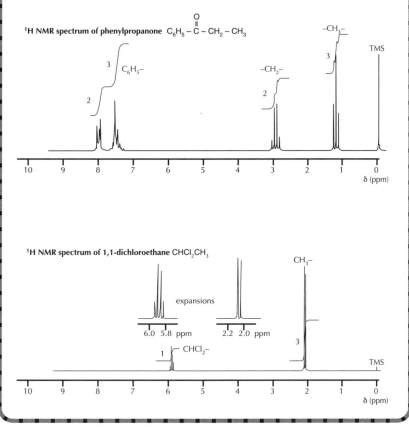

GAS–LIQUID CHROMATOGRAPHY (GLC)

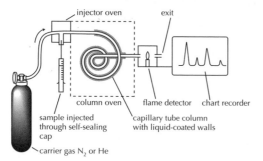

injector oven   exit

column oven

flame detector   chart recorder

sample injected through self-sealing cap

capillary tube column with liquid-coated walls

carrier gas N_2 or He

Schematic diagram of a gas–liquid chromatograph

This method is used to separate and identify the components present in mixtures of volatile liquids that do not decompose at temperatures at or near their boiling points. The stationary phase consists of a liquid (e.g. a long chain alkane) coated onto a solid support in a long, thin capillary tube. The mobile phase is an inert gas such as nitrogen or helium. The sample is injected through a self-sealing cap into an oven for vaporization. The sample is then carried by the inert gas into the column which is coiled and fitted into an oven. At the end of the column the separated components exit into a detector. This is usually a flame ionization detector connected to a chart recorder.

Each component will have a separate retention time and the area under the peak will be proportional to the amount of component present. GLC can thus detect the alcohol concentration in a sample of blood and can be used to provide evidence in court cases involving drunken driving. It is possible to programme the temperature of the oven to increase during the operation to speed up the elution of the less volatile components.

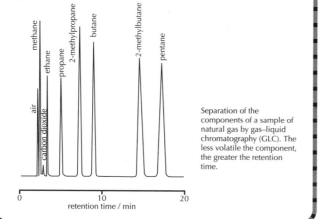

Separation of the components of a sample of natural gas by gas–liquid chromatography (GLC). The less volatile the component, the greater the retention time.

retention time / min

HIGH PERFORMANCE LIQUID CHROMATOGRAPHY (HPLC)

Column chromatography uses gravity to allow the solvent to elute. High performance liquid chromatography works in much the same way but the mobile phase is forced through under pressure. The stationary phase is commonly composed of silica particles with long chain alkanes adsorbed onto their surface. The separation is very efficient so long columns are not needed. The separated components are usually detected by ultraviolet spectroscopy. Like GLC the results are recorded onto a chart showing the different retention times. HPLC can be used for identification as well as for separation. It is particularly useful for non-volatile components or components that decompose near their boiling point. It can also be used to separate enantiomers (chiral separation) using columns containing optically active material.

Other uses include the analysis of oil pollutants, alcoholic beverages, food (antioxidants, sugars and vitamins), pharmaceuticals, polymers, biochemical and biotechnology research and the quality control of insecticides and herbicides.

GAS CHROMATOGRAPHY – MASS SPECTROMETRY (GC–MS)

GC–MS is an incredibly powerful tool used extensively in forensic science and in medicine and other analytical laboratories. Because the mass spectrometer is so sensitive it can be used to detect and measure the concentration of extremely small samples of substances. It is used to detect traces of banned drugs in athletes, toxins in food samples, and the different components in mixtures of crude oil.

The technique works by combining the operation of gas liquid chromatography with mass spectrometry. As the separate components elute from the chromatograph they pass straight into a mass spectrometer. The mass spectrometer is connected to a computer which contains a library of the spectra of all known compounds. The computer matches the spectra and gives a print-out of all the separate components and their concentrations. A similar technique (HPLC–MS) combines high performance liquid chromatography with mass spectrometry.

IB QUESTIONS – OPTION A – MODERN ANALYTICAL CHEMISTRY

1. **(a)** Infrared spectroscopy is a powerful tool for identifying organic compounds. State what occurs at the molecular level during the absorption of infrared (ir) radiation and identify the change that is necessary for ir absorption to occur. Discuss why infrared studies are particularly helpful in the characterisation of organic molecules. **[4]**

   (b) Use information in Table 18 of the IB data booklet to list the absorption regions expected for:

   (i) ethanoic acid. **[2]**

   (ii) methyl methanoate. **[2]**

   (c) Identify the absorption listed in **(b)** which could be used to distinguish between these two compounds. Explain why the other absorptions could not be used. **[2]**

   (d) Identify the absorption listed in **(b)** which has the highest energy and calculate its wavelength in cm. **[2]**

2. A Student used the technique of ascending paper chromatography in an experiment to investigate some permitted food dyes (labelled P1 – P5). The result is shown below.

   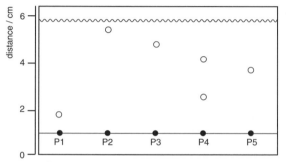

   (a) By reference to the diagram above, describe how the experiment would be carried out and explain the meaning of the terms *stationary phase, mobile phase, partition, solvent front* and *R_f value* **[8]**

   (b) Calculate the R_f value of P1. **[2]**

   (c) State, giving a reason, wheather P4 is a single substance or a mixture **[1]**

3. The infra-red spectrum of compound **A** is shown below.

   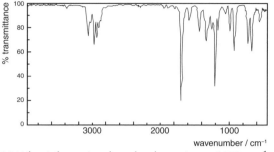

   (a) What information does the absorption at 1690 cm^{-1} give about compound **A**? **[1]**

   (b) There is a sharp absorption at 2950 cm^{-1}. What could this be due to? **[1]**

   (c) The infra-red spectrum of compound **A** does **not** show a broad absorption at about 3300 cm^{-1}. What information does give about compound **A**? **[1]**

4. This figure below depicts the visible region of the electromagnetic spectrum and the two regions nearest to it.

   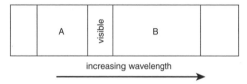

   (a) Name the regions labeled A and B, identify the atomic or molecular processes associated with each region and compare the energies of the photons involved in these processes. **[5]**

   (b) State, giving a reason, which region (A or B) could be used to obtain information about the strength of bonds. **[1]**

5. ^{1}H NMR spectroscopy can be used to distinguish between pentan-2-one and pentan-3-one. In each case state the number of peaks in their ^{1}H NMR spectra and state the ratios of the areas under each peak. **[4]**

HL

---

6. The structure of carotene is given below:

   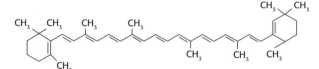

   (a) What feature of the carotene molecule accounts for the fact that it absorbs light in the visible region of the spectrum? **[2]**

   (b) The extinction coefficient of the carotene absorption spectrum is 130 000 dm^3 mol^{-1} cm^{-1} at its absorption maximum of 450 nm. Using the Beer-Lambert formula in the IB data booklet calculate the concentration of carotene required to absorb 99% of the incident light at that wavelength as it passes through a 1.0 cm cell. **[2]**

   (c) Carotene usually occurs mixed with other complex organic molecules. A technique often employed with such mixtures is to analyse them by combining gas phase chromatography with mass spectrometry. Explain briefly why these two techniques complement each other so well. **[2]**

7. The ^{1}H NMR spectrum of compound A is shown below:

   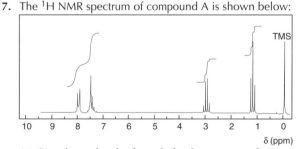

   (a) Give the molecular formula for the compound responsible for the peak at 0 ppm and suggest **two** reasons why it is used as reference. **[3]**

   (b) What information do the integration lines give about the numbers of protons with chemical shifts at 1.2, 3.0, and 7.3—8.1 ppm? **[1]**

   (c) The shifts at 1.2 and 3.0 ppm are split into a triplet and a quartet respectively. What information does this splitting pattern give about the structure of compound **A**? **[2]**

   (d) What information about compound **A** can be deduced from the peaks occurring between 7.3 and 8.1 ppm? **[1]**

Energy and amino acids

ENERGY

The daily calorific intake for a moderately active woman is about 8400 kJ (2000 kcal) per day. For an adult male undertaking physical work this increases to about 14 700 kJ (3500 kcal). Energy is provided by fats, carbohydrates, and proteins. Carbohydrates provide the main source of energy but like proteins are already partially oxidized so do not provide as much energy weight-for-weight as fats which are used to store energy.

FOOD CALORIMETRY

The energy content of a food can be found by burning the food in a food calorimeter. A known mass of the food is heated electrically and burned in a supply of oxygen. The heat produced is transferred through a copper spiral to water and the temperature increase of the water recorded. The 'water equivalent' of the whole system is calibrated and the calorific value of the food determined. Calorific values are typically recorded as kcal per 100 g or more commonly as kJ per 100 g. 1 kcal = 4.18 kJ.

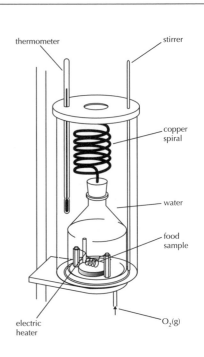

Sample calculation

A recommended daily helping of a breakfast cereal is 37.5 g. 1.00 g of the dried breakfast cereal was combusted in a food calorimeter. The temperature rise was found to be 5.50 °C. The mass of water, stirrer, thermometer and glass surroundings had a water equivalent of 626 g. Calculate the energy content of the recommended amount of the cereal.

Heat evolved = mass of water x specific heat capacity of water x temperature rise

$$= 626 \times 4.18 \times 5.50 = 14400 \text{ J} = 14.4 \text{ kJ per 1.00 g of cereal.}$$

Energy content of recommended daily helping = 37.5 x 14.4 = 540 kJ

AMINO ACIDS AND POLYPEPTIDES

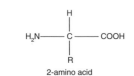

2-amino acid

Amino acids contain both an amine functional group and a carboxylic acid functional group. When they are both attached to the same carbon atom they are known as 2-amino acids (or α-amino acids).

There are about twenty 2-amino acids that occur naturally. They are the basic 'building blocks' of proteins in the body, which consist of long chains of polypeptides formed by condensation reactions between the amino acids.

When two amino acids condense they form a **dipeptide and water**. A **tripeptide** is made up from three amino acids. If the three amino acids are all different then there are six possible ways in which they can combine.

Possible tripeptides formed from alanine $H_2N–CHCH_3–COOH$, glycine $H_2N–CH_2–COOH$, and cysteine $H_2N–CH(CH_2SH)–COOH$:

$H_2N – Ala – Gly – Cys – COOH$

$HOOC – Ala – Gly – Cys – NH_2$

$H_2N – Gly – Ala – Cys – COOH$

$H_2N – Ala – Cys – Gly – COOH$

$HOOC – Ala – Cys – Gly – NH_2$

$HOOC – Gly – Ala – Cys – NH_2$

Structure and uses of proteins

STRUCTURE OF PROTEINS

Proteins are large macromolecules made up of chains of 2-amino acids. About twenty 2-amino acids occur naturally. The amino acids bond to each other through condensation reactions resulting in the formation of a polypeptide, in which the amino acid residues are joined to each other by an amide link (peptide bond).

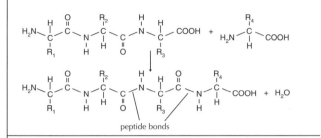

peptide bonds

Each protein contains a fixed number of amino acid residues connected to each other in strict sequence. This sequence, e.g. gly-his-ala-ala-leu- ... is known as the primary structure of proteins. The secondary structure describes the way in which the chain of amino acids folds itself due to intramolecular hydrogen bonding. The folding can either be α-helix in which the protein twists in a spiralling manner rather like a coiled spring, or β-pleated to give a sheet-like structure.

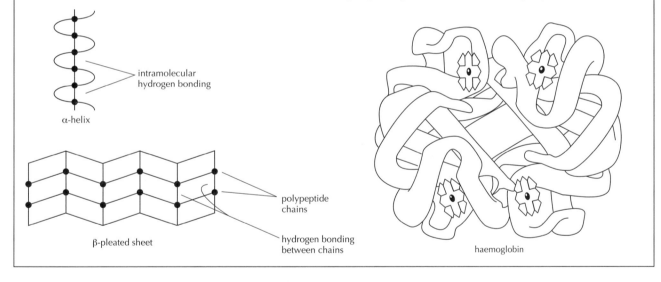

intramolecular hydrogen bonding

α-helix

β-pleated sheet

polypeptide chains

hydrogen bonding between chains

The tertiary structure describes the overall folding of the chains by interactions between distant amino acids to give the protein its three-dimensional shape. These interactions may be due to hydrogen bonds, van der Waals' attraction between non-polar side groups, and ionic attractions between polar groups. In addition two cysteine residues can form **disulfide bridges** when their sulfur atoms undergo oxidation.

Examples of interactions between side groups on polypeptide chains:

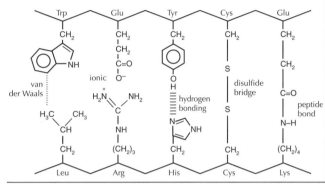

Separate polypeptide chains can interact together to give a more complex structure – this is known as the quaternary structure. Haemoglobin has a quaternary structure that includes four protein chains (two α-chains and two β-chains) grouped together around four haem groups.

haemoglobin

USES OF PROTEINS

Proteins have many different functions in the body. They can act as biological catalysts for specific reactions (enzymes). They can give structure (e.g hair and nails consist almost entirely of polypeptides coiled into α-helices), and provide a source of energy. Some hormones are proteins, e.g. FSH (follicle stimulating hormone), responsible for triggering the monthly cycle in females.

Analysis of proteins

The primary structure of proteins can be determined either by paper chromatography or by electrophoresis. In both cases the protein must first be hydrolysed by hydrochloric acid to successively release the amino acids. The three-dimensional structure of the complete protein can be confirmed by X-ray crystallography.

PAPER CHROMATOGRAPHY

A small spot of the unknown amino acid sample is placed near the bottom of a piece of chromatographic paper. Separate spots of known amino acids can be placed alongside. The paper is placed in a solvent (eluent), which then rises up the paper due to capillary action. As it meets the sample spots the different amino acids partition themselves between the eluent and the paper to different extents, and so move up the paper at different rates. When the eluent has nearly reached the top, the paper is removed from the tank, dried, and then sprayed with an organic dye (ninhydrin) to develop the chromatogram by colouring the acids. The positions of all the spots can then be compared.

If samples of known amino acids are not available the R_f value (retention factor) can be measured and compared with known values as each amino acid has a different R_f value. It is possible that two acids will have the same R_f value using the same solvent, but different values using a different solvent. If this is the case the chromatogram can be turned through 90° and run again using a second solvent.

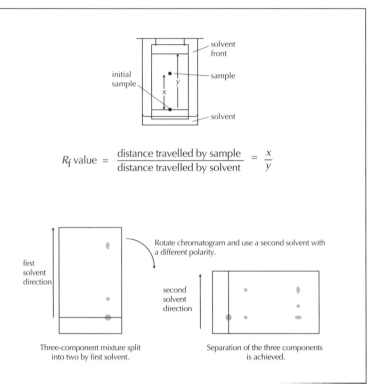

$$R_f \text{ value} = \frac{\text{distance travelled by sample}}{\text{distance travelled by solvent}} = \frac{x}{y}$$

Three-component mixture split into two by first solvent.

Rotate chromatogram and use a second solvent with a different polarity.

Separation of the three components is achieved.

ELECTROPHORESIS

The structure of amino acids alters at different pH values. At low pH (acid medium) the amine group will be protonated. At high pH (alkaline medium) the carboxylic acid group will lose a proton. This explains why amino acids can function as buffers. If H^+ ions are added they are removed as $-NH_4^+$ and if OH^- ions are added the $-COOH$ loses a proton to remove the OH^- ions as water. For each amino acid there is a unique pH value (known as the **isoelectric point**) where the acid will exist as the zwitterion.

The medium on which electrophoresis is carried out is usually a polyacrylamide gel. So the process is known as PAGE (polyacrylamide gel electrophoresis). The sample is placed in the centre of the gel and a potential difference applied across it. Depending on the pH of the buffer the different amino acids will move at different rates towards the positive and negative electrodes. At its isoelectric point a particular amino acid will not move as its charges are balanced. When separation is complete the acids can be sprayed with ninhydrin and identified by comparing the distance they have travelled with standard samples, or from a comparison of their isoelectric points.

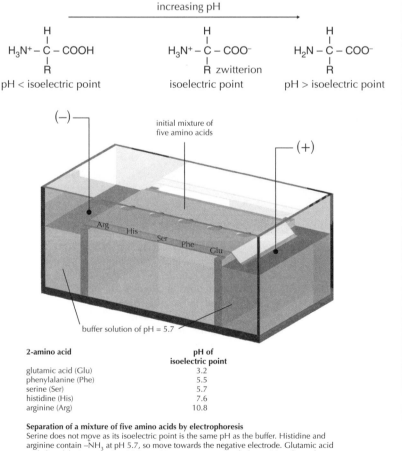

increasing pH

pH < isoelectric point

isoelectric point

pH > isoelectric point

2-amino acid	pH of isoelectric point
glutamic acid (Glu)	3.2
phenylalanine (Phe)	5.5
serine (Ser)	5.7
histidine (His)	7.6
arginine (Arg)	10.8

Separation of a mixture of five amino acids by electrophoresis
Serine does not move as its isoelectric point is the same pH as the buffer. Histidine and arginine contain $-NH_3$ at pH 5.7, so move towards the negative electrode. Glutamic acid and phenylalanine contain $-COO^-$ at pH 5.7, so move towards the positive electrode.

Carbohydrates

MONOSACCHARIDES

All monosaccharides have the empirical formula CH_2O. In addition they contain a carbonyl group ($>C=O$) and at least two –OH groups. They have between three and six carbon atoms. Monosaccharides with the general formula $C_5H_{10}O_5$ are known as pentoses (e.g. ribose) and monosaccharides with the general formula $C_6H_{12}O_6$ are known as hexoses (e.g. glucose).

Many structural isomers of monosaccharides are possible. In addition several carbon atoms are chiral (asymmetric) and give rise to optical isomerism. To make matters worse open chain structures and ring structures are possible. The form of glucose that is found in nature is known as D-glucose.

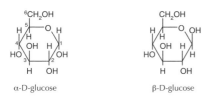

Straight chain formula of D-glucose

The ring structure of D-glucose can exist in two separate crystalline forms known as α-D-glucose and β-D-glucose. Note: the only difference is that the –OH group on the first carbon atom is inverted.

α-D-glucose   β-D-glucose

Six-membered ring monosaccharides are known as pyranoses. Hexoses can also have a furanose structure where they have a five-membered ring containing an oxygen atom.

MAJOR FUNCTIONS OF POLYSACCHARIDES IN THE BODY

Carbohydrates are used by humans:
- **to provide energy:** foods such as bread, biscuits, cakes, potatoes, and cereals are all high in carbohydrates
- **to store energy:** starch is stored in the livers of animals in the form of glycogen – also known as animal starch. Glycogen has almost the same chemical structure as amylopectin
- **as precursors** for other important biological molecules, e.g. they are components of nucleic acids and thus play an important role in the biosynthesis of proteins.
- **as dietary fibre:** Dietary fibre is mainly plant material that is not hydrolysed by enzymes secreted by the human digestive tract but may be digested by microflora in the gut. Examples include cellulose, hemicellulose, lignin and pectin. It may be helpful in preventing conditions such as diverticulosis, irritable bowel syndrome, obesity, Crohn's disease, haemerrhoids and diabetes mellitus.

POLYSACCHARIDES

Monosaccharides can undergo condensation reactions to form disaccharides and eventually polysaccharides. For example, sucrose, a disaccharide formed from the condensation of α-D-glucose in the pyranose form and β-D-fructose in the furanose form.

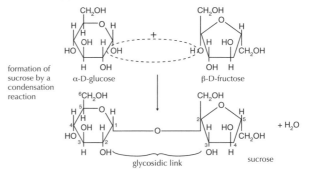

The link between the two sugars is known as a glycosidic link. In the case of sucrose the link is between the C-1 atom of glucose in the α-configuration and the C-2 atom of fructose. The link is known as an β-1,2 bond. Maltose, another disaccharide is formed from two glucose molecules condensing to form an α–1,4 bond. Lactose is a disaccharide in which the β-D-galactose is linked at the C-1 atom to the C-4 atom of β-D-glucose. This is called a β-1,4 bond.

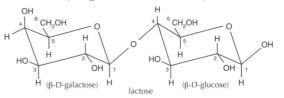

One of the most important polysaccharides is starch. Starch exists in two forms: amylose, which is water soluble, and amylopectin, which is insoluble in water.

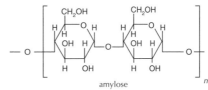

Amylose is a straight chain polymer of α-D-glucose units with α-1,4 bonds:

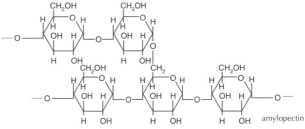

Amylopectin also consists of α-D-glucose units but it has a branched structure with both α-1,4 and α-1,6 bonds:

Most plants use starch as a store of carbohydrates and thus energy.

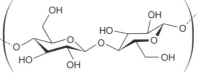

Cellulose, a polymer of β-D-glucose contains β–1,4 linkages. Cellulose, together with lignin, provides the structure to the cell walls of green plants. Most animals, including all mammals, do not have the enzyme cellulase so are unable to digest cellulose or other dietary fibre polysaccharides.

Lipids (1)

Lipids are organic molecules with long hydrocarbon chains that are soluble in non- polar solvents. They are mainly used for energy storage, insulating and protecting vital organs, forming cell membranes and, in some cases, acting as hormones. Three important types of lipids are triglycerides (fats and oils), phospholipids (lecithin) and steroids (cholesterol).

FATS AND OILS

Fats and oils are triesters (triglycerides) formed from the condensation reaction of propane-1,2,3-triol (glycerol) with long chain carboxylic acids (fatty acids).

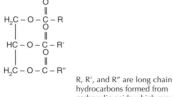

R, R', and R'' are long chain hydrocarbons formed from carboxylic acids which may be the same or different.

General formula of a fat or oil.

Fats are solid triglycerides, examples include butter, lard and tallow. Oils are liquid at room temperature and include castor oil, olive oil and linseed oil. The essential chemical difference between them is that fats contain saturated carboxylic acid groups (i.e. they do not contain C=C double bonds). Oils contain at least one C=C double bond and are said to be unsaturated. Most oils contain several C=C double bonds and are known as polyunsaturated.

PHOSPHOLIPIDS

Phospholipids form an integral part of all cell membranes. They are essentially made of four components. A backbone such as propane-1,2,3-triol (glycerol), linked by esterification to two fatty acids and a phosphate group which is itself condensed to a nitrogen containing alcohol. There are many different phospholipids. They can be exemplified by phosphatidyl choline – the major component of lecithin, present in egg yolk.

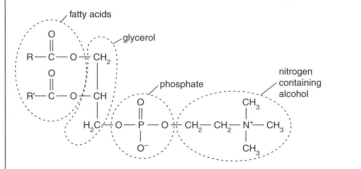

The structure of phosphatidyl choline showing the origins of the four distinct components.

CHOLESTEROL

Cholesterol has the characteristic four ring structure possessed by all steroids.

cholestrol

steroid 'backbone'

It is transported around the body by lipoproteins. **Low density lipoproteins** (LDL) are in the order of 18-25 nm and transport cholesterol to the arteries where it can line the walls of the arteries leading to cardiovascular diseases. The major source of these low density lipoproteins are saturated fats, in particular those derived from lauric (C_{12}), myristic (C_{14}) and palmitic (C_{16}) acids. Smaller lipoproteins, in the order of 8-11 nm, known as **high density lipoproteins** (HDL) can remove the cholesterol from the arteries and transport it back to the liver.

FATTY ACIDS

Stearic acid (m.pt 69.6 °C) and linoleic acid (m.pt –5.0 °C) both contain the same number of carbon atoms and have similar molar masses. However, linoleic acid contains two double bonds. Generally the more unsaturated the fatty acid the lower its melting point. The regular tetrahedral arrangement of saturated acids means that they can pack together closely, so the van der Waals' forces holding molecules together are stronger as the surface area between them is greater. As the bond angle at the C=C double bonds changes from 109.5° to 120° in unsaturated acids it produces a 'kink' in the chain. They are unable to pack so closely and the van der Waals' forces between the molecules become weaker, which results in lower melting points. This packing arrangement is similar in fats and explains why unsaturated fats (oils) have lower melting points.

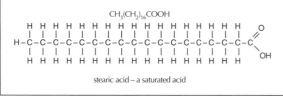

stearic acid – a saturated acid

Name		Number of C atoms per molecule	Number of C=C bonds	Melting point / °C
saturated fatty acids				
lauric acid	$CH_3(CH_2)_{10}COOH$	12	0	44.2
myristic acid	$CH_3(CH_2)_{12}COOH$	14	0	54.1
palmitic acid	$CH_3(CH_2)_{14}COOH$	16	0	62.7
stearic acid	$CH_3(CH_2)_{16}COOH$	18	0	69.6
unsaturated fatty acids				
oleic acid	$CH_3(CH_2)_7CH=CH(CH_2)_7COOH$	18	1	10.5
linoleic acid	$CH_3(CH_2)_4CH=CHCH_2CH=CH(CH_2)_7COOH$	18	2	–5.0

Lipids (2)

ESSENTIAL FATTY ACIDS

Most naturally occurring fats contain a mixture of saturated, mono-unsaturated and poly-unsaturated fatty acids and are classified according to the predominant type. Essential fatty acids are fatty acids that the body is unable to synthesize. Two essential unsaturated fatty acids are ω-6 linoleic acid and ω-3 linolenic acid. From these the body is able to synthesize longer and more unsaturated fatty acids. Green leaves are a good source of ω-3 fatty acids whereas most seeds and vegetable oils are a good source of ω-6 fatty acids.

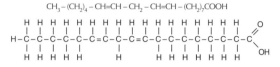

The structure of linoleic acid (*cis,cis*-9,12-octadecadienoic acid)

The double bonds in linoleic acid are on the ninth and twelfth carbon atoms of the acid but it is known as an omega(ω)-six fatty acid as the first double bond is on the sixth carbon atom from the end of the hydrocarbon chain (i.e. counting from the other end).

ω-3 fatty acids, such as linolenic acid, have the first double bond on the third carbon atom from the end of the hydrocarbon chain.

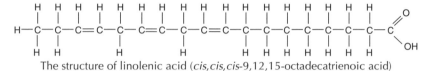

The structure of linolenic acid (*cis,cis,cis*-9,12,15-octadecatrienoic acid)

When fatty acids are made synthetically by partially hydrogenating other polyunsaturated fatty acids then the *trans*-isomers may be formed. *Trans*-fatty acids are present in fried foods such as French fries and some margarines. They increase the formation of LDL cholesterol and thus increase the risk of heart disease.

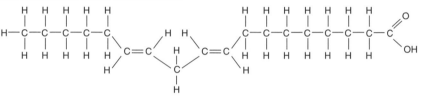

The structure of the *trans,trans*- form of linoleic acid

HYDROLYSIS OF FATS

In the body, fats and oils are hydrolyzed by enzymes, known as lipases, to glycerol and fatty acids.

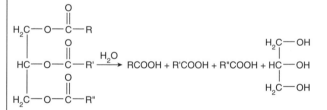

The hydrolysis of fats. (This is the reverse of the formation of fats from glycerol and fatty acids)

These in turn are broken down by a series of redox reactions to produce ultimately carbon dioxide, water and energy. Because they are essentially long chain hydrocarbons with only two oxygen atoms each on the three carboxyl atoms fats are in a less oxidized form than carbohydrates so weight for weight produce more energy.

DETERMINING THE NUMBER OF C=C BONDS IN AN UNSATURATED FAT

Unsaturated fats can undergo addition reactions. Margarine is made by the hydrogenation of vegetable oils so that it is a solid at room temperature. The addition of iodine to unsaturated fats can be used to determine the number of C=C double bonds since one mole of iodine will react quantitatively with one mole of double bonds. Iodine is coloured. As the iodine is added to the unsaturated fat the purple colour of the iodine will disappear as the addition reaction takes place. Often fats are described by their iodine number, which is the number of grams of iodine that add to 100 g of the fat.

$$\text{C=C} + I_2 \longrightarrow \text{-C-C-}$$

unsaturated fat   diiodo-addition product

THE ROLES OF LIPIDS IN THE BODY

- Energy storage. Because they contain proportionately less oxygen than carbohydrates they release more energy when oxidized.
- Insulation and protection of organs. Fats are stored in adipose tissue which provides both insulation and protection to parts of the body.
- Steroid hormones. Examples include progesterone and testosterone.
- Cell membranes. Lipids provide the structural component of cell membranes.

In addition omega-3-poly-unsaturated fatty acids are thought to reduce the risk of heart disease and polyunsaturated fats may lower the level of LDL (bad) cholesterol. The negative effects of lipids include the increase in LDL cholesterol from *trans*- fatty acids and saturated fatty acids such as lauric (C_{12}), myristic (C_{14}) and palmitic (C_{16}) acids.

Micro- and macro-nutrients

DEFINITIONS

Micro-nutrients are substances required in very small amounts (mg or µg). They mainly function as a co-factor of enzymes and include vitamins and trace minerals such as Fe, Cu, F, Zn, I, Se, Mn, Mo, Cr, Co and B.

Macro-nutrients are substances that are required in relatively large amounts (> 0.005% of body mass) and include proteins, fats, carbohydrates and minerals (Na, Mg, K, Ca, P, S and Cl).

MALNUTRITION

Malnutrition occurs when either too much food is consumed which leads to obesity or the diet is lacking in one or more essential micro- and macro-nutrients. Specific micro-nutrient deficiencies include:

- Fe – anaemia
- I – goiter
- vitamin A (retinol) – xerophthalmia, night blindness
- vitamin B_3 (niacin) – pellagra
- vitamin B_1 (thiamin) – beriberi
- vitamin C (ascorbic acid) – scurvy
- Vitamin D (calciferol) – rickets

Deficiencies in macronutrients such as protein include marasmus and kwashiorkor.

Solutions to combat malnutrition include

- eating fresh food rich in vitamins and minerals
- adding nutrients that are missing in commonly consumed foods
- genetic modification of food
- providing nutritional supplements

VITAMINS

Vitamins can be classified as fat soluble or water soluble. The structure of fat soluble vitamins is characterised by long, non-polar hydrocarbon chains or rings. These include vitamins A, D, E, F and K. They can accumulate in the fatty tissues of the body. In some cases an excess of fat soluble vitamins can be as serious as a deficiency. The molecules of water soluble vitamins, such as vitamin C and the eight B-group vitamins, contain hydrogen attached directly to electronegative oxygen or nitrogen atoms that can hydrogen bond with water molecules. They do not accumulate in the body so a regular intake is required. Vitamins containing C=C double bonds and –OH groups are readily oxidized and keeping food refrigerated slows down this process.

VITAMIN A (RETINOL)

Although it does contain one -OH group, vitamin A is fat soluble due to the long non-polar hydrocarbon chain. Unlike most other vitamins it is not broken down readily by cooking. Vitamin A is an aid to night vision.

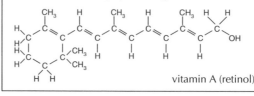

vitamin A (retinol)

VITAMIN C (ASCORBIC ACID)

Due to the large number of polar -OH groups vitamin C is soluble in water so is not retained for long by the body. The most famous disease associated with a lack of vitamin C is scorbutus ('scurvy'). The symptoms are swollen legs, rotten gums and bloody lesions. It was a common disease in sailors, who spent long periods without fresh food, until the cause was recognised.

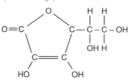
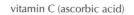

vitamin C (ascorbic acid)

VITAMIN D (CALCIFEROL)

Vitamin D is essentially a large hydrocarbon with one –OH group and is fat soluble.

A deficiency of vitamin D leads to bone softening and malformation - a condition known as rickets.

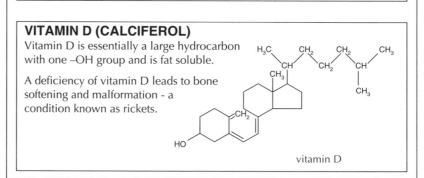

vitamin D

Hormones

Hormones are chemicals produced in endocrine glands and transported to the site of action by the blood stream. Hormones act as chemical messengers and perform a variety of different functions. Examples of specific hormones include antidiuretic hormone (ADH) which controls body water content, aldosterone (a steroid which regulates the sodium and potassium balance in the blood), adrenaline, thyroxin, insulin and the sex hormones.

ADRENALINE

Adrenaline (epinephrine) is produced in the adrenal glands – two small organs located above the kidneys. It is a stimulant closely related to the amphetamine drugs. It is released in times of excitement and causes a rapid dilation of the pupils and airways and increases heartbeat and the rate of release of sugar into the bloodstream. It is sometimes known as the 'fight or flight' hormone.

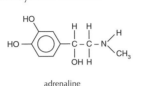

adrenaline

SEX HORMONES

The sex hormones are all steroids. Steroids contain a characteristic four-ring structure. The basic building block for the other steroids is cholesterol formed in the liver and found in all tissues, the blood, brain, and

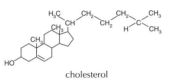

cholesterol

spinal cord. The male sex hormones are produced in the testes and comprise mainly testosterone and androsterone. They are anabolic – encouraging tissue, muscle, and bone growth – and androgenic – conferring

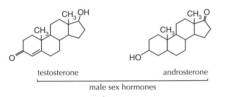

testosterone androsterone

male sex hormones

the male sexual characteristics. The female sex hormones are structurally very similar with just small changes in the functional groups attached to the steroid framework. They are produced in the ovaries from puberty until menopause. The two main female sex hormones are oestradiol and progesterone. They are responsible for sexual development and for the menstrual and reproductive cycles in women.

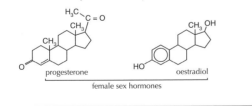

progesterone oestradiol

female sex hormones

THYROXINE

Thyroxine is produced in the thyroid gland located in the neck. It is unusual in that it contains iodine. A lack of iodine in the diet can cause the thyroid gland to swell to produce the condition known as goitre. Thyroxine regulates the body's metabolism. Low levels of thyroxine cause hypothyroidism, characterized by lethargy as well as sensitivity to cold and a dry skin. An overactive thyroid gland can cause the opposite effect. This is known as hyperthyroidism with the symptoms of anxiety, weight loss, intolerance to heat, and protruding eyes.

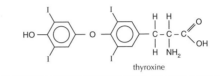

thyroxine

INSULIN

Insulin is a protein containing fifty-one amino acid residues. It is formed in the pancreas – an organ located at the back of the abdomen – and regulates blood sugar levels. In diabetics the levels of insulin are low or absent and glucose is not transferred sufficiently from the bloodstream to the tissues. This is known as hyperglycaemia and results in thirst, weight loss, lethargy, coma, and circulation problems. Long term sufferers of diabetes can suffer blindness, kidney failure, and need limbs amputated due to poor circulation. It is treated by reducing sugar intake and taking daily insulin injections. Too much insulin can cause hypoglycaemia, where the blood sugar level falls resulting in dizziness and fainting.

ORAL CONTRACEPTIVES

At the beginning of the menstrual cycle the pituitary releases follicle stimulating hormone (FSH). FSH travels to the ovaries causing the release of oestradiol, which prepares for the release of the ovum or egg and the build up of the uterine wall. After about two weeks a feedback system stops the release of FSH and triggers the release of the luteinizing hormone (LH). This travels to the ovaries and releases progesterone. The progesterone causes the egg to be transported to the uterus as well as continuing to build up the uterine wall. If the egg is fertilized it embeds itself into the uterine wall and hormone levels rise dramatically, otherwise hormone levels fall and menstruation begins.

The most common 'pill' contains a mixture of oestradiol and progesterone and mimics pregnancy by intentionally keeping the hormones at a high levels so that no more eggs are released. It is usual to take the pill for 21 days and then a placebo for 7 days so that a mild period will result, but without the risk that the hormone levels will fall and allow the unexpected release of an egg. Oestradiol and progesterone may also be given to post menopausal women as hormone replacement therapy (HRT) partly to prevent brittle bone disease (osteoporosis).

ANABOLIC STEROIDS

The anabolic steroids have similar structures to testosterone and build up muscle. They may be given to someone recuperating from a serious illness to build up muscles weakened by inactivity. Some athletes have abused these drugs as they can enhance athletic performance. Competitors are given random urine tests to detect these and other banned substances.

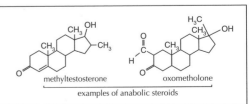

methyltestosterone oxomethalone

examples of anabolic steroids

 **Enzymes**

CATALYTIC ACTIVITY AND ACTIVE SITE

Enzymes are protein molecules that catalyse biological reactions. Each enzyme is highly specific for a particular reaction, and extremely efficient, often being able to increase the rate of reaction by a factor greater than 10^8. Enzymes work by providing an alternative pathway for the reaction with a lower activation energy, so that more of the reactant particles (substrate) will possess the necessary minimum activation energy.

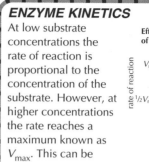

Effect of adding an enzyme on activation energy

The specificity of enzymes depends on their tertiary and quaternary structure. The part of an enzyme that reacts with the substrate is known as the active site. This is a groove or pocket in the enzyme where the substrate will bind. The site is not necessarily rigid but can alter its shape to allow for a better fit – known as the induced fit theory.

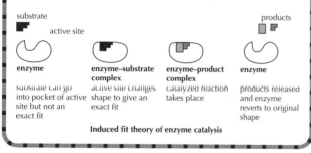

substrate can go into pocket of active site but not an exact fit

active site changes shape to give an exact fit

catalyzed reaction takes place

products released and enzyme reverts to original shape

Induced fit theory of enzyme catalysis

COMPETITIVE AND NON-COMPETITIVE INHIBITION

Inhibitors are substances that slow down the rate of enzyme-catalyzed reactions. Competitive inhibitors resemble the substrate in shape, but cannot react. They slow down the reaction because they can occupy the active site on the enzyme thus making it less accessible to the substrate. Non-competitive inhibitors also bind to the enzyme, but not on the active site. This causes the enzyme to change its shape so that the substrate cannot bind. As the substrate concentration is increased the effect of competitive inhibitors lessens, as there is increased competition for the active sites by the substrate. With non-competitive inhibitors increasing the substrate concentration has no effect, as the enzyme's shape still remains altered.

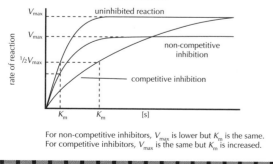

Effect of substrate concentration on inhibitors

For non-competitive inhibitors, V_{max} is lower but K_m is the same. For competitive inhibitors, V_{max} is the same but K_m is increased.

ENZYME KINETICS

At low substrate concentrations the rate of reaction is proportional to the concentration of the substrate. However, at higher concentrations the rate reaches a maximum known as V_{max}. This can be

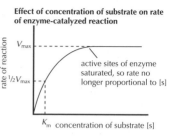

Effect of concentration of substrate on rate of enzyme-catalyzed reaction

explained in terms of enzyme saturation. At low substrate concentrations there are enough active sites present for the substrate to bind to and react. Once all the sites are used up the enzyme cannot work any faster.

The Michaelis–Menten constant K_m is the substrate concentration when the rate of the reaction is $\frac{1}{2} V_{max}$. K_m for a particular enzyme with a particular substrate will always be the same. It indicates whether the enzyme functions efficiently at low substrate concentrations, or whether high substrate concentrations are necessary for efficient catalysis.

EFFECT OF TEMPERATURE, pH, AND HEAVY METAL IONS ON ENZYME ACTIVITY

The action of an enzyme depends on its specific shape. Increasing the temperature will initially increase the rate of enzyme-catalyzed reactions, as more of the reactants will possess the minimum activation energy. The optimum temperature for most enzymes is about 40 °C. Above this temperature enzymes rapidly become denatured as the weak bonds holding the tertiary structure together break.

At different pH values the charges on the amino acid residues change affecting the bonds between them, and so altering the tertiary structure and making the enzyme ineffective. Heavy metals can poison enzymes by reacting with –SH groups replacing the hydrogen atom with a heavy metal atom, or ion so that the tertiary structure is altered.

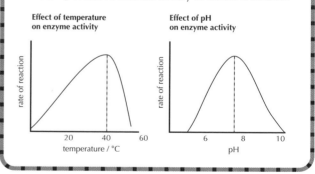

Effect of temperature on enzyme activity

Effect of pH on enzyme activity

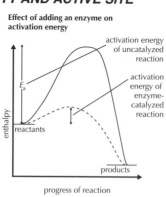

Nucleic acids

STRUCTURE OF NUCLEOTIDES AND NUCLEIC ACIDS

Almost all cells in the human body contain DNA (deoxyribonucleic acid). DNA and a related material RNA (ribonucleic acid) are macromolecules with relative molar masses of up to several million. Both nucleic acids are made up of repeating base-sugar-phosphate units called **nucleotides**. A nucleotide of DNA contains deoxyribose (a pentose sugar), a phosphate group, and one of four nitrogen-containing bases, adenine, guanine, cytosine, or thymine. RNA contains a different sugar, ribose, but also contains a phosphate group and four nitrogen-containing bases. Three of the bases are the same as those in DNA but the fourth, uracil, replaces thymine.

In DNA the polynucleotide units are wound into a helical shape with about 10 nucleotide units per complete turn. Two helices are then held together by hydrogen bonds between the bases to give the characteristic double helix structure. The hydrogen bonds are very specific. Cytosine can only hydrogen bond with guanine and adenine can only hydrogen bond with thymine (uracil in RNA).

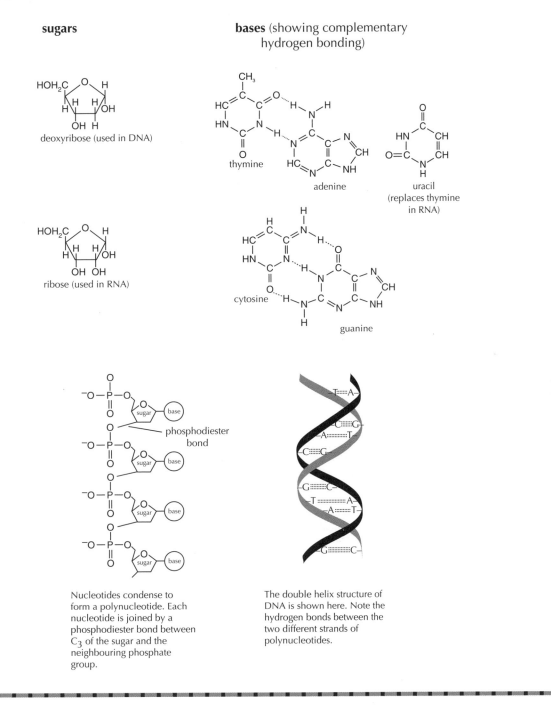

sugars

deoxyribose (used in DNA)

ribose (used in RNA)

bases (showing complementary hydrogen bonding)

thymine

adenine

uracil (replaces thymine in RNA)

cytosine

guanine

phosphodiester bond

Nucleotides condense to form a polynucleotide. Each nucleotide is joined by a phosphodiester bond between C_3 of the sugar and the neighbouring phosphate group.

The double helix structure of DNA is shown here. Note the hydrogen bonds between the two different strands of polynucleotides.

THE GENETIC CODE

When cells divide the genetic information has to be replicated intact. The genetic information is stored in chromosomes found inside the nucleus. In humans there are 23 pairs of chromosomes. Chromosomes are effectively a very long DNA sequence. The DNA in the cell starts to partly unzip as hydrogen bonds between the bases break. Sugar base units will be picked up from the aqueous solution to form a complementary new strand. Because adenosine can only hydrogen bond with thymine (A–T) and cytosine can only hydrogen bond with guanine (C–G) the new strand formed will be identical to the original.

DNA resides wholly in the nucleus whilst protein synthesis takes places in the cytoplasm. The information required to make complex proteins is passed from the DNA to messenger RNA by a similar unzipping process, known as transcription, except that the new strand of mRNA contains a different sugar and uracil in place of thymine.

The coded information held in the mRNA is then used to direct protein synthesis using a triplet code by a process known as translation. Each sequence of three bases represents one amino acid and is known as the **triplet code**. The triplet code allows for up to 64 permutations known as **codons**. This is more than sufficient to represent the 20 amino acids and several different codons may represent the same amino acid. Consecutive DNA codons of AAA, TAA, AGA, GTG, and CTT will transcribe to RNA codons of UUU, AUU, UCU, CAC, and GAA which will cause part of a strand of a protein to be formed that contains the amino acid residues - Phe-Ile-Ser-His-Glu. In 2000 the human genome – the complete sequence of bases in human DNA – was finally determined and published on the Internet.

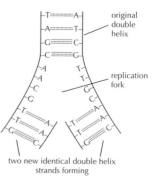

two new identical double helix strands forming

Replication of DNA

UUU	Phe	UCU	Ser	UAU	Tyr	UGU	Cys
UUC	Phe	UCC	Ser	UAC	Tyr	UGC	Cys
UUA	Leu	UCA	Ser	UAA	Terminator	UGA	Terminator
UUG	Leu	UCG	Ser	UAA	Terminator	UGG	Trp
CUU	Leu	CCU	Pro	CAU	His	CGU	Arg
CUC	Leu	CCC	Pro	CAC	His	CGC	Arg
CUA	Leu	CCA	Pro	CAA	Gln	CGA	Arg
CUG	Leu	CCG	Pro	CAG	Gln	CGG	Arg
AUU	Ile	ACU	Thr	AAU	Asn	AGU	Ser
AUC	Ile	ACC	Thr	AAC	Asn	AGC	Ser
AUA	Ile	ACA	Thr	AAA	Lys	AGA	Arg
AUG	Met	ACG	Thr	AAG	Lys	AGG	Arg
GUU	Val	GCU	Ala	GAU	Asp	GGU	Gly
GUC	Val	GCC	Ala	GAC	Asp	GGC	Gly
GUA	Val	GCA	Ala	GAA	Glu	GGA	Gly
GUG	Val	GCG	Ala	GAG	Glu	GGG	Gly

The genetic code carried by RNA

DNA PROFILING

DNA profiling (also known as DNA fingerprinting) was developed in 1985. In this process a small amount of cellular material is required. This may be obtained from blood, semen, hair, or saliva. The DNA is extracted and broken down into smaller fragments known as minisatellites using restriction enzymes. DNA not only contains coded triplets, but also has regions where there is no coded message in the base sequence. It is these regions where the splits occur to form the minisatellites which are unique to the person giving the sample (except in the case of identical twins). The fragments are separated into bands by gel electrophoresis. The resulting pattern is

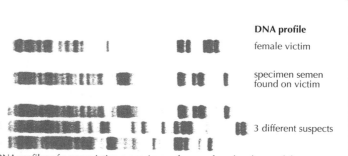

DNA profile

female victim

specimen semen found on victim

3 different suspects

DNA profiles of a rape victim, a specimen of semen found on her, and three different suspects. Which one was the rapist?

transferred to a nylon membrane and then labelled with radioactive ^{32}P which binds to particular bands of the DNA. X-ray film is then exposed to the radiation produced by the ^{32}P. The film is developed to give the characteristic 'fingerprint' of all the fragments. Use is made of this in court cases to positively identify murderers and rapists and to prove paternity. It also allows palaeontologists to map the evolutionary tree of extinct species.

AEROBIC AND ANAEROBIC RESPIRATION

When glucose burns in a plentiful supply of oxygen it forms carbon dioxide and water and releases energy.

$$C_6H_{12}O_6(s) + 6O_2(g) \longrightarrow 6CO_2(g) + 6H_2O\ (l)\quad \Delta H^\ominus = -2816\ kJ\ mol^{-1}$$

In the body one mole of glucose is broken down in a process known as glycolysis to two moles of pyruvate (2-oxopropanoate). Glycolysis produces energy by storing it in molecules of adenosine triphosphate, ATP. The pyruvate may then break down aerobically or anaerobically. In the aerobic decomposition glucose is oxidised by molecular oxygen (which is itself reduced) to form carbon dioxide and water and release energy. It can be summarized by the following two reactions:

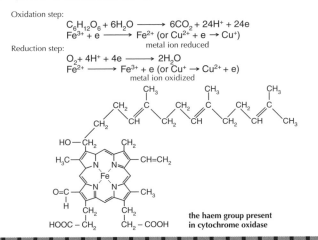

$$C_6H_{12}O_6 + O_2 \longrightarrow 2 \begin{array}{c}\text{(pyruvic acid structure)}\end{array} + 2H_2O$$

pyruvic acid
(pyruvate is the anion of the acid)

then

$$2\ C_3H_4O_3 + 5O_2 \longrightarrow 6\ CO_2 + 4H_2O$$

overall

$$2\ C_6H_{12}O_6 + 6O_2 \longrightarrow 6\ CO_2 + 6H_2O$$

Theoretically one mole of glucose can produce thirty six moles of ATP by aerobic decomposition. During anaerobic decomposition, which occurs in the absence of oxygen, only two moles of ATP are produced and the product formed in humans is lactic acid (2-hydroxypropanoic acid), $CH_3CH(OH)COOH$. However ATP is created more quickly during anaerobic decomposition so during short bursts of strenuous activity, muscle cells may also use anaerobic respiration which accounts for the build up of lactic acid in the muscles. In yeast the pyruvate is decomposed not to lactic acid but to ethanol and carbon dioxide. This is the basis of fermentation.

$$C_6H_{12}O_6 \longrightarrow 2C_2H_5OH + 2CO_2$$

ELECTRON TRANSPORT

Food does not burn in the body to produce energy, instead it is oxidized by a series of redox reactions involving the transport of electrons. The reactions take place in the mitochondria found inside cells. The oxidizing enzymes are called cytochromes. Cytochromes contain copper or iron. The metal ion in cytochrome oxidase is surrounded by a porphyrin ligand. This contains four nitrogen atoms each of which donates a pair of electrons, so that it occupies four of the sites around the metal ion and is said to be a tetradentate ligand. During each step of the the overall oxidation of glucose the Fe^{3+} ion is reduced to Fe^{2+} (or the Cu^{2+} ion is reduced to Cu^+). In the reduction stage when oxygen is reduced to water the Fe^{2+} ion is oxidized back to Fe^{3+} (or Cu^+ is oxidized to Cu^{2+}).

Oxidation step:
$$C_6H_{12}O_6 + 6H_2O \longrightarrow 6CO_2 + 24H^+ + 24e$$
$$Fe^{3+} + e \longrightarrow Fe^{2+}\ (\text{or } Cu^{2+} + e \rightarrow Cu^+)$$
metal ion reduced

Reduction step:
$$O_2 + 4H^+ + 4e \longrightarrow 2H_2O$$
$$Fe^{2+} \longrightarrow Fe^{3+} + e\ (\text{or } Cu^+ \rightarrow Cu^{2+} + e)$$
metal ion oxidized

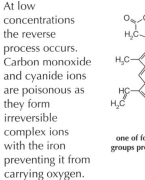

the haem group present
in cytochrome oxidase

HAEMOGLOBIN AS AN OXYGEN CARRIER

The ability of iron to form complexes is also important in haemoglobin. Haemoglobin in the blood is responsible for carrying oxygen around the body during respiration. Haemoglobin contains four large polypeptide groups, and four iron ions each surrounded by flat porphyrin ligands known as haem groups. Haem is a prosthetic group, a group essential for the protein to be able to carry out its function.

At high oxygen concentrations haemoglobin binds to oxygen molecules. In the process the oxygen bonds onto the iron in the haem group as an extra ligand.

At low concentrations the reverse process occurs. Carbon monoxide and cyanide ions are poisonous as they form irreversible complex ions with the iron preventing it from carrying oxygen.

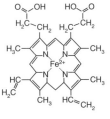

one of four haem prosthetic
groups present in haemoglobin

1. (a) The structures of two important sex hormones, testosterone and oestradiol are given in the IB data booklet. Apart from the extra methyl group, state **two** different functional groups which are present in testosterone but are absent in oestradiol. Suggest **one** simple chemical test which you could use to distinguish between oestradiol and testosterone in a school laboratory. [3]

   (b) Two synthetic steroids that are used in contraception are norethynodrel, the first commercially available oral contraceptive, and RU-486, the active ingredient of the 'morning after pill'.

   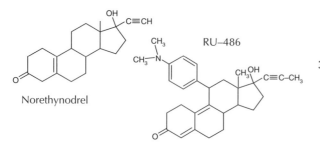

   Norethynodrel

   Name **two** different functional groups that are present in both norethynodrel and RU-486 but are absent in progesterone. Name **two** other functional groups that are present in RU-486 but absent in both progesterone and norethynodrel. Describe briefly how norethynodrel functions as an oral contraceptive. [7]

2. The diagram below shows a triacyclglycerol (triglyceride) molecule.

   $$CH_3(CH_2)_4CH=CHCH_2CH=CH(CH_2)_7CO-O-CH_2$$
   $$CH_3(CH_2)_4CH=CHCH_2CH=CH(CH_2)_7CO-O-CH$$
   $$CH_3(CH_2)_7CH=CH(CH_2)_7CO-O-CH_2$$

   (a) To which of the main dietary groups does this molecule belong? [1]

   (b) Is this molecule likely to have come from an animal or vegetable source? Explain your answer. [2]

   (c) Is this molecule likely to be a solid or an oil? Explain your answer. [2]

   (d) What are the major functions of this class of molecules in the body? [2]

3. The structures of vitamin A (retinol) and vitamin C (ascorbic acid) are given in the IB data booklet.

   (a) (i) Name **two** functional groups which are present in retinol. [2]

   (ii) By referring to the structures, classify vitamin A and vitamin C as water or fat soluble and account for the difference on the molecular level. [3]

   (iii) State **one** physical symptom of each of vitamin A and vitamin C deficiency. State the common name given to vitamin C deficiency. [3]

   (b) 0.014 moles of a particular oil was found to react exactly with 14.2g of iodine. Calculate the number of moles of iodine that reacted and state what can be deduced about the structure of the oil from this information. [3]

HL

4. (a) Iron combines reversibly with oxygen in haemoglobin. Apart from the ability to use different oxidation states, give **two** typical characteristics of transition metals that are shown by iron in haemoglobin. [2]

   (b) The ability of haemoglobin to carry oxygen at body temperature depends on the concentration of oxygen, the concentration of carbon dioxide and on the pH. The graph shows how the percentage saturation of haemoglobin with oxygen changes with pH at different partial pressures of oxygen.

   (i) The partial pressure of oxygen in active muscle is shown by the dotted line at 2.8 kPa. What is the difference in the percentage saturation of haemoglobin with oxygen in active muscle when the pH changes from 7.6 to 7.2? [1]

   (ii) When the cells in muscles respire they excrete carbon dioxide and sometimes lactic acid as waste products. Explain how this affects the ability of haemoglobin to carry oxygen. [2]

   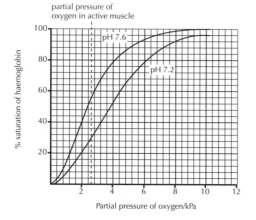

Iron

THE BLAST FURNACE

Iron is produced by reducing iron ores in a blast furnace. Traditionally the reducing agents are carbon monoxide and carbon. A modern blast furnace, which uses carbon monoxide, carbon, and hydrogen as the reducing agents, is capable of producing 10 000 tonnes of molten iron per day. Most of the molten 'pig' iron produced is converted directly into steel but some is cooled to make cast iron goods, such as engine cylinder blocks.

Raw materials

The solid charge is fed through the hopper via a conveyer belt. It consists of:

1. Iron ore (mainly haematite Fe_2O_3, magnetite Fe_3O_4, or hydrated oxides, e.g. geothite FeOH.OH and iron oxide obtained by roasting iron sulfides, e.g. iron pyrites, FeS_2) or scrap (recycled) iron.
2. Coke made by heating coal in the absence of air.
3. Limestone $CaCO_3$ to remove high melting point impurities by forming 'slag'.

In addition preheated air is blown in through nozzles, known as *tuyères*, at the bottom of the furnace. This air is enriched with oxygen and may also include hydrocarbons, such as oil or natural gas, to replace up to 40% of the coke.

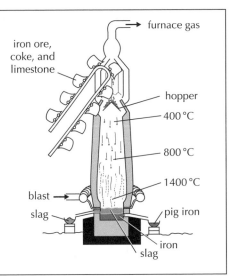

ESSENTIAL PROCESSES

Coke burns to form carbon monoxide:

$$2C(s) + O_2(g) \rightarrow 2CO(g)$$

In the reducing conditions incomplete combustion of the added hydrocarbons occurs:

e.g. $CH_4(g) + \frac{1}{2}O_2 \rightarrow CO(g) + 2H_2(g)$

The reducing gases pass up the furnace where they reduce the iron oxides in a series of stages depending on the temperature and composition of the gas. Examples of overall reactions taking place include:

$$Fe_2O_3(s) + 3CO(g) \rightarrow 2Fe(l) + 3CO_2(g)$$
$$Fe_3O_4(s) + 4H_2(g) \rightarrow 3Fe(l) + 4H_2O(g)$$
$$FeO(s) + CO(g) \rightarrow Fe(l) + CO_2(g)$$

In addition coke itself can reduce iron ore,

e.g. $Fe_2O_3(s) + 3C(s) \rightarrow 2Fe(l) + 3CO(g)$

The partially oxidized gases (furnace gas) which emerge from the top of the furnace are used as a fuel to preheat the air blasted in through the *tuyères*.

At high temperatures the limestone decomposes:

$$CaCO_3(s) \rightarrow CaO(s) + CO_2(g)$$

The carbon dioxide reacts with coke to produce carbon monoxide

$$CO_2(g) + C(s) \rightarrow 2CO(g)$$

and the coke can also react with water vapour to produce more carbon monoxide and hydrogen:

$$H_2O(g) + C(s) \rightarrow H_2(g) + CO(g)$$

The calcium oxide reacts with high melting point impurities to form a complex aluminosilicate 'slag' that also contains most of the sulfur impurities,

e.g. $CaO(s) + SiO_2(s) \rightarrow CaSiO_3(l)$

At the very high temperatures at the bottom of the furnace the molten iron and liquid slag separate into two layers with the less dense slag on top. Both are tapped off as more charge is added to the furnace in a continuous process. The molten iron (known as pig iron) contains phosphorus and sulfur together with small amounts of other elements, such as manganese and silicon and about 4 to 5% carbon. The slag is used for road making or is treated to make by-products, such as cement and thermal insulation.

Steel

CONVERSION OF IRON INTO STEEL

Most steel is made by the basic oxygen process. Molten iron is added to a vessel known as an oxygen converter. Preheated oxygen is injected at high pressure into the vessel and the impurities are oxidized,

e.g.
$$C + O_2 \rightarrow CO_2$$
$$P_4 + 5O_2 \rightarrow P_4O_{10}$$
$$Si + O_2 \rightarrow SiO_2$$

Apart from carbon dioxide the products react with added lime to form a slag. As the reactions are highly exothermic the temperature is controlled by adding scrap steel. The dissolved oxygen in the steel must be removed by adding controlled amounts of aluminium or silicon before the steel is suitable for casting or rolling. During this process other elements, such as chromium and nickel, are also added to form the precise alloy required.

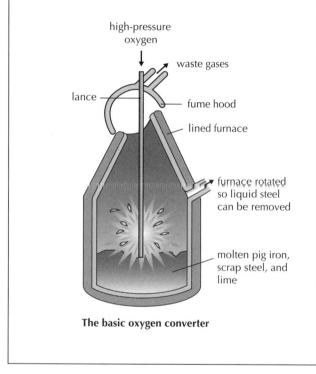

The basic oxygen converter

ALLOYS

Alloys are a homogeneous mixture of metals, or a mixture of metals and a non-metal. Transition metals often form alloys with each other as their atoms have similar atomic radii and the crystal structure is not seriously disrupted. The addition of other elements into the metallic structure can alter the physical properties. Some properties such as density, and electrical conductivity may not differ much from the constituent elements. However others, such as tensile strength and melting point, may differ considerably. For example, solder, an alloy of tin and lead, is a good conductor of electricity but has a lower melting point than either of its constituent elements. It is used to permanently join electrical contacts and to join two lead pipes.

PROPERTIES AND USES OF STEEL ALLOYS

Steel is an alloy of iron, carbon, and other metallic and non-metallic elements. It has a wide range of uses and by adjusting its composition it can be tailor-made with specific properties. For example, chromium increases the resistance of steel to corrosion. Stainless steel, used for kitchen knives and sinks, etc., contains about 18% chromium and 8% nickel. Toughened steel for use in drill bits, which need to retain a sharp cutting edge at high temperatures, contains up to 20% molybdenum. The mechanical properties of steel are affected by heat treatment. This involves the controlled heating and cooling of the steel to change the nature of the crystals in the steel. It does not alter the shape of the steel. Steel can be made less brittle by **tempering** which involves heating the steel to about 400 – 600 °C and allowing it to cool slowly. To make the steel more ductile a process known as **annealing** is used. This involves heating the steel to a higher temperature (about 1040 °C) followed by slow cooling. If hard steel is required then after annealing the steel must be **quenched** by rapid cooling.

RECYCLING OF IRON AND STEEL

The worldwide annual production of steel is about 750 million tonnes. The process of iron and steel making is very energy intensive. This is expensive and uses up precious energy resources. Damage to the environment is caused by extraction of iron ores and by the iron and steel making processes.

To conserve both energy and the environment efficient schemes are now in operation to recycle used or unwanted scrap iron or steel. Currently about 310 million tonnes of the steel produced worldwide comes from recycled steel. This represents an energy saving equivalent to about 160 million tonnes of coal and the conservation of about 200 million tonnes of iron ore.

Aluminium

PRODUCTION OF ALUMINIUM

Aluminium is primarily made by the electrolytic reduction of aluminium oxide.

The main ore of aluminium is bauxite. The aluminium is mainly in the form of the hydroxide $Al(OH)_3$ and the principal impurites are iron(III) oxide and titanium hydroxide. The impurities are removed by heating powdered bauxite with sodium hydroxide solution. The aluminium hydroxide dissolves because it is amphoteric.

$$Al(OH)_3(s) + NaOH(aq) \rightarrow NaAlO_2(aq) + 2H_2O(l)$$

The aluminate solution is filtered leaving the impurities behind. Seeding with aluminium hydroxide then reverses the reaction. The pure recrystallized aluminium hydroxide is then heated to produce aluminium oxide (alumina).

$$2Al(OH)_3(s) \rightarrow Al_2O_3(s) + 3H_2O(l)$$

In a separate process hydrogen fluoride is added to the aluminate solution followed by sodium carbonate to precipitate cryolite (sodium hexafluoroaluminate(III), Na_3AlF_6).

$$NaAlO_2(aq) + 6HF(g) + Na_2CO_3(aq) \rightarrow Na_3AlF_6(s) + 3H_2O(l) + CO_2(g)$$

The electrolysis of molten alumina takes places in an open-topped steel container lined with graphite. Alumina has a melting point of 2045 °C so it is mixed with cryolite. The resulting solution melts at about 950 °C so that much less energy is required. The aluminium is produced on the graphite lining which acts as the negative electrode (cathode). Molten aluminium is more dense than cryolite so it collects at the bottom of the cell where it can be syphoned off periodically.

$$Al^{3+}(l) + 3e^- \rightarrow Al(l)$$

The positive electrode is made of blocks of graphite. As the oxide ions are oxidized some of the oxygen formed reacts with the graphite blocks so that they have to be renewed regularly.

$$2O^{2-}(l) \rightarrow O_2(g) + 4e^-$$
$$C(s) + O_2(g) \rightarrow CO_2(g)$$

A modern cell can produce up to two tonnes of aluminium per day.

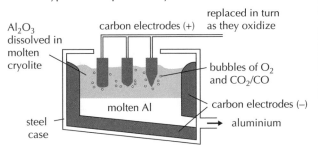

Production of aluminium by electrolysis

PROPERTIES AND USES

The aluminium produced from the electrolysis is more than 99% pure with the main impurities being small amounts of iron and silicon. Aluminium is malleable and can be shaped easily. It is an excellent conductor of heat and electricity. It has a much lower density than iron and yet can form alloys that are stronger than steel. Unlike iron it is resistant to corrosion as it forms a protective layer of aluminium oxide. The thickness of the oxide layer can be further increased by a process known as anodizing.

Some common uses of aluminium

Use	Examples
Transport	Superstructures of trains, ships and aeroplanes. Alloy engines for cars
Construction	Window frames, doors, roofing
Power transmission	Overhead electricity cables, capacitor foil
Kitchen utensils	Kettles, saucepans
Packaging	Drink cans, foil wrapping
Chemical industry	$Al(OH)_3$ – flame retarder, paper making
	$Al_2(SO_4)_3$ – flocculant in sewage treatment and to precipitate PO_4^{3-}
	Al_2O_3 – catalyst and catalytic support material, abrasive

ENVIRONMENTAL IMPACT OF ALUMINIUM PRODUCTION

The production of aluminium requires very large quantities of electricity. A cheap source of electricity is therefore essential and aluminium plants are usually located either where hydroelectric power is readily available or near sources of coal or natural gas that can be utilized for electricity generation. As with iron, the recycling of aluminium is an important method of saving energy and minimizing the damage to the environment. Iron and steel can easily be separated from aluminium by using a magnet. The cost of recycling aluminium is low as it requires only about 5% of the energy that would be needed to produce the same amount of aluminium from bauxite. Recycling aluminium also reduces global warming as less carbon dioxide (from the oxidized positive electrodes) is released into the atmosphere.

The oil industry

IMPORTANCE OF OIL AS A CHEMICAL FEEDSTOCK

Crude oil is a mixture of many different hydrocarbons, most of which are alkanes. It was formed millions of years ago from the decay of marine organisms. It is often found together with natural gas. The economy of the modern world depends to a large extent on oil. Its main use is as a fuel to supply the world's energy demands but about 10% of it is used, after refining, as chemical feedstock. This makes it the most significant source of organic chemicals including plastics, pesticides, food additives, pharmaceuticals, detergents, cosmetics, dyes, and solvents. It is a limited resource and future generations may question why most of it is being burned with all the attendant problems of pollution rather than being used to make useful products.

FRACTIONAL DISTILLATION OF OIL

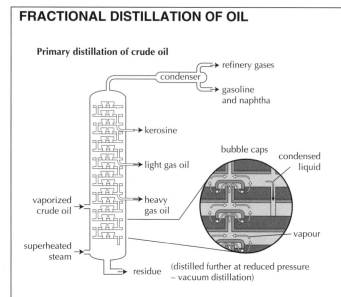

Primary distillation of crude oil

A single substance with a low boiling point can be separated from involatile impurities by simple distillation. When a mixture of volatile liquids with different boiling points is heated the vapour will be richer in the more volatile substances. By condensing the vapour then revaporizing the condensed liquid the new vapour will be even richer in the more volatile substances. By repeating the process many times the mixture can be separated into its different components according to their boiling point ranges – a process known as **fractional distillation**.

Crude oil is heated to about 450 °C and the vaporized hydrocarbons move up a fractionating column. Superheated steam is added to maintain the temperature and ensure the vapour moves up through the column. Bubble caps in the column allow the hot rising vapour to pass through the liquid that has condensed at each level. At each level the condensed higher boiling point components are tapped off or re-routed to lower levels.

REMOVAL OF SULFUR

Crude oil contains appreciable amounts of sulfur, mainly in the form of hydrogen sulfide, caused by the anaerobic decay of sulfur-containing amino acids in the original marine organisms. This must be removed before the oil can be refined as it can poison catalysts by blocking their active sites. It is also removed to prevent acid rain pollution caused by the production of sulfur dioxide when the oil is burned.

CRACKING

Cracking is the process conducted at high temperatures whereby large hydrocarbons are broken down into smaller, more useful molecules. The products are usually alkanes and alkenes. For example, decane can be broken down to form octane and ethene.

$$C_{10}H_{22}(g) \rightarrow C_8H_{18}(g) + C_2H_4(g)$$

The alkanes are usually branched isomers (e.g. 2,2,4-trimethylpentane) and are added to gasoline (petrol) to improve the octane rating. The alkenes are used to make other chemicals, particularly addition polymers.

Steam cracking

The feedstock is preheated, vaporized, and mixed with steam and then converted to low molecular mass alkenes at 1000–1150 K.

Catalytic cracking

The use of a silica/alumina catalyst enables the cracking to take place at the relatively lower temperature of about 750 K.

THE COMPOSITION AND CHARACTERISTICS OF CRUDE OIL FRACTIONS

The number of carbon atoms, boiling ranges, and the uses of the different fractions are summarized in the table:

Fraction	Carbon chain length	Boiling range / °C	Main uses
Refinery gas	1–4	<30	Used as fuel on site, gaseous cooking fuel, and as feedstock for chemicals, e.g. methane is used to provide hydrogen gas for the Haber process.
Gasoline and naphtha	5–10	40–180	Gasoline (petrol) for cars. Feedstock for organic chemicals (by steam cracking).
Kerosine	11–12	160–250	Fuel for jet engines; domestic heating; cracked to provide extra gasoline (petrol).
Gas oil (diesel oil)	13–25	220–350	Diesel engines and industrial heating; cracked to produce extra gasoline (petrol).
Residue	>20	>350	Fuel for large furnaces; vacuum distilled to make lubricating oils and waxes. Residue of bitumen and asphalt used to surface roads and waterproof roofs.

Addition polymers

EXAMPLES OF ADDITION POLYMERS

Addition polymers are formed by addition reactions of alkenes.

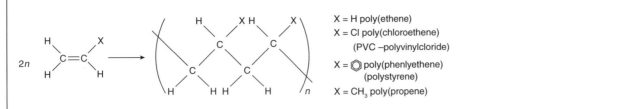

X = H poly(ethene)

X = Cl poly(chloroethene)
    (PVC –polyvinylcloride)

X = ⬡ poly(phenlyethene)
    (polystyrene)

X = CH₃ poly(propene)

RELATIONSHIP BETWEEN STRUCTURAL FEATURES AND PROPERTIES OF POLYMERS

Polymers can also be divided into thermoplastics and thermosets. Most alkenes polymerize to form thermoplastics which can be remoulded each time they are heated. Thermosets cannot be softened or remoulded once they are formed and will be permanently destroyed if heated to a high temperature. Generally the longer the chain length of a polymer the higher the melting point and strength. Apart from the length of the chain there are two other main structural features of addition polymers that affect their particular properties.

Branching

Depending on the reaction conditions ethene can form high density or low density polythene. In high density poly(ethene), HDPE there is little branching. This gives long chains that can fit together closely making the polymer stronger, denser, and more rigid than low density poly(ethene), LDPE. The presence of side chains in low density poly(ethene) results in a more resilient and flexible structure making it ideal for the production of film products, such as food wrappings.

Orientation of alkyl groups

In poly(propene) the methyl groups can all have the same orientation along the polymer chain – **isotactic**. Due to the regular structure isotactic polymers are more crystalline and tough. Isotactic poly(propene) is a thermoplastic and can be moulded into objects, such as car bumpers, and drawn into fibres for clothes and carpets. In **atactic** poly(propene) the chains are more loosely held so the polymer is soft and flexible, making it suitable for sealants and roofing materials.

isotactic poly(propene) – all methyl groups orientated in same direction

atactic poly(propene) – methyl groups arranged randomly

MODIFICATIONS

Plasticizers

Plasticizers are small molecules that can fit between the long polymer chains. They act as lubricants and weaken the attraction between the chains, making the plastic more flexible. By varying the amount of plasticizer added PVC can form a complete range of polymers from rigid to fully pliable.

Volatile hydrocarbons

If pentane is added during the formation of polystyrene and the product heated in steam the pentane vaporizes producing expanded polystyrene. This light material is a good thermal insulator and is also used as packaging as it has good shock-absorbing properties.

ADVANTAGES AND DISADVANTAGES OF POLYMER USE

The examples above illustrate how polymers can be tailor-made to perform a variety of functions based on properties such as strength, density, thermal and electrical insulation, flexibility, and lack of reactivity. There are, however, some disadvantages.

1. **Depletion of natural resources** The majority of polymers are carbon based. Currently oil is the major source of carbon although in the past it was coal. Both are fossil fuels and are in limited supply.
2. **Disposal** Because of their lack of reactivity plastics are not easily disposed of. Some, particularly PVC and poly(propene), can be recycled and others (e.g. nylon) are weakened and eventually decomposed by ultraviolet light. Plastics can be burned but if the temperature is not high enough poisonous dioxins can be produced along with toxic gases, such as hydrogen cyanide and hydrogen chloride.
3. **Biodegradability** Most plastics do not occur naturally and are not degraded by micro-organisms. By incorporating natural polymers, such as starch, into plastics they can be made more biodegradeable. However, in the anaerobic conditions present in landfills biodegradation is very slow or will not occur at all.

Catalysts

HOMOGENEOUS AND HETEROGENEOUS CATALYSIS

Catalysts function by providing an alternative reaction pathway with a lower activation energy so that more of the reactant particles will possess the necessary minimum energy to react when they collide. This has been covered in Topic 6 – Kinetics. Catalysts may be in the same phase as the reactants (usually the same phase as the product(s) too) in which case they are known as **homogeneous catalysts**. A phase is similar to a state except that there is a physically distinct boundary between two phases. It is possible to have a single state but two phases. An example is two immiscible liquids such as oil and water. If the catalyst is in a different phase to the reactants then it is functioning as a **heterogeneous catalyst**.

Examples of homogenous catalysis:

Sulphuric acid in esterification:

$$CH_3COOH(l) + C_2H_5OH(l) \underset{}{\overset{H^+(aq)}{\rightleftharpoons}} CH_3COOC_2H_5(l) + H_2O(l)$$

Nitrogen monoxide in the oxidation of sulfur dioxide

$$2SO_2(g) + O_2(g) \overset{NO(g)}{\rightleftharpoons} 2SO_3(g)$$

Iron(II) ions in the reaction between peroxodisulfate(VI) ions and iodide ions

$$S_2O_8^{2-}(aq) + 2I^-(aq) \overset{Fe^{2+}(aq)}{\rightleftharpoons} 2SO_4^{2-}(aq) + I_2(aq)$$

Examples of heterogeneous catalysis:

Iron in the Haber process to manufacture ammonia

$$N_2(g) + 3H_2(g) \overset{Fe(s)}{\rightleftharpoons} 2NH_3(g)$$

Nickel in the hydrogenation of alkenes

$$C_2H_4(g) + H_2(g) \overset{Ni(s)}{\rightleftharpoons} C_2H_6(g)$$

Vanadium(V) oxide in the Contact process

$$2SO_2(g) + O_2(g) \overset{V_2O_5(s)}{\rightleftharpoons} 2SO_3(g)$$

MODES OF ACTION OF CATALYSTS

Although homogenous catalysts can be recovered chemically unchanged at the end of the reaction they can form intermediate compounds during the reaction.

For example when nitrogen monoxide catalyses the oxidation of sulfur dioxide it is thought that nitrogen dioxide is formed as an intermediate compound.

$$2NO(g) + O_2(g) \rightarrow 2NO_2(g)$$

This is a redox reaction and the oxidation number of nitrogen has increased from +2 to +4. In the second step the nitrogen dioxide is reduced back to nitrogen monoxide by the sulfur dioxide

$$2SO_2(g) + 2NO_2(g) \rightarrow 2SO_3(g) + 2NO(g)$$

So that nitrogen monoxide is unchanged at the end of the reaction and the overall equation is

$$2SO_2(g) + O_2(g) \rightarrow 2SO_3(g)$$

Because transition metal ions can exist in more than one oxidation state they tend to make good homogenous catalysts. In the reaction between peroxodisulfate(VI) ions and iodide ions it is thought that the $Fe^{2+}(aq)$ ions are oxidized to $Fe^{3+}(aq)$ ions and then back to $Fe^{2+}(aq)$ ions.

$$S_2O_8^{2-}(aq) + 2Fe2+(aq) \rightarrow 2SO_4^{2-}(aq) + 2Fe^{3+}(aq)$$

$$2I^-(aq) + 2Fe^{3+}(aq) \rightarrow 2Fe^{2+}(aq) + I_2(aq)$$

Heterogeneous catalysts tend to function by adsorbing reactant molecules onto the surface of the catalyst (the active site) and bringing them into close contact with each other in the correct orientation. Many transition metals and their compounds have the ability to physically adsorb large amounts of gases on their surface which makes them particularly good heterogeneous catalysts. For example nickel or palladium can adsorb ethene and hydrogen so that they can react to form ethane.

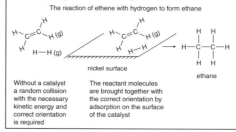

The reaction of ethene with hydrogen to form ethane

nickel surface

Without a catalyst a random collision with the necessary kinetic energy and correct orientation is required

The reactant molecules are brought together with the correct orientation by adsorption on the surface of the catalyst

ethane

CHOICE OF CATALYST

The advantage of a homogeneous catalyst compared to a heterogeneous catalyst is that all of the catalyst is exposed to the reactants whereas in heterogeneous catalysis the efficiency of the catalyst is dependent upon the surface area. The disadvantage of a homogenous catalyst is that it is usually harder to remove the catalyst after the reaction whereas a heterogeneous catalyst can be relatively easily removed by filtration. The most efficient catalysts of all are enzymes – biological catalysts – but these are very specific for a particular biological reaction due to the shape of the active site. When choosing a catalyst for a chemical reaction the following factors should be considered.

- Selectivity – Will the catalyst produce only the desired product?

- Efficiency – Will the catalyst cause a considerable increase in the rate?
  Will it continue to work well under severe conditions, such as those experienced by catalytic converters in cars, as well as mild conditions?

- Environmental impact – Will it be easy to dispose of the catalyst without causing harm to the environment? Many transition metals are classed as heavy metals and can cause problems if they enter the soil or ground water.

- Potential for poisoning – Catalysts rely on reactants occupying the active site reversibly. A poison will occupy the active site irreversibly so blocking access to reactants. Poisons include carbon monoxide, cyanide ions and sulfur.

- Cost – Industry is profit-based. Many transition metals such as rhodium, platinum and palladium are expensive and the cost to benefit ratio needs to be carefully calculated.

Fuel cells and rechargeable batteries

FUEL CELLS

A fuel cell utilizes the reaction between oxygen and hydrogen to produce water. Unlike combustion the energy is given out not as heat but as electricity. As reactants are used up more are added so a fuel cell can give a continuous supply of electricity. They are used in spacecraft as they do not need recharging. The electrolyte is aqueous sodium hydroxide. It is contained within the cell using porous electrodes which allow the passage of water, hydrogen, and oxygen.

Oxidation (– electrode)

$H_2 + 2OH^- \rightarrow 2H_2O + 2e^-$

Reduction (+ electrode)

$O_2 + 2H_2O + 4e^- \rightarrow 4OH^-$

Overall equation: $2H_2 + O_2 \rightarrow 2H_2O$

hydrogen in →

← oxygen in

$H_2O \rightarrow$

porous negative carbon elecrode with Pd or Pt catalyst

Na^+

$\leftarrow OH^-$

porous positive carbon electrode with Pt or Ag catalyst

water

NaOH solution

water out

hydrogen–oxygen fuel cell

RECHARGEABLE BATTERIES VERSUS FUEL CELLS

A battery is a general term for a voltaic or electrochemical cell in which chemical energy is converted into electrical energy (see Topic 9 – Oxidation and reduction). The electrons transferred in the spontaneous redox reaction taking place in the voltaic cell produce the electricity. Batteries are a useful way to store and transport relatively small amounts of energy. Some batteries (primary cells) can only be used once whereas secondary cells can be recharged. The main advantage of a fuel cell over a rechargeable battery is that it is non-polluting as water is the only product. As a fuel cell produces a continuous supply of electricity by the addition of more raw materials it does not need to be electrically recharged and the electrodes degrade less. However fuel cells are currently considerably more expensive than rechargeable batteries to produce.

NICKEL-CADMIUM AND LITHIUM-ION BATTERIES

Rechargeable nickel-cadmium (NiCd) batteries are used in electronics and toys. They have a cell potential of 1.2 V. The positive electrode is made of nickel hydroxide which is separated from the negative electrode made of cadmium hydroxide. The electrolyte is potassium hydroxide. During discharge the following reaction occurs:

$2NiO(OH) + Cd + 2H_2O \rightarrow 2Ni(OH)_2 + Cd(OH)_2$

This process is reversed during charging. One of the disadvantages of NiCd batteries is that cadmium is an extremely toxic heavy metal so the batteries need to be disposed of responsibly. Laptops, cell phones and other handheld devices often use lithium-ion batteries. These contain lithium atoms complexed to other ions, e.g. Li_xCoO_2, in the positive electrode and it is these ions rather than lithium itself which undergo the redox reactions. The negative electrode is made of graphite. Lithium-ion batteries are much lighter than NiCd batteries and produce a higher voltage, 3.6 – 4.2 V, but they do not have such a long lifespan.

LEAD–ACID BATTERY

The lead–acid battery is used in automobiles and is an example of a secondary cell. Usually it consists of six cells in series producing a total voltage of 12 V. The electrolyte is an aqueous solution of sulfuric acid. The negative electrodes are made of lead and the positive electrodes are made of lead(IV) oxide.

Oxidation (– electrode) $Pb + SO_4^{2-} \rightarrow PbSO_4 + 2e^-$

Reduction (+ electrode)

$PbO_2 + 4H^+ + SO_4^{2-} + 2e^- \rightarrow PbSO_4 + 2H_2O$

The overall reaction taking place is thus:

$Pb + PbO_2 + 4H^+ + 2SO_4^{2-} \rightarrow 2PbSO_4 + 2H_2O$

The reverse reaction takes place during charging. This can be done using a battery charger or through the alternator as the automobile is being driven. As sulfuric acid is used up during discharging the density of the electrolyte can be measured using a hydrometer to give an indication of the state of the battery. The disadvantages of lead–acid batteries are that they are heavy and both lead and sulfuric acid are potentially polluting.

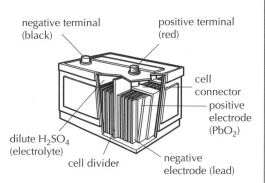

negative terminal (black)

positive terminal (red)

cell connector

positive electrode (PbO₂)

dilute H₂SO₄ (electrolyte)

cell divider

negative electrode (lead)

Lead–acid battery – during recharging hydrogen and oxygen are evolved from the electrolysis of dilute H_2SO_4 so it needs topping up occasionally with distilled water.

Liquid crystals (1)

LIQUID CRYSTALS

Liquid crystals are a phase or state of matter that lies between the solid and liquid state. In a liquid crystal the molecules tend to retain their orientation as in a solid but they can also move to different positions as in a liquid. The physical properties of liquid crystals (such as electrical conductivity, optical activity and elasticity) depend upon the orientation of the molecules relative to some fixed axis in the material. Examples of substances which can behave as liquid crystals under certain conditions include DNA, soap solution, graphite and cellulose together with some more specialized substances such as biphenyl nitriles.

An example of a biphenyl nitrile is 4-pentyl-4'-cyanobiphenyl (known as 5CB)

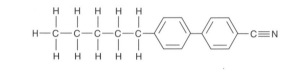

LYOTROPIC AND THERMOTROPIC LIQUID CRYSTALS

Liquid crystals only show liquid crystal properties under certain conditions. They are very sensitive to temperature and concentration. Essentially liquid crystals can be divided into two main types.

Lyotropic liquid crystals are solutions which show the liquid crystal phase at certain concentrations. An example of lyotropic liquid crystals is soap solution. At low dilution the polar soap molecules behave randomly but at higher concentrations they group together into larger units called micelles which in the liquid crystal phase are ordered in their orientation.

Thermotropic liquid crystals are pure substances that show liquid crystals behaviour over a range of temperature between the solid and liquid states. Examples of a thermotropic liquid crystal are biphenyl nitriles used in liquid crystal displays (LCDs).

The use of thermotropic liquid crystals in a calculator screen

Within the thermotropic liquid crystal phase the rod-shaped molecules which are typically about 2.5×10^{-9} metres in length exist in groups or domains. The molecules can flow and are randomly distributed as in a liquid but within each domain they all point in the same direction. This is known as the **nematic phase**. As the temperature increases the orientation becomes increasingly more disrupted until eventually the directional order is lost and the normal liquid phase is formed.

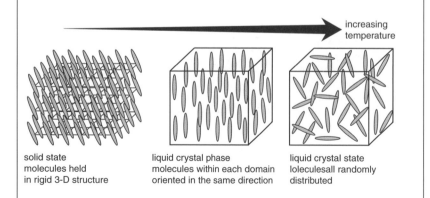

increasing temperature

solid state
molecules held
in rigid 3-D structure

liquid crystal phase
molecules within each domain
oriented in the same direction

liquid crystal state
loleculesall randomly
distributed

PRINCIPLES OF LCD DEVICES

Since liquid crystal molecules are polar their orientation can be controlled by an applied electric field. The orientation of the molecules affects the ability of the liquid crystal molecules to transmit light. In liquid crystal displays used in digital watches, calculators and laptops a small voltage is applied across a thin film of the material.

The flat screens used for computer monitors or televisions use liquid crystals

This controls the areas of the display that are light and dark and hence gives the characteristic readings of the pictures or letters. The great advantage of LCDs over other types of electronic display is that they use extremely small electric currents. The disadvantage is that they only work within a certain temperature range which explains why a digital watch or laptop screen may give a strange display in very hot or cold temperatures. Hence for use in an LCD a liquid crystal should:

- be a chemically stable compound
- contain polar molecules
- remain stable in the liquid crystal phase over a suitable range of temperature
- be able to orientate quickly (rapid switching speed).

Nanotechnology

NANOTECHNOLOGY

Nanotechnology is defined as the research and technology of compounds within the range of one to one hundred nanometres (1.0×10^{-9} m to 1.0×10^{-7} m) in length, i.e. on the atomic scale. It creates and uses structures which have novel properties based on their small size. Nanotechnology covers many separate scientific disciplines. At one extreme it can be seen as an extension of the existing sciences which have been reduced to the nano-scale or at the other extreme it can be seen as a completely new discipline of science and technology. There are two main approaches. The "bottom-up" approach involves building materials and devices from individual atoms, molecules or components. The "top-down" approach involves constructing nano-objects from larger entities. Sometimes physical techniques are used which allow atoms to be manipulated and positioned to specific requirements. For example, a process known as dip pen nanolithography can be used to place atoms in specific positions using an atomic force microscope. It is also possible to use chemical reactions such as in DNA nanotechnology where the specific base-pairing due to hydrogen bonding can be utilized to build desired molecules and structures.

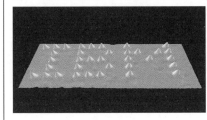

An image IBM 'written' with xenon atoms on nickel using dip pen nanolithography

NANOTUBES

Nanotubes are tubes with a diameter in the region of just a few nanometres. They are made using only carbon atoms. The basic building block is buckminsterfullerene, an allotrope of carbon, containing sixty carbon atoms arranged into hexagonal rings and pentagonal rings. By using a hemisphere of buckminsterfullerene to close the ends, tubes up to 4 cm in length have been made.

Both single and multi-walled tubes, made from concentric nanotubes, have been formed. A wide variety of different materials, including elements, metal oxides and even small proteins have been inserted inside the tubes. Because they have a greatly increased ratio of surface area to volume there is the possibility of them acting as extremely efficient and highly selective catalysts. They also have huge tensile strength – it is estimated that bundles of nanotubes should be in the region of fifty times stronger than steel and yet much lighter in mass. Because they are so small in size the quantum effect is appreciable which results in an altering of the electronic and optical properties traditionally associated with the different forms of carbon such as graphite. As the behaviour of electrons depends on the length of the tube some nanotubes are conductors and others are semiconductors. Research is currently underway to use carbon nanotubes to create transistors and other electronic devices much smaller than can be created using silicon chips. By inserting silver halides into single-walled nanotubes then decomposing them to silver, metallic nanowires of pure silver with a diameter of 2.0×10^{-8} m have been made which are the thinnest electrical 'wires' in existence.

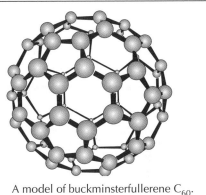

A model of buckminsterfullerene C_{60}.

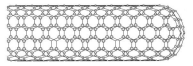

A diagram of part of a nanotube

IMPLICATIONS OF NANOTECHNOLOGY

Nanotechnology is concerned with the ability to control or manipulate on the atomic scale. It has the potential to solve many problems such as increase food production, prevent, monitor and cure diseases and improve information and communication technology, although most of these benefits still probably lie somewhere far in the future. The 2004 United Nations Task Force on Science, Technology and Innovation noted that some of the advantages of nanotechnology include production using little manual labour, land, or maintenance, high productivity, low cost, and modest requirements for materials. As a consequence many developing countries are investing large sums of money in research and development. However little is known about the potential risks associated with developing this technology. The hazards associated with small airborne particles are not properly known or covered by current toxicity regulations. The human immune system may be defenceless against new nano-scale products. There may also be social problems too as poorer societies may suffer as established technologies become redundant and demands for commodities change rapidly.

Condensation polymers

FORMATION OF CONDENSATION POLYMERS

Addition polymers are usually made from alkenes whereas condensation polymers are made from monomers that contain at least two reactive functional groups (or the same functional group at least twice). When the monomers condense to form the polymer small molecules such as water or hydrogen chloride are also produced. The formation of polyamides such as nylon and polyesters such as polyethylene terephthalate (PET) made from benzene–1, 4-dicarboxylic acid and ethane-1,2-diol have already been covered in Topic 10 – Organic chemistry. Other examples include phenol-methanal plastics and polyurethanes.

Phenol-methanal plastics are made by adding acid or alkali to a mixture of phenol and methanal. The methanal is first substituted in the 2- or 4- position in the phenol and then the product undergoes a condensation reaction with another molecule of phenol with the elimination of water.

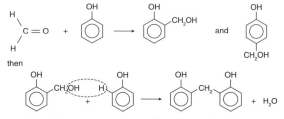

then

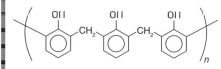

Further polymerization then takes place to build up a long chain, followed by covalent cross linking to form a three dimensional structure.

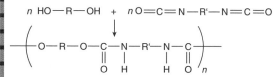

Polyurethanes are produced from the reaction of polyhydric alcohols (e.g. diols or triols) with compounds containing more than one isocyanate functional group, -NCO.

In general:

$$n\ HO-R-OH\ +\ n\ O{=}C{=}N-R'-N{=}C{=}O$$

$$\left(\!\!\begin{array}{c} O-R-O-\underset{\underset{O}{\parallel}}{C}-\underset{\underset{H}{\mid}}{N}-R'-\underset{\underset{H}{\mid}}{N}-\underset{\underset{O}{\parallel}}{C} \end{array}\!\!\right)_{\!n}$$

In one sense polyurethanes are addition polymers as no small molecules such as water are released in the reaction but unlike other addition polymers there are no urethane monomers which can add to each other. They are usually made *in situ* and moulded into the shape required.

STRUCTURAL FEATURES

The properties of condensation polymers depend upon their structural features.

In phenol-methanal plastics (commonly known as Bakelite) the benzene ring bonds in more than one position to form a 3-D cross-linked structure.

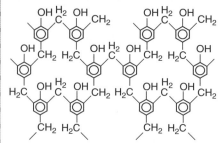

Part of the three-dimensional cross-linking structure of a phenol-methanal plastic

This gives it considerable strength and such a high melting point that it decomposes before melting (an example of a thermoset plastic). It also makes it very unreactive.

Due to its high electrical resistance it was used as the casing for early radios and is now used as one of the ingredients in worktops and printed circuit-board insulation.

Another example of cross-linking in a polymer is Kevlar, the material from which lightweight bullet-proof vests, composites for motor-cycle helmets and armour are made. Kevlar is a polyamide made by condensing 1,4-diaminobenzene (para-phenylenediamine) with benzene-1,4-dicarbonyl chloride (terephthaloyl dichloride).

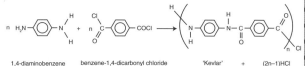

| 1,4-diaminobenzene | benzene-1,4-dicarbonyl chloride | 'Kevlar' | + | (2n–1)HCl |

It forms a strong three-dimensional structure due to hydrogen bonding between the long rigid chains.

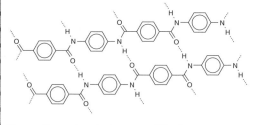

Three-dimensional structure of Kevlar

MODIFICATION OF POLYMERS

The addition of plasticizers and volatile hydrocarbons to addition polymers has already been covered. Air can be blown into polyurethanes to make polyurethane foams for use as cushions and thermal insulation. The fibres of polyesters can be blended with other manufactured or natural fibres for making clothes which are dye-fast and more comfortable than pure polyesters. When ethyne is polymerized the *cis*- form of poly(ethyne) is found to be an electrical conductor due to the delocalization of the π electrons.

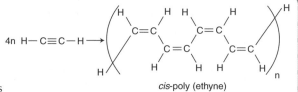

cis-poly (ethyne)

The conductivity can be dramatically increased by adding chemicals such as iodine which can remove electrons or alkali metals which can add electrons. This process is known as doping (the doping of silicon is covered on the next page). Like addition polymers the disadvantages of condensation polymers include their disposal and the effect on the environment. Condensation polymers are not generally biodegradable and when polyurethanes are burned they can release hydrogen cyanide gas.

Organic mechanisms / silicon and photovoltaic cells

MANUFACTURE OF POLY(ETHENE)

Low density poly(ethene), LDPE

The manufacture of low density poly(ethene) is carried out at very high pressures (1000–3000 atm) at a temperature of about 500 K. An initiator, such as an organic peroxide or a trace of oxygen, is added. Under these conditions free radicals are formed.

$$R–O–O–R \rightarrow 2RO^{\bullet}$$ free radical formation

$$RO^{\bullet} + H_2C{=}CH_2 \rightarrow R–O–CH_2–CH_2^{\bullet}$$ propagation

$$R–O–CH_2–CH_2^{\bullet} + H_2C{=}CH_2 \rightarrow R–O–CH_2–CH_2–CH_2–CH_2^{\bullet}$$ etc.

Termination takes place when two radicals combine. The average polymer molecule contains between about 4×10^3 and 4×10^4 carbon atoms with many short branches. The branches affect both the degree of crystallinity and the density of the material. LDPE generally has a density of about 0.92 g cm^{-3} and is used mostly for packaging.

High density poly(ethene), HDPE

High density poly(ethene) is manufactured by polymerising ethene at a low temperature (about 350 K) and pressure (1–50 atm) using a Ziegler–Natta catalyst. The catalyst is a suspension of titanium(III) or titanium(IV) chloride together with an alkyl–aluminium compound (e.g. triethylaluminium $Al(C_2H_5)_3$). The mechanism is complex and still not thoroughly understood. Essentially it involves the insertion of the monomer between the catalyst and the growing polymer chain. This is known as co-ordination polymerization (sometimes described as anionic polymerization). The titanium atom is attached to one end of the growing hydrocarbon chain and uses its empty d orbitals to form a co-ordination complex with the π electrons of the new incoming ethene molecule.

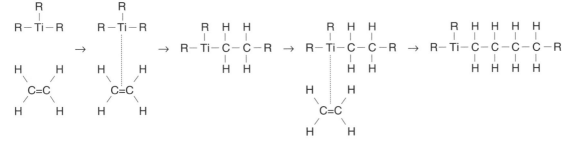

A simplified reaction sequence of the Ziegler–Natta catalysed polymerization of ethene is shown above.

The resulting polymer consists mainly of linear chains with very little branching. This gives it a higher density (0.96 g cm^{-3}) and a more rigid structure as the chains can fit together more closely. It is used to make containers and pipes.

SILICON AND PHOTOVOLTAIC CELLS

Metals conduct electricity because they contain delocalized electrons. Non-metals (apart from graphite) are poor conductors because the electrons are normally held in fixed positions. Silicon is a semiconductor.

A crystal of silicon contains a lattice of silicon atoms bonded to each other by shared pairs of electrons. These electrons are in fixed positions so silicon is a poor conductor under normal conditions. However, the energy required to excite an electron and free it from its bonding position is equivalent to the energy of light with a wavelength of 1.1×10^{-6} m. Visible light has a shorter wavelength in the range of $4–7 \times 10^{-7}$ m. This is higher in energy and so sunlight is able to excite an electron in silicon. The electron is then free to move through the crystal lattice making it an electrical conductor. This is the basis of the photoelectric effect and is the theory behind photoelectric cells. In practice the process is not very efficient and the cost of purifying the silicon is high. However solar cells are not polluting and do not use up valuable fossil fuel reserves.

Normal silicon with all electrons in fixed pairs.

Sunlight provides sufficient energy to release an electron from a fixed

One method of improving the efficiency of the photoelectric effect is by doping. This process involves adding very small amount of atoms of other elements usually from group 3 (Al, Ga, or In) or from group 5 (P or As). When a group 5 element is added the extra electron can move easily throughout the crystal lattice making it a better conductor compared with pure silicon. Such doping produces an n-type semiconductor because the conductivity is due to negative electrons. When a group 3 element is added the element now has one fewer electron than silicon. This produces a 'hole' in the lattice. When a free electron moves into this hole it produces a new hole where the electron was formerly located. The hole can be regarded as a positive carrier so the semiconductor is known as a p-type.

STRUCTURAL FEATURES OF BIPHENYL NITRILES

When the structure of 4-pentyl-4'-cyanobiphenyl is compared with other compounds which have good liquid crystal properties it can be seen that they have several features in common.

- The nitrile group in the biphenyl nitrile (and the $-NN^+(O^-)-$ and $-C=N-$ groups in the other two molecules) is polar. This ensures that the intermolecular forces are strong enough to align in a common direction.

- The two benzene rings in the molecules ensure that the molecules are rigid and therefore more rod-shaped.

- The long alkane chain group on the end of the molecule ensures that the molecules cannot pack so closely together and so helps to maintain the liquid-crystal state.

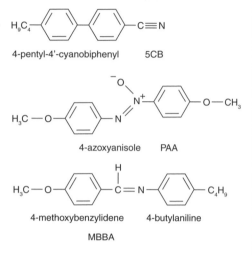

4-pentyl-4'-cyanobiphenyl   5CB

4-azoxyanisole   PAA

4-methoxybenzylidene   4-butylaniline

MBBA

THE WORKINGS OF THE LCD DEVICE

In a liquid-crystal display each pixel contains a liquid crystal film sandwiched between two glass plates. The plates have many very fine scratches at right angles to each other and have the property of polarizing light. The liquid crystal molecules in contact with the glass line up with the scratches and form a twisted arrangement between the plates due to intermolecular forces. This is known as **twisted nematic geometry**. The property utilized by the liquid crystals is their ability to interact with plane polarized light (see page 73) which is rotated through 90° by the molecules as it passes through the film. When the two polarizers are aligned with the scratches, light will pass through the film and the pixel will appear bright. When a potential difference is applied across the film, the polar molecules will align with the film thus losing their twisted structure and ability to interact with the light. Plane polarized light will now no longer be rotated so that the pixel appears dark.

the operation of the twisted nematic liquid crystal display

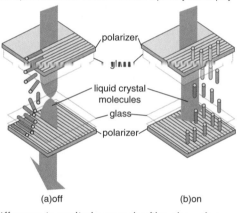

(a)off   (b)on

KEVLAR

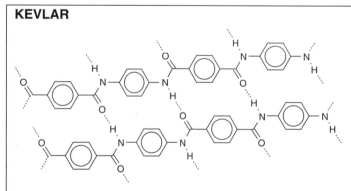

Kevlar consists of rigid, rod-shaped polar molecules with strong intramolecular hydrogen bonding between the chains. This gives a very ordered and strong structure hence its use as a protective material.

In a solution of concentrated sulfuric acid it can act as a lyotropic liquid crystal. Kevlar is lyotropic because the alignment of its molecules depends upon the concentration of the solution. In strong acid solution the oxygen and nitrogen atoms in the amide linkage become protonated and this breaks the hydrogen bonding between the chains. In very strong acidic solution the structure is broken down completely and Kevlar loses its liquid-crystal properties and also its protective properties so equipment made of Kevlar should be stored well away from acids.

The chlor-alkali industry

ELECTROLYSIS OF SODIUM CHLORIDE

Chlorine is a powerful oxidizing agent with a standard electrode potential of +1.36 V. Apart from fluorine, very few chemical oxidizing agents are powerful enough to oxidize chloride ions to chlorine so the manufacture of chlorine gas depends on using electrons themselves.

Chlorine gas is formed during the electrolysis of molten sodium chloride in the industrial production of sodium metal; however, the main source of chlorine is the electrolysis of aqueous sodium chloride (brine) as this requires less energy. Sodium chloride is a cheap raw material, which is readily available, and the process also produces sodium hydroxide and hydrogen, both of which are important industrial products. There are two main methods by which the electrolysis is achieved – the mercury cell and the diaphragm cell.

THE MERCURY CELL

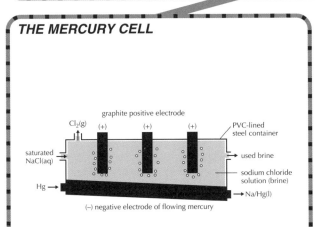

The negative electrode (cathode) is made of flowing mercury. Although sodium is above hydrogen in the electrochemical series sodium is preferentially discharged as it forms an alloy (known as an amalgam) with the mercury.

$$Na^+(aq) + e^- + Hg(l) \rightarrow Na/Hg(l)$$

The mercury then flows out of the electrolysis cell into a separate chamber where it reacts with water to produce hydrogen and sodium hydroxide solution. The mercury is recycled back into the electrolytic cell.

$$Na/Hg(l) + H_2O(l) \rightarrow Na^+(aq) + OH^-(aq) + \tfrac{1}{2}H_2(g) + Hg(l)$$

The cell itself is made of PVC-lined steel and the positive electrode (anode) where the chlorine is formed is made of graphite.

$$2Cl^-(aq) \rightarrow Cl_2(g) + 2e^-$$

THE DIAPHRAGM CELL

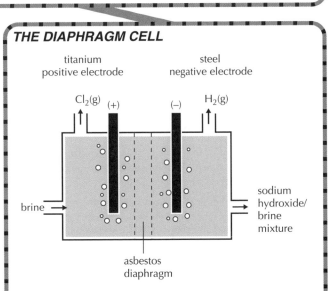

In the diaphragm cell the positive electrode is made of titanium and the negative electrode is made of steel. Hydrogen is formed at the negative electrode and chlorine at the positive electrode.

negative electrode $2H_2O(l) + 2e^- \rightarrow H_2(g) + 2OH^-(aq)$
positive electrode $2Cl^-(aq) \rightarrow Cl_2(g) + 2e^-$

The diaphragm, made of asbestos, allows the sodium chloride solution to flow between the electrodes but separates the chlorine and hydrogen gas and helps to prevent the OH^- ions flowing towards the positive electrode. The sodium hydroxide solution formed accumulates in the cathode compartment and is piped off. The resulting solution contains about 10% sodium hydroxide and 15% unused sodium chloride by mass. It is concentrated by evaporation and the sodium chloride crystallizes out leaving a 50% solution of sodium hydroxide.

A more modern version of the diaphragm cell (known as an ion exchange membrane cell) uses a partially permeable ion exchange membrane rather than asbestos. The membrane is made of a fluorinated polymer and is permeable to positive ions but not negative ions.

ENVIRONMENTAL IMPACT OF THE CHLOR-ALKALI INDUSTRY

In many parts of the world the mercury cell has been replaced by the diaphragm or ion exchange membrane cell. The ion exchange membrane cell is much cheaper to run due to the development of modern polymers. The main reason why the use of the mercury cell has been discontinued is environmental. In theory all the mercury is recycled but in practice some leaks into the environment and can build up in the food chain to toxic levels.

The chlorine produced has many important industrial uses, among them chlorinated organic solvents, water purification, pesticides, feedstock for inorganic chemicals (e.g. hydrochloric acid), and bleaching paper. Concern is mounting over the use of chlorinated organic compounds. Several have been shown to be carcinogenic and the C–Cl bond can break homolytically in the presence of ultraviolet light at higher altitudes to form chlorine radicals, which can contribute to ozone depletion.

IB QUESTIONS – OPTION C –
CHEMISTRY IN INDUSTRY AND TECHNOLOGY

1. **(a)** Aluminium is manufactured by the electrolysis of alumina dissolved in molten cryolite.

    (i) Explain the function of the cryolite. [1]

    (ii) Give an ionic equation for the reaction at the positive electrode (anode) during the electrolysis. [1]

    (iii) Explain with the aid of an equation why the positive electrode slowly disappears. [1]

    (b) Explain how the production of pure alumina from bauxite takes advantage of the amphoteric nature of aluminium oxide. [2]

    (c) Give **two** properties **and** related uses which make aluminium an important metal in today's world. [2]

    (d) Despite aluminium being the most abundant metal in the earth's crust, it is frequently recycled. Give **two** reasons which favour recycling. [2]

2. **(a)** Explain why crude oil contains small amounts of sulfur. [1]

    (b) Why must the sulfur be removed from crude oil before further refining takes place? [1]

    (c) One of the chemical processes used in the refining of crude oil is *cracking*. Give a balanced equation for the thermal cracking of $C_{10}H_{22}$ and explain why the process is important. [2]

    (d) Discuss the disadvantages of using crude oil as an energy source rather than as a chemical feedstock. [4]

3. **(a)** Explain what is meant by *nanotechnology*. [1]

    (b) Explain what is meant by *heterogeneous catalysis* [1]

    (c) Suggest two reasons why carbon nanotubes have the potential to be good heterogeneous catalysts. [2]

    (d) State two reasons why the research and development of nanotechnology may cause health concerns. [2]

4. Poly(ethene) is the most commonly used synthetic polymer. It is produced in low-density and high-density forms.

    (a) Identify which form has the higher melting point. Explain by reference to its structure and bonding. [4]

    (b) Describe how the properties of **two** named polymers can be modified by adding a different substance in each case. [3]

    (c) After use, discarded polymers may be burned or dumped. Discuss the problems associated with

    (i) burying polymers on landfill sites

    (ii) dumping polymers into the sea

    (iii) burning polymers. [6]

HL

---

5. The structures of two important polymers are given below:

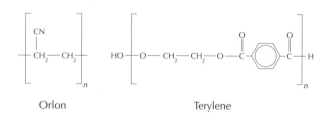

Orlon   Terylene

    (a) Deduce and draw the structural formulas of the monomers used to make these polymers. [3]

    (b) State the conditions necessary for the manufacture of high density polythene. How does the mechanism for the manufacture of high density polythene differ from the mechanism for the manufacture of low density polythene? [3]

6. 4-octyl-4′-cyanobiphenyl is an example of a thermotropic liquid crystal.

    (a) Explain what is meant by the term *thermotropic liquid crystal*. [2]

    (b) 4-octyl-4′-cyanobiphenyl is a typical example of a biphenyl nitrile. Explain the importance of the following structural features when accounting for its liquid crystal properties

    (i) the nitrile group.

    (ii) the biphenyl group.

    (iii) the octyl group. [6]

    (c) Outline the principles of the liquid crystal display device. [4]

Pharmaceutical products

THE EFFECTS OF DRUGS AND MEDICINES

For centuries man has used natural materials to provide relief from pain, heal injuries, and cure disease. Many of these folk remedies have been shown to be very effective and the active ingredients isolated and identified.

Morphine was extracted from the poppy *Papaver somniferum* early in the nineteenth century and later salicylic acid, the precursor of aspirin, was isolated from willow bark. The words 'drug' or 'medicine' are commonly applied to these substances, but they have different connotations in different countries and are difficult to define precisely. For example, the pineal gland in humans, a small lump of tissue that resides at the base of the brain, produces a substance called melatonin. This substance is known to bring on the onset of sleep and is often taken by people suffering from 'jet-lag'. As it occurs naturally in very low amounts in many foods it is classed as a food in America and can readily be bought. However, it is unavailable in Europe since it is classed as a drug because of its potential to modify physiological functions in humans. Generally a drug or medicine is any chemical (natural or man-made), which does one or more of the following:
- alters incoming sensory sensations
- alters mood or emotions
- alters the physiological state (including consciousness, activity level, or co-ordination).

Drugs and medicines are commonly (but not always) taken to improve health. They accomplish this by assisting the body in its natural healing process. The mechanism of drug action is still not fully understood, and there is evidence that the body can be 'fooled' into healing itself through the 'placebo' effect.

RESEARCH, DEVELOPMENT, AND TESTING OF NEW PRODUCTS

The research and development of new drugs is a long and expensive process. Traditionally a new product is isolated from an existing species, or synthesized chemically and then subjected to thorough laboratory and clinical pharmacological studies to demonstrate its effectiveness. Before studies are allowed on humans it must be tested on animals to determine the **lethal dose** required to kill fifty percent of the animal population, known as the LD_{50}. The effective dose required to bring about a noticeable effect in 50% of the population is also obtained. This is known as the ED_{50}. A factor known as the *therapeutic index* (or therapeutic window) is then calculated.

$$\text{Therapeutic index} = \frac{LD_{50}}{ED_{50}}$$

This relates the therapeutic effects of the drug to its toxic effects. If the therapeutic index is 100 then $LD_{50}/ED_{50} = 100:1$. This means that a hundredfold increase in the dose corresponding to the ED_{50} would result in a 50% death rate.

Once the therapeutic index has been established the drug can then be used in an initial clinical trial on humans. This is usually on volunteers as well as on patients half of whom are given a placebo. This initial

METHODS OF ADMINISTERING DRUGS

In order to reach the site where their effects are needed the majority of drugs must be absorbed into the bloodstream. The method of administering the drug determines the route taken and the speed with which it is absorbed into the blood. The four main methods are: by mouth (oral); inhalation; through the anus (rectal), and by injection (parenteral).

Drugs may also be applied topically so that the effect is limited mainly to the site of the disorder, such as the surface of the skin. Such drugs may come in the form of creams, ointments, sprays, and drops.

The three different methods of injection

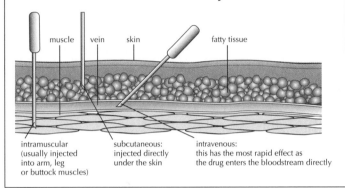

muscle   vein   skin   fatty tissue

intramuscular (usually injected into arm, leg or buttock muscles)

subcutaneous: injected directly under the skin

intravenous: this has the most rapid effect as the drug enters the bloodstream directly

THALIDOMIDE – HOW IT CAN ALL GO WRONG

In 1958 a German pharmaceutical company launched a massive publicity campaign for a new tranquilizer to combat 'morning sickness' in pregnant women. The drug was sold world-wide under brand names such as Thalidomide and Contergan. In many countries it was sold without prescription and marketed as completely innocuous. Reports of severe adverse side-effects began to appear in 1959, and it later transpired that as early as 1956 clinical trials by the company itself had revealed problems. Nevertheless because it was so profitable the company continued to market the drug heavily and sales kept increasing until it was withdrawn in 1961. By that time many children had been born with absent or severely malformed limbs.

trial is closely monitored to establish the drug's safety and possible side effects.

Drugs usually have unwanted **side effects**, for example aspirin can cause bleeding of the stomach and morphine, which is normally used for pain relief, can cause constipation. Side effects can be relative depending on why the drug is taken. People with diarrhoea are sometimes given a kaolin and morphine mixture, and people who have suffered from a heart attack are advised to take aspirin as it is effective as an anti-clotting agent. The severity of the complaint will determine an acceptable **risk-to-benefit ratio**. If an effective treatment is found for a life threatening disease then a high risk from side effects will be more acceptable.

The **tolerance** of the drug is also determined. Drug tolerance occurs as the body adapts to the action of the drug. A person taking the drug needs larger and larger doses to achieve the original effect. The danger with tolerance is that as the dose increases so do the risks of dependence and the possibility of reaching the lethal dose.

If the drug passes the initial clinical trial it will then go through a rigorous series of further phases, where its use is gradually widened in a variety of clinical situations. If it passes all these trials it will eventually be approved by the drug administration of a particular country, for use either as an OTC (over the counter) drug or for use only through prescription by a doctor.

Antacids

DIGESTION

The process of digestion involves the breakdown of food into molecules that can be utilized by individual cells in the body. The process starts in the mouth through the mechanical action of chewing and by the action of the digestive enzyme, amylase, present in saliva. Much of the digestive process, however, takes place in the stomach, a collapsible muscular bag which can hold between 2 and 4 litres of food.

The walls of the stomach are lined with a layer of cells which secrete mucus, pepsinogen (a precursor for the enzyme pepsin that breaks down proteins into peptides), and hydrochloric acid, collectively known as gastric juices. The hydrogen ion concentration of the hydrochloric acid normally lies between 3×10^{-2} mol dm^{-3} and 3×10^{-3} mol dm^{-3} giving a pH value between 1.5 and 2.5. The wall of the stomach is protected from the action of the acid by a lining of mucus. Problems can arise if the stomach lining is damaged or when too much acid is produced, which can eat away at the mucus lining.

TREATMENT OF INDIGESTION

The discomfort caused by excess acid is known as indigestion and may result from overeating, alcohol, smoking, anxiety, or in some people from eating certain types of food. Some drugs, such as aspirin, can also irritate the stomach lining and can result in ulceration of the stomach walls by the gastric acid. Antacids are used to combat excess stomach acid. They are most effective if taken between one and three hours after eating, as food typically remains in the stomach for up to four hours after a meal.

Antacids are essentially simple bases, such as metal oxides, hydroxides, carbonates, or hydrogencarbonates. They work by neutralizing the acid, preventing inflammation, relieving pain and discomfort, and allow the mucus layer and stomach lining to mend. When used in the treatment of ulcers they prevent acid from attacking the damaged stomach lining and so allow the ulcer to heal. Common examples include $Al(OH)_3$, $NaHCO_3$, $CaCO_3$, and 'milk of magnesia', which is a mixture of MgO and $Mg(OH)_2$. Typical neutralization reactions are:

$$NaHCO_3(s) + HCl(aq) \rightarrow NaCl(aq) + CO_2(g) + H_2O(l)$$

$$MgO(s) + 2HCl(aq) \rightarrow MgCl_2(aq) + H_2O(l)$$

$$Al(OH)_3(s) + 3HCl(aq) \rightarrow AlCl_3(aq) + 3H_2O(l)$$

FAST EFFECTIVE RELIEF FROM STOMACH UPSETS

'MILK OF MAGNESIA' LIQUID REGD.

ORIGINAL FLAVOUR

Sterling Health

SIDE EFFECTS

Although relatively harmless, antacids can have side effects. Magnesium compounds can cause diarrhoea, whereas aluminium compounds can cause constipation. Aluminium compounds can interfere with the absorption of phosphate from the diet causing possible bone damage if taken in high doses over a long period. Sodium hydrogen carbonate produces carbon dioxide gas, which may cause bloating and belching.

Antacids are commonly combined with alginates and anti-foaming agents. Alginates float on the contents of the stomach to produce a neutralizing layer. This prevents heartburn, which is caused when the stomach acid rises up the oesophagus. Anti-foaming agents are used to help prevent flatulence. The most usual anti-foaming agent is dimethicone.

WORKED EXAMPLE

Which would be the most effective in combating indigestion – a spoonful of liquid containing 1.00 g of magnesium hydroxide, or a spoonful of liquid containing 1.00 g of aluminium hydroxide?

M_r for $Mg(OH)_2 = 24.30 + (2 \times 17.01) = 58.33$

M_r for $Al(OH)_3 = 26.98 + (3 \times 17.01) = 78.01$

Amount of $Mg(OH)_2$ in 1.00 g $= \frac{1.00}{58.33} = 0.0171$ moles

Amount of $Al(OH)_3$ in 1.00 g $= \frac{1.00}{78.01} = 0.0128$ moles

$$Mg(OH)_2(s) + 2HCl(aq) \rightarrow MgCl_2(aq) + 2H_2O(l)$$

Therefore amount of HCl neutralized by 1.00 g of $Mg(OH)_2$
$= 2 \times 0.0171 = 0.0342$ moles

$$Al(OH)_3(s) + 3HCl(aq) \rightarrow AlCl_3(aq) + 3H_2O(l)$$

Therefore amount of HCl neutralized by 1.00 g of $Al(OH)_3$
$= 3 \times 0.0128 = 0.0384$ moles

Therefore the aluminium hydroxide would be slightly more effective.

Analgesics

MILD ANALGESICS
Aspirin

For a long time the bark of the willow tree (*Salix alba*) was used as a traditional medicine to relieve the fever symptoms of malaria. In the 1860s chemists showed that the active ingredient in willow bark is salicylic acid (2-hydroxybenzoic acid) and by 1870 salicylic acid was in wide use as a pain killer (analgesic) and fever depressant (antipyretic). However, salicylic acid has the undesirable side effect of irritating and damaging the mouth, oesophagus, and stomach membranes. In 1899 the Bayer Company of Germany introduced the ethanoate ester of salicylic acid, naming it 'Aspirin'.

Aspirin is thought to work by preventing a particular enzyme, prostaglandin synthase, being formed at the site of the injury or pain. This enzyme is involved in the synthesis of prostaglandins, which produce fever and swelling, and the transmission of pain from the site of an injury to the brain. Because of its anti-inflammatory properties aspirin can also be taken for arthritis and rheumatism. Aspirin also has an ability to prevent blood clotting and is sometimes taken to prevent strokes or the recurrence of heart attacks. Recent research suggests that aspirin may also be effective in preventing prostate cancer.

Aspirin can, however, have side effects. The most common side effect is that it causes bleeding in the lining of the stomach. A few people are allergic to aspirin with just one or two tablets leading to bronchial asthma. The taking of aspirin by children under twelve has been linked to Reye's disease – a potentially fatal liver and brain disorder with the symptoms of vomiting, lethargy, irritability, and confusion. Exceeding the safe dose of aspirin can be fatal as the salicylic acid leads to acidosis due to a lowering of the pH of the blood.

OTHER MILD ANALGESICS

Paracetamol (known as acetaminophen in the USA) is often preferred to aspirin as a mild pain reliever, particularly for young children, as its side-effects are less problematical although in rare cases it can cause kidney damage and blood disorders. Serious problems can arise, however, if an overdose is taken. Even if the overdose does not result in death it can cause brain damage and permanent damage to the liver and kidneys.

STRONG ANALGESICS

Strong analgesics are only available on prescription and are given to relieve the severe pain caused by injury, surgery, heart attack, or chronic diseases, such as cancer. They work by interacting temporarily with receptor sites in the brain, with the result that pain signals within the brain and spinal cord are blocked. The most important naturally occurring strong analgesics are morphine and codeine found in the opium poppy. The active part of the morphine molecule has been identified. Codeine and semi-synthetic (obtained by simple structural modifications to morphine) opiates, such as heroin, and totally synthetic compounds (e.g. 'demerol') all possess this basic structure and function as strong analgesics.

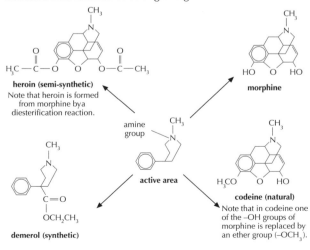

Effects of opiates

Diethanoylmorphine (heroin) is a more powerful painkiller than morphine but also more addictive. All opiates cause addiction and lead to tolerance, where there is a danger of reaching the lethal limit.

Short term effects
- Induce a feeling of euphoria (sense of well-being)
- Dulling of pain
- Depress nervous system
- Slow breathing and heart rate
- Cough reflex inhibited
- Nausea and vomiting (first time users)
- High doses lead to coma and/or death

Long term effects
- Constipation
- Loss of sex drive
- Disrupts menstrual cycle
- Poor eating habits
- Risk of AIDS, hepatitis, etc. through shared needles
- Social problems, e.g. theft, prostitution

Withdrawal symptoms occur within 6 to 24 hours for addicts if the supply of the drug is stopped. These include hot and cold sweats, diarrhoea, anxiety, and cramps. One treatment to wean addicts off their addiction is to use methadone as a replacement for heroin. Methadone is also an amine and functions as an analgesic but does not produce the euphoria craved by addicts.

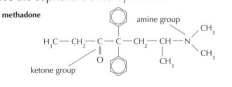

Depressants

EFFECTS OF DEPRESSANTS

Depressants are drugs which depress the central nervous system by interfering with the transmission of nerve impulses in the nerve cells (neurones). Depressants slow down the functions of the body including mental activity. In low doses they induce a feeling of calm and relieve anxiety and may induce sleep, but in larger doses they can cause loss of consciousness, coma, and death. The most commonly taken depressant is alcohol (ethanol). Rather confusingly depressants are sometimes described as anti-depressants, because they relieve the symptoms of mental depression.

USE AND ABUSE OF ALCOHOL (ETHANOL)

Medically alcohol is used as an antiseptic before injections and is also used to harden the skin. Drinking large and regular amounts of alcohol can cause psychological and physical dependence, known as alcoholism. The social costs of road accidents, violent behaviour, and family breakdowns due to alcohol consumption are huge.

Short term effects In moderate quantities it gives the drinker a feeling of relaxation and confidence and increases sociability. It dilates small blood vessels leading to flushing and a feeling of warmth. With increasing amounts judgement and concentration become progressively impaired. Violent behaviour is possible. Speech becomes slurred and loss of balance occurs. Loss of consciousness may follow at high concentrations and there is a risk of death from inhalation of vomit or stoppage of breathing.

Long term effects Long term heavy drinking can lead to severe liver disease including cirrhosis and liver cancer. It is linked with coronary heart disease, high blood pressure, strokes and an increasing risk of dementia. During pregnancy it can cause miscarriage and lead to fetal abnormalities. Sudden discontinuation of alcohol by heavy users can lead to delirium tremens (the 'DTs'), which includes severe shaking that can last up to four days.

SYNERGISTIC EFFECTS OF ALCOHOL

Ethanol can interact with, and considerably enhance the effect of other drugs because it depresses the central nervous system itself. This synergistic effect can be fatal, particularly when alcohol is taken together with benzodiazepines, narcotics, barbiturates, and solvents. With aspirin it increases the risk of stomach bleeding.

OTHER DEPRESSANTS

Other depressants commonly prescribed to reduce anxiety and relieve stress or to help insomnia include the benzodiazepines and prozac. They do not however remove the causes and are usually only prescribed for a limited period while counselling or psychotherapy are put in place, as they can induce dependence. They are also used as a premedication in hospitals before general surgery.

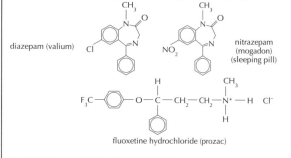

diazepam (valium)

nitrazepam (mogadon) (sleeping pill)

fluoxetine hydrochloride (prozac)

LEGAL LIMITS FOR DRIVING

In many countries the legal limit for driving is a blood alcohol concentration (BAC) of 80 mg of ethanol per 100 cm^3 of blood. In some countries it is even lower. After drinking, the concentration of alcohol in the blood increases for some time as the ethanol is absorbed, then it slowly decreases as it is metabolized and excreted. A unit of alcohol is roughly equal to:

• 280 cm^3 ($\frac{1}{2}$ pint) of beer or lager
• a glass of wine
• a measure of spirits.

Estimate of the alcohol consumption needed to exceed a legal limit of 80 mg of ethanol per 100 cm^3 of blood for a man of average weight (for a woman the quantities need to be reduced by about 30%)

Number of units	Time taken drinking / hours
4.0	1
5.0	2
6.0	3
6.5	4
7.5	5

DETECTION OF ALCOHOL IN BREATH, BLOOD, AND URINE

At the roadside a motorist may be asked to blow into a breathalyser. This may involve acidified potassium (or sodium) dichromate(VI) crystals turning green as they are reduced by the alcohol to Cr^{3+}, or the use of a fuel cell where the ethanol is oxidized to produce electricity. None of these are accurate enough to be used in court. At the police station a blood or urine sample may be taken and sent to a forensic science laboratory for analysis using gas liquid chromatography.

infrared intoximeter

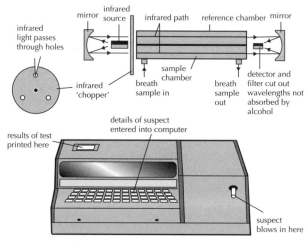

Modern intoximeters can now be used in a police station to accurately measure the amount of alcohol in the breath. They are based on the principle that the C–H bonds in ethanol absorb infrared radiation of a particular wavelength –3.39 micrometres (µm). The suspect blows a sample of breath into a chamber. Infrared radiation from a heated source passes through a 'chopper' (a rotating slotted disc), which makes the beam then pass alternately through the sample chamber and a chamber containing no breath sample. The intensities of the two emerging beams are compared and the amount of radiation absorbed by the sample is then converted into micrograms of ethanol per 100 cm^3 of breath.

Stimulants

Stimulants are drugs that increase a person's state of mental alertness.

AMPHETAMINES

Amphetamine is chemically related to adrenaline, the 'flight or fight' hormone. It is a sympathomimetic drug, that is, one which mimics the effect of stimulation on the sympathetic nervous system. This is the part of the nervous system which deals with subconscious nerve responses, such as speeding up the heart and increasing sweat production.

Amphetamines were initially used to treat narcolepsy (an uncontrollable desire for sleep) and were issued to airmen in World War II to combat fatigue. In the 1950s and 1960s they were used as anti-depressants and slimming pills. Regular use can lead to both tolerance and dependence. Short-term effects include increase in heart rate and breathing, dilation of the pupils, decrease in appetite followed by fatigue and possible depression as the effects wear off. Long-term effects include weight loss, constipation, and emotional instability.

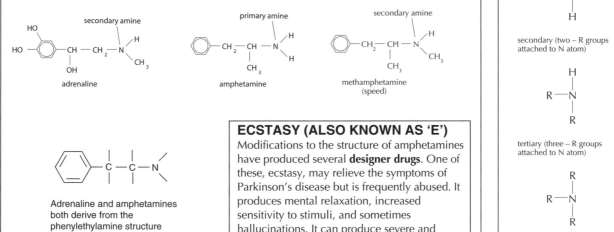

Adrenaline and amphetamines both derive from the phenylethylamine structure

ECSTASY (ALSO KNOWN AS 'E')

Modifications to the structure of amphetamines have produced several **designer drugs**. One of these, ecstasy, may relieve the symptoms of Parkinson's disease but is frequently abused. It produces mental relaxation, increased sensitivity to stimuli, and sometimes hallucinations. It can produce severe and sometimes fatal effects even after a single dose.

CLASSIFICATION OF AMINES

primary (one – R group attached to N atom)

secondary (two – R groups attached to N atom)

tertiary (three – R groups attached to N atom)

CAFFEINE

Caffeine is the most widely used stimulant in the world. It is present in coffee, tea, chocolate, and cola drinks and is also found in some pain killers and other medicines. There is evidence that consuming 400 mg of caffeine a day, or more, can cause dependence and physical side effects.

Caffeine content of different products

cup of ground coffee	80–120 mg
cup of instant coffee	65 mg
cup of tea	40 mg
can of cola	40 mg
bar (100 g) of plain chocolate	80 mg

Caffeine is a diuretic (causes frequent urination) and increases alertness, concentration, and restlessness. It is included in

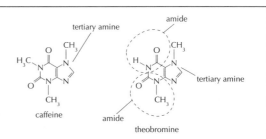

many common painkillers as it speeds up their effects. Like nicotine, morphine, codeine, and cocaine, caffeine is an alkaloid. Alkaloids are nitrogen-containing compounds of plant origin containing heterocyclic rings (rings containing other atoms as well as carbon) and a tertiary amine group. A compound with a similar structure to caffeine, which is also found in chocolate, is theobromine (although note that it contains no bromine!).

NICOTINE

It is the nicotine in tobacco that is largely responsible for causing approximately one third of the world's population to be addicted to smoking. Stopping smoking can produce temporary withdrawal symptoms that include a craving for tobacco, nausea, weight gain, insomnia, irritability, and depression. Like amphetamines nicotine is sympathomimetic. It increases concentration and relieves tension and the physical effects include increased heart rate and blood pressure, and reduction in urine output. The long term effects include increased risk of heart disease and coronary thrombosis. Its stimulatory effects may also lead to excess production of stomach acid leading to an increased risk of

peptic ulcers. The other well-known risks of smoking include chronic lung diseases, adverse effects on pregnancy, and cancers of the lung, mouth, and throat.

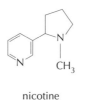

nicotine

Antibacterials

INFECTIOUS ORGANISMS

Many different types of micro-organism can cause disease, among them bacteria, viruses, fungi, yeasts, and protozoa. A typical bacterium consists of a single cell with a protective wall made up of a complex mixture of proteins, sugars, and lipids. Inside the cell wall is the cytoplasm, which may contain granules of glycogen, lipids, and other food reserves. Each bacterial cell contains a single chromosome consisting of a strand of deoxyribonucleic acid (DNA). Some bacteria are aerobic – that is they require oxygen and are more likely to infect surface areas, such as the skin or respiratory tract. Others are anaerobic and multiply in oxygen-free or low oxygen surroundings, such as the bowel. Not all bacteria cause disease and some are beneficial. Many exist on the skin or in the bowel without causing ill effects and some cannot live either in or on the body.

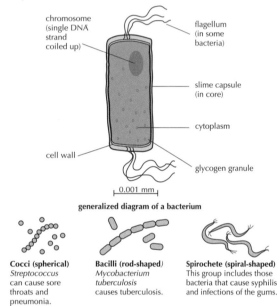

generalized diagram of a bacterium

Cocci (spherical)
Streptococcus can cause sore throats and pneumonia.

Bacilli (rod-shaped)
Mycobacterium tuberculosis causes tuberculosis.

Spirochete (spiral-shaped)
This group includes those bacteria that cause syphilis and infections of the gums.

DISCOVERY OF PENICILLIN

Antibacterials are chemicals, which prevent the growth and multiplication of bacteria. The first effective antibacterial, the dye trypan red, was developed by Paul Ehrlich to cure sleeping sickness. In 1910 Ehrlich also developed an arsenic containing compound, salvarsan, which was effective against syphilis. In 1935 the first 'sulfa dug' prontosil, which was effective against streptococcal bacteria, was developed.

However, the real discovery in the fight against bacteria was made by Alexander Fleming in 1928. Fleming, a bacteriologist, was working with cultures of *Staphylococcus aureus*, a bacterium that causes boils and other types of infections. He left an open petri dish containing one of the cultures in the laboratory, while he went away on holiday. Upon his return he noticed that a mould had developed and had inhibited growth of the bacterium. He deduced that the mould (*Penicillium notatum*) produced a compound (which he called penicillin), which inhibited the growth of bacteria. Although he published his results Fleming did not pursue his discovery. It was Howard Florey and Ernest Chain who overcame the problems associated with isolating and purifying penicillin. In 1941 they used penicillin on a policeman who was dying of septicaemia. They recorded a dramatic improvement in his condition, but unfortunately their meagre supply ran out before the policeman was cured and he relapsed and died. The search was on to produce penicillin in bulk. It was solved in America by growing strains of the penicillin mould in large tanks containing corn-steep liquor. In the 1950s the structure of penicillin was determined and this enabled chemists to synthesize different types of penicillin and other antibiotics (antibacterials originating from moulds) in the laboratory without recourse to moulds.

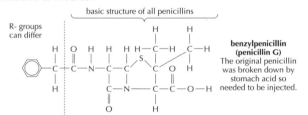

basic structure of all penicillins

R- groups can differ

benzylpenicillin (penicillin G)
The original penicillin was broken down by stomach acid so needed to be injected.

NARROW AND BROAD SPECTRUM ANTIBIOTICS

The penicillins are **narrow spectrum antibiotics**, effective against only certain types of bacteria. Other types of antibiotics, such as the tetracyclines (e.g. aureomycin and terramycin), are effective against a much broader range of bacteria (broad spectrum antibiotics). When a doctor is confronted with a patient suspected to be suffering from a bacterial infection the organism needs to be identified by taking blood, sputum, urine, pus, or stool samples. This takes time so initially a broad spectrum antibiotic may be prescribed. Once the bacterium is known the treatment may be switched to a narrow spectrum antibiotic, which is the recommended treatment for the identified organism.

MECHANISM OF ACTION OF ANTIBIOTICS

There are two main mechanisms by which antibiotics destroy bacteria. Penicillins and the cephalosporins prevent bacteria from making normal cell walls. Other antibiotics act inside the bacteria interfering with the chemical activities essential to their life function.

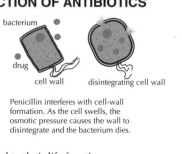

bacterium

drug

cell wall

disintegrating cell wall

Penicillin interferes with cell-wall formation. As the cell swells, the osmotic pressure causes the wall to disintegrate and the bacterium dies.

OVERUSE AND RESISTANCE TO PENICILLINS

When penicillin became readily available to doctors it was often given to cure minor illnesses, such as a sore throat. Certain bacteria were resistant to penicillin and were able to multiply. Their resistance was due to the presence of an enzyme called penicillinase, which could deactivate the original penicillin, penicillin G. To combat this chemists developed other penicillins whereby the active part of the molecule is retained but the side chain is modified. However, as bacteria multiply and mutate so fast it is a continual battle to find new antibiotics, which are effective against an ever more resistant breed of 'super bugs'. These include the methicillin-resistant *Staphylococcus aureus* (MRSA) and strains of *Mycobacterium tuberculosis* which cause tuberculosis, TB. To treat these infections a strict adherence to a treatment regime often involving a 'cocktail' of different antibiotics is required to prevent the risk of further resistance developing

The use of antibiotics in animal feedstocks has also contributed to this problem. Healthy animals are given antibiotics to prevent risk of disease, but the antibiotics are passed on through the meat and dairy products to humans, increasing the development of resistant bacteria.

Antivirals

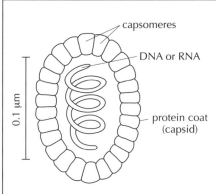

capsomeres

DNA or RNA

protein coat (capsid)

0.1 μm

STRUCTURE OF A VIRUS

There are many different types of virus with varying shape and structure. All viruses, however, have a central core of DNA or RNA (ribonucleic acid) surrounded by a coat (capsid) of regularly packed protein units (capsomeres), each containing many protein molecules. Unlike bacteria they have no nucleus or cytoplasm and are therefore not cells. They do not feed, excrete, or grow and they can only reproduce inside the cells of living organisms using materials provided by the host cell.

MULTIPLICATION OF VIRUSES

Athough viruses can survive outside the host they can only replicate by penetrating the living host cell and injecting their DNA or RNA into the cell's cytoplasm. The virus then 'takes over' the biochemical machinery inside the cell. This causes the cell to die or become seriously altered and causes the symptoms of the viral infection. The cell is made to produce new DNA or RNA and forms large numbers of new viruses. These are then released and move on to infect other healthy cells.

MODE OF ACTION OF ANTIVIRAL DRUGS

Common viral infections include the common cold, influenza, and childhood diseases, such as mumps and chicken pox. Fortunately the body's own defence mechanism is usually strong enough to overcome infections such as these and drugs are given more to remove the associated pain, fight the fever or to counteract secondary infections. One difficulty in treating viral infections is the speed with which the virus multiplies. By the time the symptoms have appeared the viruses are so numerous that antiviral drugs will have little effect. During the past few years some drugs have been developed to fight specific viral infections. They can work in different ways. Some work by altering the cell's genetic material so that the virus cannot use it to multiply. An example of this is acyclovir, which is applied topically to treat cold sores caused by the herpes virus. Its structure is similar to deoxyguanosine, one of the building blocks of DNA. It tricks the viral enzymes into using it as a building block for the viral DNA and thus prevents the virus from multiplying. However, it is difficult to eliminate the virus completely so the infection may flare up again at a later date.

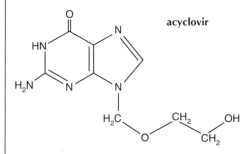

acyclovir

Others work by preventing the new viruses formed from leaving the cell. One such drug is amantadine, which is active against the influenza virus. One of the enzymes used by all influenza viruses to stick to the host cell wall as it leaves is called neuraminidase, and the drug works by inhibiting the active site on this enzyme. One of the problems with developing antiviral drugs is that the viruses themselves are regularly mutating – this is particularly true with the Human Immunodeficiency Virus (HIV).

AIDS

AIDS (Acquired Immune Deficiency Syndrome) is caused by a retrovirus – that is, it contains RNA rather than DNA. The virus invades certain types of white blood cells, known as T helper-cells, which normally activate other cells in the immune system, which leaves the body unable to fight infection. Once it invades a healthy cell its first task is to make viral-DNA from the RNA template using an enzyme called reverse transcriptase. This is opposite to the process that takes place in normal cells in which RNA is made from a DNA template using transcriptase as the enzyme.

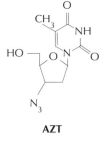

AZT

There are various ways in which a drug may be developed to eradicate the HIV virus. One drug that has met with some success is AZT (zidovudine). This combines with the enzyme that the HIV virus uses to build DNA from RNA and clogs up its active site. It is therefore a reverse-transcriptase inhibitor. Since it is only retroviruses that use this enzyme AZT does not affect normal cells.

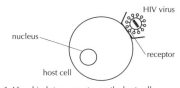

nucleus

HIV virus

receptor

host cell

1. Virus binds to a receptor on the host cell.
Possible action of antiviral
Binding site could be altered to prevent virus attaching to host cell.

2. Virus enters host cell.
Possible action of antiviral
Cell wall could be altered to prevent virus entering cell.

viral RNA

reverse transcriptase

altered form of RNA

3. Virus loses its protective coat and releases RNA and reverse transcriptase.
Possible action of antiviral
Drugs might be developed which would prevent the virus from losing its protective coat.

4. The reverse transcriptase converts the viral RNA into a form which can enter the nucleus of the host cell so it can integrate with the cell's DNA.
Possible action of antiviral
AZT works by blocking the action of reverse transcriptase.

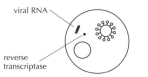

viral RNA

protein

new HIV viruses

dead host cell

5. The host cell produces new viral RNA and protein.
Possible action of antiviral
May be able to inhibit the production of new viral RNA and proteins by altering the genetic material of the virus.

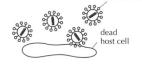

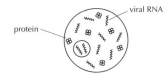

6.The new RNA and proteins form new viruses which then leave the host cell. The host cell dies.
Possible action of antiviral
Develop a drug which prevents the new viruses from leaving the cell. This is how amantadine works against the influenza virus.

GEOMETRIC ISOMERISM

Geometric isomerism can exist in inorganic compounds as well as in organic compounds, e.g. the square planar compounds of diamminedichloroplatinum(II) can exist in both *cis*- and *trans*- forms.

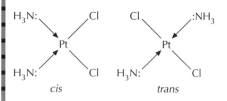

The *cis*- form (known as cisplatin) is highly effective in the treatment of testicular and ovarian cancers as well as other forms of cancer. Transplatin is not an effective anti-cancer drug. Cisplatin has no overall charge so can diffuse through the cancer cell membrane. Inside the cell it exchanges a chloride ion for a molecule of water to form $[Pt(NH_3)_2Cl(H_2O)]^+$. This complex ion enters the cell nucleus, where it binds to the DNA by exchanging another chloride ion to form $[Pt(NH_3)_2(DNA)]^{2+}$. This alters the cancer cell's DNA, so that when the cell tries to replicate it cannot be copied correctly and the cell dies.

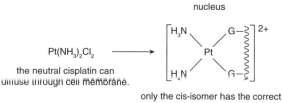

the neutral cisplatin can diffuse through cell membrane.

only the cis-isomer has the correct stereochemistry to bond to two guanine bases on the DNA

RING STRAIN

All penicillins contain a four-membered beta-lactam ring. This ring contains two carbon atoms that are sp^3 hybridized, a nitrogen atom that is sp^3 hybridized and a carbon atom that is sp^2 hybridized. Because of the restrictions of the ring the normal bond angles of 109.5° and 120° are not able to be obtained. This makes the amide group in the ring highly reactive as the ring can readily break due to the strained angles. The group containing the beta-lactam ring is very similar to a combination of the two amino acids cysteine and valine. When the ring opens these parts of the penicillin become covalently bonded to the enzyme that synthesizes the cell walls of the bacterium, thus blocking its action.

CHIRALITY

Asymmetric or chiral carbon atoms form two different optically active forms. Because of their different stereochemistry the two enantiomers can behave in totally different ways in the body. It is now realized that one of the enantiomers of thalidomide gives the benefits associated with the drug and it is the other enantiomer that is thought to be responsible for causing the fetal deformities. When new drugs are synthesized nowadays the pharmacological activity of both forms is studied separately.

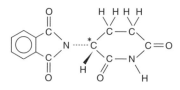

The structure of thalidomide. The chiral carbon atom is marked with an asterisk, *

POLARITY

Heroin is much more potent and produces a much greater feeling of euphoria than morphine. This can be explained by the difference in the polarity of the two substances. Morphine molecules contain two polar -OH groups. When morphine is converted into heroin these are replaced by much less polar ethanoate groups. This makes heroin much more soluble in lipids which are non polar. Heroin is thus able to rapidly penetrate the lipid-based blood-brain barrier and reach the brain in higher concentrations than morphine.

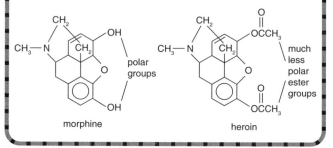

morphine   heroin

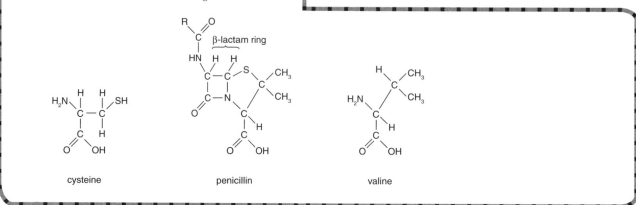

cysteine   penicillin   valine

COMBINATORIAL CHEMISTRY AND PARALLEL SYNTHESIS

Drug companies possess 'libraries' of compounds which have been screened for drug activity. In the past it could take many years to build up such a library as the substances were synthesized, purified, and screened by traditional methods. The advent of very sensitive techniques, such as mass spectrometry, to identify extremely small amounts of substances has lead to the development of 'combinatorial libraries'. In combinatorial chemistry very large numbers of related compounds can be prepared quickly. One way is to use computer controlled syringes to carry out repetitive chemical techniques but much use is now made of solid phase organic chemistry.

The starting material for the reaction is covalently bonded to very small beads (100 micrometres diameter) of polystyrene-based resin. A process called 'mix and split' is then used. Imagine the process for just three amino acids. After the first coupling, all the resin beads are then split into individual portions for the next step so that when reacted again all the nine possible combinations of dipeptides are formed. After another step all 27 possible combinations of tripeptides have been formed. If the process is scaled up for all the 20 naturally occurring amino acids then each cycle produces 400 dipeptides, 8000 tripeptides, and 160 000 tetrapeptides, etc. By using a large excess of the second and subsequent amino acids the reaction can be made to give a high yield. The final products can be purified easily by filtering off the beads from the reaction mixture and washing. Preliminary screening for drug activity can then take place either *in vitro* or *in vivo* by measuring the ability of a compound to affect enzymes and bind to receptor cells.

This process was done first for amino acids but has been extended to cover many other types of active molecules, such as those containing the benzodiazepine group in depressants to form very large combinatorial libraries. Once a particular substance is identified as potentially useful it can be made on a larger scale. In the future it may become unnecessary to actually make the initial compounds. By knowing the exact shape both of the active site on the enzyme or receptor and the shapes of the active groups within a potential drug the use of virtual computer modelling can produce a virtual library.

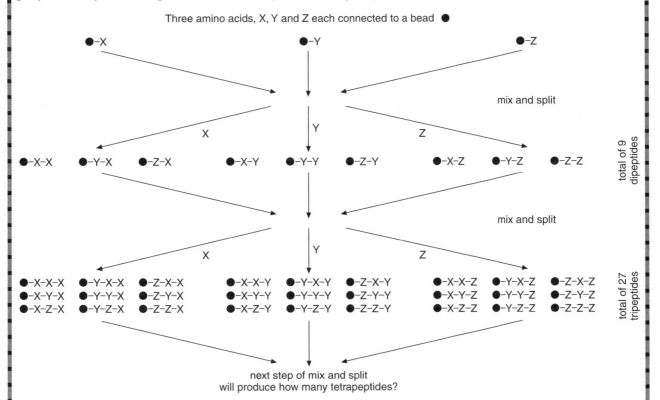

Three amino acids, X, Y and Z each connected to a bead ●

An alternative technique to complement combinatorial chemistry is parallel synthesis. This still uses solid-phase chemistry but on a slightly larger scale than combinatorial chemistry and has the advantage that all intermediates and products are generated separately. Parallel synthesis involves the synthesis of a highly reactive intermediate via a series of simple steps. Sets of individual compounds are then prepared simultaneously by reacting with a number of different reagents in arrays of physically separate reaction vessels or micro-compartments without interchange of intermediates during the assembly process. This gives a smaller more focussed library than that obtained with combinatorial chemistry. Essentially it is the first step towards producing larger yields of materials identified from screening tests so that they can be fully characterized without the need for laborious identification procedures. It can then be scaled up to production quantity levels.

Drug design (2)

MODIFYING THE POLARITY

Many medicines and drugs are fairly complex organic molecules with low polarity. This means that they tend to be insoluble in water and other polar environments in the body. To increase their solubility so that they can be more easily transported around the body they can be administered as an ionic salt.

If they contain amines then they can be converted into their hydrochloride salt. This is analogous to the reaction of ammonia with hydrochloric acid.

$$NH_3 + HCl \rightarrow NH_4^+Cl^-$$

Examples of drugs which are administered in an ionic form as their hydrochloride include the selective serotonin re-uptake inhibitor (SSRI) fluoxetine. This is prescribed under the name of Prozac® for people suffering with depression. The opiates too contain an amine group and are often administered as their hydrochloride salt. The white crystalline form of heroin, known as diamorphine, is in fact diacetylmorphine hydrochloride and in this form it is soluble in water and so can be injected.

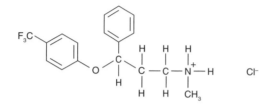

fluoxetine hydrochloride (prozac)

Drugs that contain a carboxylic acid group can similarly be made polar by converting them into their anion and administering them as their sodium or calcium salt.

$$RCOOH + NaOH \rightarrow RCOO^-Na^+ + H_2O$$

This is the case for soluble aspirin. Once the aspirin anion reaches the strongly acidic environment of the stomach it reverts back to the unionized form.

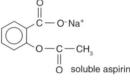

soluble aspirin

USE OF CHIRAL AUXILLIARIES

The traditional synthesis of an optically active compound normally produces a racemic mixture (50:50 mixture) of the two enantiomers which then has to be separated into the two isomers by using chromatography together with an another optically active compound. A more recent technique using **chiral auxilliaries** now makes it possible to synthesize just the desired isomer. Attaching an auxilliary, which is itself optically active, to the starting material creates the stereochemical conditions necessary for the reaction to form only one enantiomer. After the desired product has been formed the auxilliary is removed and recycled.

One drug for which this technique has been used to great effect is the anti-cancer drug taxol. Although it does occur naturally in yew trees the commercial semi-synthesis of taxol provides the necessary quantity to meet the demand.

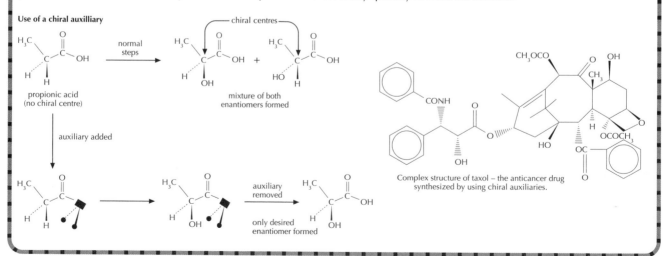

Complex structure of taxol – the anticancer drug synthesized by using chiral auxiliaries.

Mind altering drugs

THE INDOLE RING STRUCTURE

The hallucinogenic drugs LSD, psilocybin, and psilocin are all amines and contain the indole ring structure. Mescaline is also an amine and hallucinogenic and contains a structure similar to indole, except that the five-membered nitrogen ring is not closed.

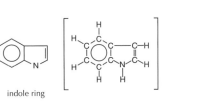

indole ring

LYSERGIC ACID DIETHYLAMIDE, LSD

LSD does not occur naturally but is derived from ergot, a fungus that grows on wheat. Only 0.028 mg of LSD will produce a noticeable effect in most people. The short term effects, which last for about 12 hours, include restlessness, dizziness, the desire to laugh, and distortions in sound and visual perceptions. In some cases unpleasant hallucinations lead to feelings of despair and suicide may be attempted. LSD takers may believe they are able to fly and driving under the influence can be extremely dangerous.

The long term risks include severe depression and the possibility of recurrences of LSD effects ('flashbacks') months or even years afterwards. There appears to be no lasting physical ill effects but LSD may cause psychological dependence. LSD has been used medically in psychotherapy, but is no longer used legitimately as it may lead to psychosis in susceptible patients. It is thought that LSD works by blocking the action of serotonin, one of the compounds responsible for transmitting impulses across synapses in the brain.

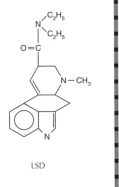

LSD

PSILOCYBIN

Psilocybin and the structurally related compound psilocin are found in the liberty cap mushroom *Psilocybe semilanceata* ('magic mushroom'). They are only mildly hallucinogenic causing slight alteration of perception and distortion of the senses together with feelings of exhilaration and insight. Although tolerance develops they are not addictive and it is not known if they can cause long term harm. The biggest dangers lie in the inability to recognize the mushrooms correctly, as similar-looking fungi are poisonous.

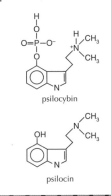

psilocybin

psilocin

MESCALINE

Mescaline is one of the oldest known hallucinogens. It comes from the peyote cactus found in South and Central America. Spanish explorers described the effects on the Mexican Indians in 1560 as 'terrible or ludicrous visions which lasted for two or three days'. Mescaline alters visual and auditory perceptions and appetite is reduced. It may cause unpleasant mental effects with people who are already anxious or depressed. Depending on how it is taken, other active substances in the plant may cause nausea and trembling. The effects of mescalin can be considerably enhanced when taken with alcohol. The long term effects are not clearly known but there is evidence that it could cause liver damage.

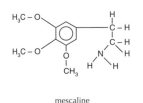

mescaline

MARIJUANA

Marijuana (also known as cannabis) is a mild hallucinogen made from the flowering tops, stems, leaves, and seeds of the hemp plant *Cannabis sativa*. Hashish, made from the resin of the plant is about five times stronger than marijuana. The active ingredient in marijuana is tetrahydrocannabinol (THC). The short term effects include a feeling of relaxation and enhanced auditory and visual perception. Loss of the sense of time, confusion, and emotional distress can also result. Hallucinations may occur in rare cases. It can have a synergistic effect and increase the risk of sedation with depressants. The long term effects include apathy and lethargy, and a lowering of fertility together with all the risks associated with tobacco smoking.

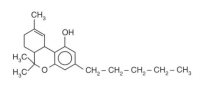

tetrahydrocannabinol (THC)

SHOULD MARIJUANA BE LEGALIZED?

Arguments for include:
- Medicinal use: research on the medical uses of cannabis is continuing but there is evidence that it is useful to prevent vomiting (anti-emetic), as a pain killer in cancer chemotherapy and AIDS, and for the treatment of glaucoma, epilepsy, Parkinson's disease, and Huntington's chorea.
- Lower crime rate – police freed to concentrate on other criminal activities.
- Freedom of the individual.

Arguments against include:
- Cost to society due to increased risk of heart disease and cancer through smoking.
- Claims by some that cannabis can lead on to harder drugs.
- Increased risk of dangerous driving while under the influence.

IB QUESTIONS – OPTION D – MEDICINES AND DRUGS

1. Viral infections are common but they are very difficult to treat by the use of drugs. One antiviral drug that has had some success, particularly in counteracting the effects of cold sores and shingles, is Acyclovir. Its structure is given below.

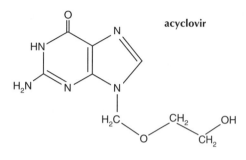

acyclovir

(a) Describe very briefly the basic structure of a virus. **[1]**

(b) Give **two** reasons why viruses are so difficult to deal with. **[2]**

(c) Identify the amide group in the structure of Acyclovir given above, by drawing a circle around it. **[1]**

(d) Explain why Acyclovir has a greater solubility in dilute acids than it has in water. **[2]**

(e) Many drugs, including Acyclovir, can be taken by mouth. However some drugs have to be administered by injection. Suggest **two** reasons why such drugs cannot be taken orally. **[2]**

2. The general structure of penicillin is given below.

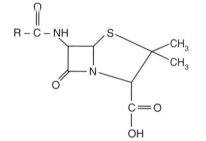

(a) The letter R in the above structure represents a side chain. State **one** reason why there are a number of different modifications of this side chain. **[1]**

(b) State why a prescribed course of penicillin should be completed. **[1]**

(c) Apart from the cost, what is an effect of the over prescription of penicillins? **[1]**

(d) What do you understand by the term 'broad based spectrum' antibiotics? **[1]**

(e) State which group of micro-organisms penicillins kill and briefly explain how they do this. **[3]**

3. The mild analgesic, aspirin, can be made from 2-hydroxybenzoic acid by reaction with ethanoyl chloride. The balanced equation for the reaction is,

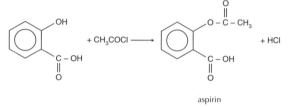

aspirin

(a) Write the molecular formula of 2-hydroxybenzoic acid. **[1]**

(b) Draw a circle around **one** of the functional groups in the structure of aspirin above. What is the name of the group and what is its effect on pH indicator solution? **[3]**

(c) Paracetamol can also be made from a simpler organic compound by reaction with ethanoyl chloride. Draw the structure of this compound. (The structure of paracetamol is shown in the IB data booklet) **[2]**

(d) Both aspirin and paracetamol are important mild analgesics, but like all drugs they have some disadvantages. Give **one** disadvantage in the use of each drug. **[2]**

4. (a) Cisplatin $Pt(NH_3)_2Cl_2$ is an effective anticancer drug. It bonds with the guanine in the DNA present in cancer cells and prevents it from replicating.

   (i) Draw the structure of *trans*-platin. **[1]**

   (ii) Describe the feature of guanine that enables it to bond with cisplatin and name the type of reaction that occurs when the bonds are formed. **[2]**

(b) Explain why it is important to carry out clinical trials on all the different enantiomers of a new drug. **[2]**

(c) Most reactions to form chiral compounds give a racemic mixture which then has to be separated into the two different enantiomers. Describe how a chiral auxilliary can be used to isolate the desired enantiomer of a particular drug. **[3]**

(d) The anticancer drug taxol can be synthesised using chiral auxilliaries. Part of its structure is shown below. Identify with an asterisk * **two** chiral centres. **[2]**

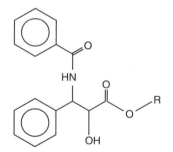

Air pollution

MAIN POLLUTANTS

Unpolluted air contains about 78% nitrogen, 21% oxygen, 1.0% argon, 0.03% carbon dioxide, and trace amounts of other gases together with up to 4% water vapour. An air pollutant can be described as a substance that is not normally present in air or a substance that is normally present but in excess amounts. The main *primary* air pollutants are carbon monoxide, oxides of nitrogen, sulfur dioxide, particulates, and hydrocarbons. *Secondary* pollutants are compounds formed when primary pollutants react in the air.

Pollutant	Natural source	Man-made source	Effect on health	Methods of reduction
Carbon monoxide (CO)	Incomplete oxidation of methane $CH_4 + 1\frac{1}{2}O_2 \rightarrow CO + 2H_2O$	Incomplete combustion of fossil fuels, e.g. $C_8H_{18} + 8\frac{1}{2}O_2 \rightarrow 8CO + 9H_2O$	Prevents haemoglobin from carrying oxygen by forming carboxyhaemoglobin	Use of lean burn engine, thermal exhaust reactor, or catalytic converter
Oxides of nitrogen (NO_x), e.g. N_2O, NO and NO_2	Electrical storms and biological processes	At high temperatures inside internal combustion engines $N_2 + O_2 \rightarrow 2NO$	Respiratory irritant leading to respiratory tract infections	Use of lean burn engine, recirculation of exhaust gases or catalytic converter
Sulfur dioxide (SO_2) (can be oxidized in the air to also form SO_3)	Oxidation of H_2S produced by volcanoes, and decay of organic matter	Combustion of sulfur-containing coal and smelting of sulfide ores $S + O_2 \rightarrow SO_2$	Respiratory irritant leading to respiratory tract infections	Removal of sulfur from fossil fuels before combustion. Alkaline scrubbing. Fluidized bed combustion
Particulates	Soot, ash, dust asbestos, sand, smoke, pollen, bacterial and fungal spores	Burning of fossil fuels, particularly coal and diesel	Can affect the respiratory system and cause lung diseases, such as emphysema, bronchitis, and cancer	Sedimentation chambers. Electrostatic precipitation
Volatile organic compounds (VOCs) (C_xH_y or R–H)	Plants, e.g. rice. Many plants emit unsaturated hydrocarbons called terpenes	Unburned or partially burned gasoline and other fuels; solvents	Some (e.g. benzene) are carcinogenic. Can form toxic secondary pollutants, e.g. PANs (peroxyacylnitrates)	Catalytic converter

METHODS OF REMOVAL

Thermal exhaust reactor

Exhaust from the car engine is combined with more air and reacts due to the heat of the exhaust gases. Carbon monoxide is converted into carbon dioxide and unburned hydrocarbons are also combusted.

$$2CO(g) + O_2(g) \rightarrow 2CO_2(g)$$

Lean burn engines

By adjusting the carburettor the ratio air:fuel can be altered. The higher the ratio the less carbon monoxide emitted as more complete combustion occurs. Unfortunately this produces higher temperatures so more NO_x is produced. At lower ratios less NO_x but more carbon monoxide will be emitted.

Catalytic converter

The hot exhaust gases are passed over a catalyst of platinum, rhodium, or palladium. These fully oxidize carbon monoxide and unburned volatile organic compounds, VOCs, and also catalyse the reaction between carbon monoxide and nitrogen oxide.

$$2CO(g) + 2NO(g) \rightarrow 2CO_2(g) + N_2(g)$$

Sulfur dioxide

Some sulfur is present in coal as metal sulfides (e.g. FeS) and can be removed physically by crushing the coal and mixing with water. The more dense sulfides sink to the bottom and the cleaned coal can be skimmed off. Sulfur is also removed from oil before it is refined by converting it into hydrogen sulfide.

Sulfur dioxide can be removed from the exhaust of coal burning plants by 'scrubbing' with an alkaline slurry of limestone ($CaCO_3$) and lime (CaO). The resulting sludge is used for landfill or as gypsum ($CaSO_4.2H_2O$) to make plasterboard.

$$CaCO_3(s) + SO_2(g) \rightarrow CaSO_3(s) + CO_2(g)$$

$$CaO(s) + SO_2(g) \rightarrow CaSO_3(s)$$

$$2CaSO_3(s) + O_2(g) + 4H_2O(g) \rightarrow 2CaSO_4.2H_2O(s)$$

A more modern method known as fluidized bed combustion involves burning the coal on a bed of limestone which removes the sulfur as $CaSO_3$ or $CaSO_4$ as the coal burns.

wet alkaline scrubber

output of clean gas

slurry of $CaCO_3$/CaO

input of gas containing SO_2

sludge removed

Electrostatic precipitation

Particulates are solid or liquid particles suspended in the air. Larger particles can be allowed to settle under the influence of gravity in sedimentation chambers. For smaller particles an electrostatic precipitation chamber can be used. The charged particulates are attracted to the oppositely charged electrodes, which are shaken periodically so that the aggregated particulates fall to the bottom of the precipitator where they can be removed.

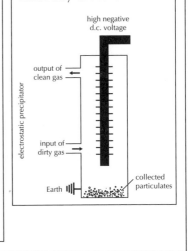

high negative d.c. voltage

output of clean gas

electrostatic precipitator

input of dirty gas

Earth

collected particulates

Acid deposition (1)

OXIDES OF SULFUR SO$_x$

Sulfur dioxide occurs naturally from volcanoes. It is produced industrially from the combustion of sulfur-containing fossil fuels and the smelting of sulfide ores.

$$S(s) + O_2(g) \rightarrow SO_2(g)$$

In the presence of sunlight sulfur dioxide is oxidized to sulfur trioxide.

$$SO_2(g) + \tfrac{1}{2}O_2(g) \rightarrow SO_3(g)$$

The oxides can react with water in the air to form sulfurous acid and sulfuric acid:

$$SO_2(g) + H_2O(l) \rightarrow$$
$$H_2SO_3(aq)$$
and
$$SO_3(g) + H_2O(l) \rightarrow$$
$$H_2SO_4(aq)$$

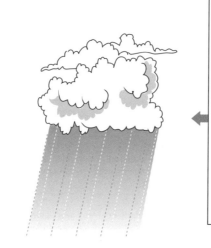

OXIDES OF NITROGEN NO$_x$

Nitrogen oxides occur naturally from electrical storms and bacterial action. Nitrogen monoxide is produced in the internal combustion engine and in jet engines.

$$N_2(g) + O_2(g) \rightarrow 2NO(g)$$

Oxidation to nitrogen dioxide occurs in the air.

$$2NO(g) + O_2(g) \rightarrow 2NO_2(g)$$

The nitrogen dioxide then reacts with water to form nitric acid and nitrous acid:

$$2NO_2(g) + H_2O(l) \rightarrow$$
$$HNO_3(aq) + HNO_2(aq)$$

or is oxidized directly to nitric acid by oxygen in the presence of water:

$$4NO_2(g) + O_2(g) + 2H_2O(l) \rightarrow 4HNO_3(aq)$$

ACID DEPOSITION

Pure rainwater is naturally acidic with a pH of 5.65 due to the presence of dissolved carbon dioxide. Carbon dioxide itself is *not* responsible for acid rain since acid rain is defined as rain with a pH less than 5.6. It is the oxides of sulfur and nitrogen present in the atmosphere which are responsible for **acid deposition** - the process by which acidic particles, gases and precipitation leave the atmosphere. Wet deposition, due to the acidic oxides dissolving and reacting with water in the air, is known as 'acid rain' and includes fog, snow and dew as well as rain. Dry deposition includes acidic gases and particles.

VEGETATION

Increased acidity in the soil leaches important nutrients, such as Ca^{2+}, Mg^{2+}, and K^+. Reduction in Mg^{2+} can cause reduction in chlorophyll and consequently lowers the ability of plants to photosynthesize. Many trees have been seriously affected by acid rain. Symptoms include stunted growth, thinning of tree tops, and yellowing and loss of leaves. The main cause is the aluminium leached from rocks into the soil water. The Al^{3+} ion damages the roots and prevents the tree from taking up enough water and nutrients to survive.

LAKES AND RIVERS

Increased levels of aluminium ions in water can kill fish. Aquatic life is also highly sensitive to pH. Below pH 6 the number of sensitive fish, such as salmon and minnow, decline as do insect larvae and algae. Snails cannot survive a pH less than 5.2 and below pH 5.0 many microscopic animal species disappear. Below pH 4.0 lakes are effectively dead. The nitrates present in acid rain can also lead to eutrophication.

BUILDINGS

Stone, such as marble, that contains calcium carbonate is eroded by acid rain. With sulphuric acid the calcium carbonate reacts to form calcium sulfate, which can be washed away by rainwater thus exposing more stone to corrosion. Salts can also form within the stone that can cause the stone to crack and disintegrate.

$$CaCO_3(s) + H_2SO_4(aq) \rightarrow$$
$$CaSO_4(aq) + CO_2(g) + H_2O(l)$$

HUMAN HEALTH

The acids formed when NO$_x$ and SO$_x$ dissolve in water irritate the mucus membranes and increase the risk of respiratory illnesses, such as asthma, bronchitis, and emphysema. In acidic water there is more probability of poisonous ions, such as Cu^{2+} and Pb^{2+}, leaching from pipes and high levels of aluminium in water may be linked to Alzheimer's disease.

METHODS TO LOWER OR COUNTERACT THE EFFECTS OF ACID RAIN

1. Lower the amounts of NO$_x$ and SO$_x$ formed, e.g. by improved engine design, the use of catalytic converters, and removing sulfur before, during, and after combustion of sulfur-containing fuels.

2. Switch to alternative methods of energy (e.g. wind and solar power) and reducing the amount of fuel burned, e.g. by reducing private transport and increasing public transport and designing more efficient power stations.

3. Liming of lakes – adding calcium oxide or calcium hydroxide (lime) neutralizes the acidity, increases the amount of calcium ions and precipitates aluminium from solution. This has been shown to be effective in many, but not all, lakes where it has been tried.

Greenhouse effect and global warming

THE GREENHOUSE EFFECT

A steady state equilibrium exists between the energy reaching the Earth from the Sun and the energy reflected by the Earth back into space. This regulates the mean average temperature of the Earth's surface. The incoming radiation is short wave ultraviolet and visible radiation. Some is reflected back into space and some is absorbed by the atmosphere before it reaches the surface. The energy radiated back from the Earth's surface is longer wavelength infrared radiation. Not all of the radiation escapes. Greenhouse gases in the atmosphere allow the passage of the incoming short wave radiation but absorb some of the reflected infrared radiation and re-radiate it back to the Earth's surface. Because of its abundance water is the main greenhouse gas.

The bonds in the carbon dioxide molecule absorb radiation of a different wavelength than the bonds in water molecules. Although it only constitutes 0.03% of the atmosphere, carbon dioxide therefore plays a key role in keeping the average global temperature at about 15 °C.

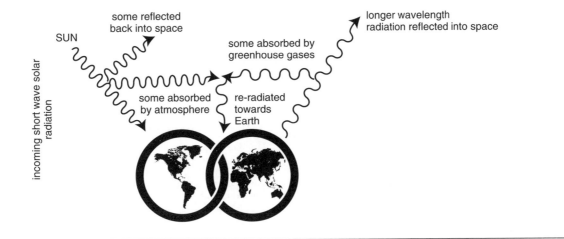

GREENHOUSE GASES

The contribution to global warming made by different greenhouse gases will depend both on their concentration in the atmosphere (abundance) and on their ability to absorb infrared radiation. Apart from water, carbon dioxide contributes about 50% to global warming. CFCs are thousands of times better at absorbing heat than CO_2 but because their concentrations are so low their effect is limited to about 14%.

Gas	Main source	Heat trapping effectiveness compared with CO_2	Overall contribution to increased global warming
H_2O	Evaporation of oceans and lakes	0.1	–
CO_2	Combustion of fossil fuels and biomass	1	50%
CH_4	Anaerobic decay of organic matter caused by intensive farming	30	18%
N_2O	Artificial fertilizers and combustion of biomass	150	6%
O_3	Secondary pollutant in photochemical smogs	2000	12%
CFCs	Refrigerants, propellants, foaming agents, solvents	10 000–25 000	14%

INFLUENCE OF GREENHOUSE GASES ON GLOBAL WARMING

Most of the greenhouse gases have natural as well as man-made sources. As man has burned more fossil fuels the concentration of carbon dioxide in the air has risen steadily. Readings taken from Mauna Loa in Hawaii show a steady increase in the concentration of carbon dioxide by about 1 ppm (0.0001%) each year for the past fifty years. During the same period measurements of the Earth's mean temperature also show a general increase. During the past 100 years the mean temperature of the Earth has increased by about 1 °C although there were some years where the mean temperature fell rather than rose. Evidence from ice core samples in Greenland shows that there have also been large fluctuations in global temperature in the past; however, most scientists now accept that the current global warming is a direct consequence of the increased emission of greenhouse gases.

The predicted consequences of global warming are complex and there is not always agreement. The two most likely effects are:

1. Changes in agriculture and biodistribution as the climate changes.

2. Rising sea-levels due to thermal expansion and the melting of polar ice caps and glaciers.

PARTICULATES

Particulates, such as soot and volcanic dust, can have the opposite effect to greenhouse gases. They cool the Earth by scattering the short wave radiation from the Sun and reflecting it back into space. The lowering of mean global temperatures during the 1940s and 1960s has been attributed to the increased volcanic activity during these periods.

Ozone depletion (1)

FORMATION AND DEPLETION OF OZONE IN THE STRATOSPHERE

The ozone layer occurs in the stratosphere between about 12 km and 50 km above the surface of the Earth. Stratospheric ozone is in dynamic equilibrium with oxygen and is continually being formed and decomposed. The strong double bond in oxygen is broken by high energy ultraviolet light from the Sun to form atoms. These oxygen atoms are called radicals as they possess an unpaired electron and are very reactive. One oxygen radical can then react with an oxygen molecule to form ozone.

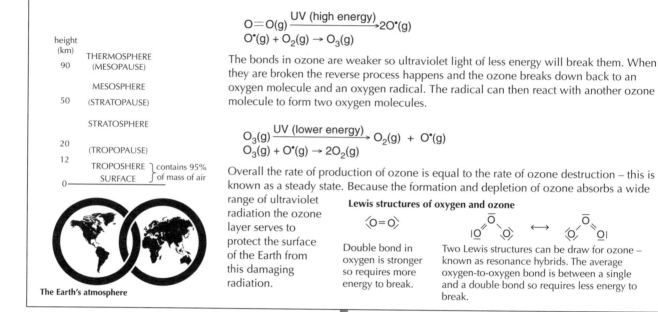

$$O{=}O(g) \xrightarrow{\text{UV (high energy)}} 2O^{\bullet}(g)$$
$$O^{\bullet}(g) + O_2(g) \rightarrow O_3(g)$$

height (km)	
90	THERMOSPHERE (MESOPAUSE)
	MESOSPHERE
50	(STRATOPAUSE)
	STRATOSPHERE
20	(TROPOPAUSE)
12	
	TROPOSHERE ⎤ contains 95%
0	SURFACE ⎦ of mass of air

The Earth's atmosphere

The bonds in ozone are weaker so ultraviolet light of less energy will break them. When they are broken the reverse process happens and the ozone breaks down back to an oxygen molecule and an oxygen radical. The radical can then react with another ozone molecule to form two oxygen molecules.

$$O_3(g) \xrightarrow{\text{UV (lower energy)}} O_2(g) + O^{\bullet}(g)$$
$$O_3(g) + O^{\bullet}(g) \rightarrow 2O_2(g)$$

Overall the rate of production of ozone is equal to the rate of ozone destruction – this is known as a steady state. Because the formation and depletion of ozone absorbs a wide range of ultraviolet radiation the ozone layer serves to protect the surface of the Earth from this damaging radiation.

Lewis structures of oxygen and ozone

Double bond in oxygen is stronger so requires more energy to break.

Two Lewis structures can be draw for ozone – known as resonance hybrids. The average oxygen-to-oxygen bond is between a single and a double bond so requires less energy to break.

LOWERING OF OZONE CONCENTRATION IN THE STRATOSPHERE

Measurements of the concentration of ozone in the stratosphere have shown that the amount of ozone in the 'ozone layer' has been decreasing. This is particularly true over both the south and north poles. The concentration above both the Antarctic and Arctic is seasonal. The biggest 'holes' occur during the winter and early spring. This decrease is due to particular chemicals produced and released by man. The main culprits are the CFCs – chlorofluorocarbons, the most common being dichlorodifluoromethane CCl_2F_2 also known as freon or CFC–12. CFCs were developed and used for refrigerants, propellants for aerosols, foaming agents for expanding plastics, and cleaning solvents. Other substances that also damage the ozone layer are oxides of nitrogen NO_x formed from internal combustion engines, power stations, and jet aeroplanes.

dichlorodifluoromethane, a chlorofluorocarbon (CFC)

ENVIRONMENTAL EFFECTS OF OZONE DEPLETION

Ultraviolet light has sufficient energy to damage biological molecules, such as amino acids, proteins, and nucleic acids. Because of ozone depletion more ultraviolet light has been reaching the Earth's surface.

Effect on humans
- Increase in sunburn.
- Increase in melanoma and non-melanoma skin cancer.
- Increase in eye cataracts and blindness.

Effect on marine ecosystems
- Marine phytoplankton loss. Since they produce much biomass this provides less food for other marine organisms and loss of the CO_2 'sink'.

Effect on plants
- More susceptible to disease.
- Growth inhibited.
- Photosynthesis inhibited.

Effect on weather
- Stratospheric convection currents altered. This affects wind and oceanic circulation.

ALTERNATIVES TO CFCs

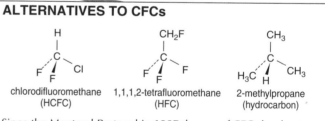

chlorodifluoromethane (HCFC)

1,1,1,2-tetrafluoromethane (HFC)

2-methylpropane (hydrocarbon)

Since the Montreal Protocol in 1987 the use of CFCs has been phased out. Even so, because of their low reactivity they are expected to remain in the atmosphere for at least the next eighty years. They destroy the ozone layer because the ultraviolet light breaks the relatively weak C–Cl bond. Substitutes for CFCs must have similar properties but not contain a bond that can be broken in ultraviolet light to form radicals. The properties required are: low reactivity, low toxicity, and low inflammability, as well as no weak C–Cl bonds. They also should not absorb infrared radiation otherwise they will act as greenhouse gases.

The most immediate replacements are HCFCs – hydrochlorofluorocarbons, such as CHF_2Cl, as they decompose more readily and do not build up in the stratosphere. Other alternatives actively being investigated are HFCs – hydrofluorocarbons, such as CF_3CH_2F, and hydrocarbons such as 2-methylpropane C_4H_{10}, for refrigerants, but they suffer from being flammable and also contribute to global warming.

Dissolved oxygen in water

IMPORTANCE OF DISSOLVED OXYGEN IN WATER

At a pressure of one atm and at a temperature of 20 °C the maximum solubility of oxygen in water is only about 9 ppm (i.e 0.009 g dm^{-3}). Although this value is small it is crucial since most aquatic plants and animals require oxygen for aerobic respiration. Fish require the highest levels and bacteria require the lowest levels. Fish need at least 3 ppm in order to be able to survive but to maintain a balanced and diversified aquatic community the oxygen content should not be less than 6 ppm.

BIOLOGICAL OXYGEN DEMAND (BOD)

The biological (or biochemical) oxygen demand is a measure of the dissolved oxygen (in ppm) required to decompose the organic matter in water biologically. It is often measured over a set time period of five days. Water that has a high BOD without the means of replenishing oxygen will rapidly not sustain aquatic life. A fast flowing river can recover its purity as the water becomes oxygenated through the mechanical action of its flow. Lakes have relatively little flow and reoxygenation is much slower or will not happen at all. Pure water has a BOD of less than 1 ppm, water with a BOD above about 5 ppm is regarded as polluted.

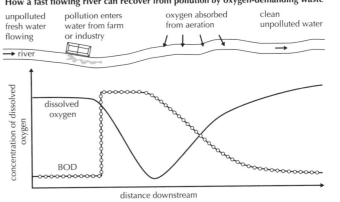

How a fast flowing river can recover from pollution by oxygen-demanding waste

MEASUREMENT OF BOD

The BOD of a sample of water can be determined by the Winkler method. The sample of the water is saturated with oxygen so the initial concentration of dissolved oxygen is known. A measured volume of the sample is then incubated at a fixed temperature for five days while micro-organisms in the water oxidize the organic material. An excess of a manganese(II) salt is then added to the sample. Under alkaline conditions manganese(II) ions are oxidized to manganese(IV) oxide by the remaining oxygen.

$$2Mn^{2+}(aq) + 4OH^-(aq) + O_2(aq) \rightarrow 2MnO_2(s) + 2H_2O(l)$$

Potassium iodide is then added which is oxidized by the manganese(IV) oxide in acidic solution to form iodine.

$$MnO_2(s) + 2I^-(aq) + 4H^+(aq) \rightarrow Mn^{2+}(aq) + I_2(aq) + 2H_2O(l)$$

The iodine released is then titrated with standard sodium thiosulfate solution.

$$I_2(aq) + 2S_2O_3^{2-}(aq) \rightarrow S_4O_6^{2-}(aq) + 2I^-(aq)$$

By knowing the number of moles of iodine produced the amount of oxygen present in the sample of water can be calculated and hence its concentration.

EUTROPHICATION

Nitrates from intensive animal farming and excess use of artificial fertilizers together with phosphates from artificial fertilizers and detergents accumulate in lakes. They act as nutrients and increase the growth of plants and algae. This can also happen in slow moving areas of sea water.

Normally when plants and algae die they decompose aerobically and form carbon dioxide and water. However, if the growth is excessive and the dissolved oxygen is not sufficient to cope anaerobic decomposition will occur. The hydrides formed, such as ammonia, hydrogen sulfide, and phosphine, not only smell foul but they poison the water. More species will die resulting in more anaerobic decay and the lake itself becomes devoid of life – a process known as **eutrophication**.

AEROBIC AND ANAEROBIC DECOMPOSITION

If sufficient oxygen is present organic material will decay aerobically and oxides or oxyanions are produced. Anaerobic decay involves organisms which do not require oxygen. The products are in the reduced form and are often foul smelling and toxic.

Element	Aerobic decay product	Anaerobic decay product
Carbon	CO_2	CH_4 (marsh gas)
Nitrogen	NO_3^-	NH_3 and amines
Hydrogen	H_2O	CH_4, NH_3, H_2S, and H_2O
Sulfur	SO_4^{2-}	H_2S ('rotten eggs' gas)
Phosphorus	PO_4^{3-}	PH_3 (phosphine)

THERMAL POLLUTION

The solubility of oxygen in water is temperature dependent. As the temperature is increased the solubility drops. At the same time the metabolic rate of fish and other organisms increases so the demand for oxygen increases. Many industries use water as a coolant and the careless discharge of heated water into rivers can cause considerable thermal pollution.

Solubility of oxygen in water at different temperatures

Temperature / °C	Solubility of O_2 in fresh water / ppm	Solubility of O_2 in sea water / ppm
0	14.71	11.71
10	11.42	9.28
20	9.14	7.50
30	7.50	6.21

Water pollutants and fresh water from sea water

PRIMARY POLLUTANTS

Heavy metals

Heavy metals that exist as ions in polluted water include cadmium, mercury, lead, chromium, nickel, copper and zinc. Their sources are varied. Cadmium is found in the effluent near zinc mining and in batteries and paints. Mercury is used a fungicide in seed dressings and used to be used as an electrode in the electrolysis of sodium chloride. Lead used to be used in paints and leaded gasoline and is still used in car batteries and roofing material. Other heavy metals originate from specific industrial processes (e.g. Cr used to be used for tanning leather), in batteries (e.g. NiCd cells) and biocides (e.g. Cu in woodworm treatment).

Pesticides

These include insecticides, such as DDT, herbicides, such as paraquat, and fungicides.

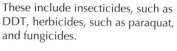
paraquat

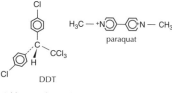
DDT

1,1,1-trichloro-2,2-bis (4-chlorophenyl) ethane
(former name **d**ichloro**d**iphenyl**t**richloroethane)

Dioxin

This is formed when waste materials containing organochlorine compounds are not incinerated at high enough temperatures. It is very persistent in the environment and extremely toxic as it accumulates in fat and liver cells. It can also cause malformations in fetuses. It was one of the herbicides present in Agent Orange used during the Vietnam war.

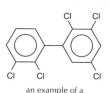
dioxin

Polychlorinated biphenyls, PCBs

These contain from one to ten chlorine atoms attached to a biphenyl molecule. They are chemically stable and have high electrical resistance so are used in transformers and capacitors. Like dioxin they persist in the environment and accumulate in fatty tissue. They affect reproductive efficiency, impair learning ability in children and are thought to be carcinogenic.

an example of a
polychlorinated biphenyl, PCB

Nitrates

Nitrates enter drinking water from intensive animal farming and from the use of artificial fertilizers. They also enter the water from acid rain. Because all nitrates are soluble they are not easily removed during treatment of sewage and so tend to accumulate. The maximum limit of nitrates in drinking water has been set at 50 mg dm^{-3} by the World Health Organisation. At higher levels there is a risk to very small babies of infantile methaemoglobinaemia (blue baby syndrome). This is because there is less acid in babies' stomachs so more of the nitrate is converted to nitrite by bacteria. The nitrite ion will oxidize the iron in haemoglobin so that it can no longer bind with oxygen and the baby suffers from oxygen starvation. Even in adults some of the nitrates are converted into nitrites which in turn are converted into nitrosamines. Nitrosamines are known to be carcinogenic.

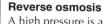
(nitrosamine)

METHODS OF OBTAINING FRESH WATER FROM SEA WATER

Distillation

Many hot countries distil sea water to obtain fresh water. One of the largest plants in Israel produces over 3 million litres a day. The sea water is heated in a series of coiled pipes and then introduced into a partially evacuated chamber. Under the reduced pressure some of the sea water boils instantly. The water vapour produced is condensed by contact with cold water pipes carrying sea water. In this way the heat released when the water condenses is used to preheat more sea water.

Reverse osmosis

A high pressure is applied to the solution side of a partially permeable membrane made of cellulose ethanoate. Water is forced out of the salt solution through the membrane leaving the salts behind. The challenge is to make a membrane that will withstand the high pressure and permit a reasonable flow of water. Commercial plants use pressures of up to 70 atm and one cubic metre of membrane can produce about 250 000 litres of fresh water each day.

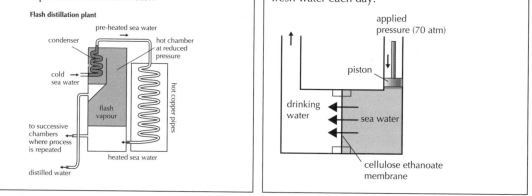

Waste water treatment

RAW SEWAGE
In some countries raw sewage is still discharged untreated into rivers and the sea. It is eventually degraded by micro-organisms but the process takes time and can cause pollution of beaches and swimming areas. In remote areas sewage is often discharged into cesspits or septic tanks where it is broken down before leaching into the ground. Waste water treated at sewage works contains floating matter, suspended matter, colloidal matter, and a range of micro-organisms. The aim of the treatment is to remove this additional material and recycle the fresh water.

PRIMARY TREATMENT
Primary treatment effectively removes about 60% of the solid material and about a third of the oxygen-demanding wastes. The incoming sewage is passed through coarse mechanical filters to remove large objects like sticks, paper, condoms, and rags. It is then passed into a grit chamber where sand and small objects settle. From there it passes into a sedimentation tank where suspended solids settle out as sludge. A mixture of calcium hydroxide and aluminium sulfate is added to aid this process.

$$Al_2(SO_4)_3(aq) + 3Ca(OH)_2(aq) \rightarrow 2Al(OH)_3(s) + 3CaSO_4(aq)$$

The two chemicals combine to form aluminium hydroxide which precipitates carrying with it suspended dirt particles – a process known as **flocculation**. Grease is removed from the surface tank by skimming. The effluent is then discharged into a waterway or passed on for secondary treatment.

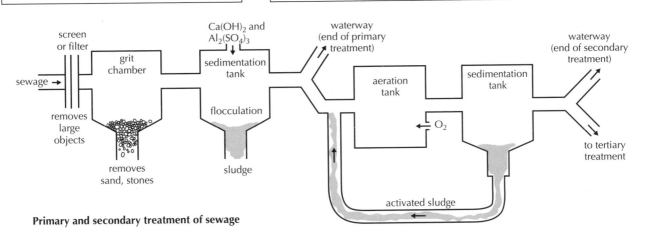

Primary and secondary treatment of sewage

SECONDARY TREATMENT
Secondary treatment removes up to 90% of the oxygen-demanding wastes (organic material and dangerous bacteria). The principle is to degrade the waste aerobically using oxygen and bacteria. One method uses trickle filters. The organic material is degraded by bacteria as the waste water trickles through a bed of stones. A more effective method is the activated sludge process. The sewage is aerated with pure oxygen in a sedimentation tank. The sludge that settles out contains active micro-organisms that digest organic waste and some of it is recycled. The water that emerges is then discharged into a waterway where it will be disinfected with chlorine or ozone before being fit for drinking.

TERTIARY TREATMENT
Although it is expensive there is an increasing need for tertiary treatment. Primary and secondary treatments do not remove heavy metal ions, nitrates and phosphates, or residual amounts of organic compounds.

Precipitation
Heavy metal ions and phosphates can be removed by precipitation.

Aluminium sulfate or calcium oxide can be used to precipitate phosphates.

$$Al^{3+}(aq) + PO_4^{3-}(aq) \rightarrow AlPO_4(s)$$

$$3Ca^{2+}(aq) + 2PO_4^{3-}(aq) \rightarrow Ca_3(PO_4)_2(s)$$

Heavy metal ions can be precipitated as insoluble hydroxides or basic salts by the addition of calcium hydroxide or sodium carbonate,

e.g. $Cr^{3+}(aq) + 3OH^-(aq) \rightarrow Cr(OH)_3(s)$

However, some metal hydroxides redissolve due to the formation of soluble complexes so some heavy metal ions, such as Zn^{2+}, Hg^{2+}, and Cd^{2+}, are precipitated as insoluble sulfides by bubbling hydrogen sulfide into the water,

e.g. $Cd^{2+}(aq) + H_2S(g) \rightarrow CdS(s) + 2H^+(aq)$

Activated carbon beds
The remaining dissolved organic material can be removed by allowing the water to flow down a bed of activated carbon. To reactivate the carbon it is periodically heated to high temperatures to oxidize the adsorbed organic compounds to carbon dioxide and water and regenerate the carbon surface.

Removal of nitrates
Nitrates are difficult to remove by chemical means as all nitrates are soluble so precipitation cannot be used. Ion exchange columns made of zeolites can be used to exchange hydroxide ions for nitrate ions but this is very expensive for large volumes of water. They are more readily removed by biological methods. Anaerobic denitrifying bacteria can reduce them to nitrogen or the water can be passed through algal ponds where the algae utilize the nitrate as a nutrient.

Soil

SOIL DEGRADATION

The composition of soil varies considerably but essentially it is a complex mixture of inorganic and organic matter including living organisms. When the quality of the soil has been affected in some way so that crop production is lowered it is known as **soil degradation**. This can occur naturally due to changes in weather patterns or can be caused by a variety of manmade factors such as acidification, desertification, contamination, erosion and salinization. Because of timber cutting, overgrazing and industrialization there is considerable concern about the current rate of soil degradation in many parts of the world.

Salinization

Salinization is caused by constant or excess irrigation. The water used for irrigation contains dissolved salts. When the water evaporates it leaves these dissolved salts behind. If the soil is poorly drained then the salts will accumulate in the fertile topsoil. These salts have two main effects both of which lead to the death of crops. Either the plants die because the salts eventually build up to toxic levels or they die of dehydration because their roots are no longer able to take up water from the salty soil.

Nutrient depletion

Plants need minerals and nutrients for healthy growth. When crops are harvested they remove the nutrients and minerals they have absorbed from the soil while growing. In the past small-scale farmers practiced crop rotation and left the soil fallow for some periods as well as adding organic material such as manure and compost to maintain the levels of nutrients. Much farming is now intensive. If proper management practices are not followed these nutrients are not being replaced.

Soil pollution

Soil pollution can be caused by a variety of factors. Industrial discharge and improper dumping of toxic waste material can cause long term soil pollution. Organic soil pollution such as oil from transport and illegal dumping of spent engine oil has more of a short term effect. The use of pesticides and fertilizers can disrupt the food chain by destroying species and thus reducing the soil's biodiversity and ultimately ruin the soil. The pollutants can also run into surface waters and move through the soil to pollute groundwater.

SOIL ORGANIC MATTER (SOM)

Although healthy plants require essential elements such as phosphorus, nitrogen, sulfur and potassium soil soon becomes degraded if only artificial fertilizers such as ammonium sulfate are applied. It is well known that animal dung and well-rotted compost made from organic watse are good nutrients for the soil, providing not only minerals but also organic matter. **Soil organic matter, SOM**, is the term generally used to represent the organic constituents of the soil. This includes plant and animal tissues, their partial decomposition products and the soil biomass. The decomposition of plants produces high molecular mass organic materials, such as polysaccharides and proteins, and simpler substances, such as sugars, amino acids and other small molecules – all of contribute towards the SOM. The resulting complex mixture is known as humus which is therefore a mixture of simple and more complex organic chemicals of plant, animal, or microbial origin.

The functions of SOM can be broadly classified into three main headings.

- Biological: Humus provides a source of energy and a source of the essential nutrient elements phosphorus, nitrogen and sulfur and thus helps to sustain healthy growth.
- Physical: Humus helps the soil to retain moisture and therefore increases the soil's capacity to withstand drought conditions, and it encourages the formation of good soil structure. Because of its dark colour humus absorbs heat and thus helps to warm the cold soil during spring.
- Chemical: Humus can act rather like clay with its **cation exchange capacity**. It contains active sites which enable it to bind to nutrient cations. This not only makes them more available to plants but also prevents them from being washed away by rain or during irrigation. Toxic cations, such as those from heavy metals, can also bind to the humus preventing them from entering the wider ecosystem. It also enhances the ability of the soil to maintain a constant pH by acting as an acid-base buffer.

COMMON ORGANIC SOIL POLLUTANTS

Pollutant	Source
Hydrocarbons and other VOCs including semi-volatile organic compounds, SVOCs.	Transport, solvents and industrial processes
Agrichemicals	Pesticides, herbicides and fungicides
Polyaromatic hydrocarbons	Incomplete combustion of coal, oil, gas, wood and garbage
Polychlorinated biphenyls, PCBs	Use as a coolant and insulator in electrical equipment such as transformers and generators.
Organotin compounds	Bactericides and fungicides used in paper, wood, textile and anti-fouling paint

Waste

METHODS OF WASTE DISPOSAL

In the past the waste from affluent societies was just dumped, burned or placed in landfill sites whereas less affluent societies recycled out of economic necessity. Incineration and landfill are still much used but now there is increasing pressure on everyone to recycle as much waste as possible. Even better is to reuse materials such as glass, metal and plastic wherever possible.

Method	Advantages	Disadvantages
Landfill	Efficient method to deal with large volumes. Filled land can be used for building or other community purposes.	Local residents may object to new sites. Once filled needs time to settle and may require maintenance as methane released.
Open dumping	Convenient and inexpensive.	Causes air and ground water pollution. Health hazard - encourages rodents and insects. Unsightly.
Ocean dumping	Source of nutrients. Convenient and inexpensive.	Danger to marine animals. Pollutes the sea.
Incineration	Reduces volume. Requires minimal space. Produces stable, odourless residue. Can be used a source of energy.	Expensive to build and operate. Can cause pollutants e.g. dioxins, if inefficiently burned. Requires energy.
Recycling	Provides a sustainable environment.	Expensive. Difficulty in separating different materials - not possible in all cases.

Recycling

Metals: Mainly aluminium and steel. The metals are sorted then melted and either re-used directly or added to the purification stage of metals formed from their ores. This is particularly important for metals such as aluminium which require large amounts of energy to produce direct from ore.

Paper: Sorted into grades. Washed to remove inks etc., made into a slurry to form new types of paper such as newspaper and toilet rolls. However energy is required to transport the paper and it may be more efficient to compost.

Glass: Sorted by colour, washed, crushed then melted and moulded into new products. Glass is not degraded during the recycling process so can be recycled many times.

Plastics: After sorting plastics are degraded to monomers by pyrolysis, hydrogenation, gasification and thermal cracking then repolymerized. Recycling causes less pollutants and uses less energy that producing new plastics from crude oil but sorting the plastics can be problematic.

TYPES OF NUCLEAR WASTE

There are various types of nuclear waste. Some are radioisotopes from research laboratories or hospitals. Some are spent fuel rods from nuclear power stations and some are materials that have come into contact with radioactive material and become contaminated themselves. Essentially nuclear waste can be divided into **high level waste** and **low level** waste. Low level waste includes items such as rubber gloves, paper towels, and protective clothing that have been used in areas where radioactive materials are handled. The level of activity is low and the half-lives of the radioactive isotopes are generally short. High level waste has high activity and generally the isotopes have long half-lives so the waste will remain active for a long period. Most high level waste comes from spent fuel rods or the reprocessing of spent nuclear fuel.

STORAGE AND DISPOSAL OF NUCLEAR WASTE

Low level waste Different methods are used to dispose of low level waste. Although many governments have now banned the practice some is simply discharged straight into the sea where it becomes diluted. Since the decay produces heat it is better to store it in vast tanks of cooled water called 'ponds' where it can lose much of its activity. Before it is then discharged into the sea it is filtered through an ion exchange resin which removes strontium and caesium, the two elements responsible for much of the radioactivity. Other methods of disposal include keeping the waste in steel containers inside concrete-lined vaults.

High level waste During reprocessing of spent fuel about 96% of the uranium is recovered for re-use. About 1% is plutonium which is a valuable fuel. The remaining 3% is high level liquid waste. One method used to treat this is to vitrify it. The liquid waste is dried in a furnace and then fed into a melting pot together with glass making material. The molten material is then poured into stainless steel tubes where it solidifies. Air flows round the containers to keep them cool. Because of the high activity and long half-lives some of the waste will remain active for hundreds if not thousands of years. The problem is how to store it safely for this length of time. Currently the best solution seems to be burying it in deep remote places that are geologically stable such as disused mines or in granite rock. The concern is that the radioactive material may eventually leach into the water table and then into drinking water.

WAVELENGTH OF ULTRAVIOLET LIGHT NECESSARY FOR OZONE AND OXYGEN DISSOCIATION

In the earlier section on ozone depletion it was stated that the ozone layer absorbs a wide range of ultraviolet light. The average bond enthalpy at 298 K for the $O{=}O$ bond is given in the IB Data Book as 496 kJ mol^{-1}. For just one double bond this equates to 8.235×10^{-19} J. The wavelength of light that corresponds to this enthalpy value (E) can be calculated by combining the expressions $E = hf$ and $c = \lambda f$ to give

$\lambda = \dfrac{hc}{E}$ where h is Planck's constant, and c is the velocity of light.

$$\lambda = \frac{6.626 \times 10^{-34} \,(J\,s)\;\; 2.998 \times 10^{8} \,(m\,s^{-1})}{8.235 \times 10^{-19} \,(J)} = 241 \text{ nm}$$

This is in the high energy region of the ultraviolet spectrum. Ozone can be described as two resonance hydrids. An alternative bonding model is to consider the π electrons to be delocalized over all three oxygen atoms. In both models the bond order is 1.5, i.e. between an O—O single bond ($\Delta H = 146$ kJ mol^{-1}) and an $O{=}O$ double bond so ultraviolet light with a longer wavelength (lower energy) is absorbed in breaking the ozone bond. The actual wavelength required is 330 nm. Working backwards this gives the strength of the O—O bond in ozone as 362 kJ mol^{-1}.

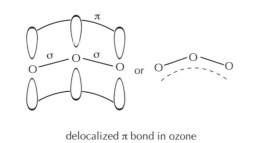

delocalized π bond in ozone

CATALYSIS OF OZONE DESTRUCTION BY CFCs AND NO$_x$

CFCs catalyze the destruction of ozone because the high energy ultraviolet light in the stratosphere causes the homolytic fission of the C—Cl bond to produce chlorine radicals. Note that it is the C—Cl bond that breaks, not the C—F bond, as the C—Cl bond strength is weaker. These radicals then break down ozone molecules and regenerate more radicals so that the process continues until the radicals eventually escape or terminate. It has been estimated that one molecule of a CFC can catalyze the breakdown of up to 100 000 molecules of ozone.

$CCl_2F_2(g) \longrightarrow CClF_2^{\bullet}(g) + Cl^{\bullet}(g)$ (radical initiation)

$Cl^{\bullet}(g) + O_3(g) \longrightarrow ClO^{\bullet}(g) + O_2(g)$ (propagation

$ClO^{\bullet}(g) + O^{\bullet}(g) \longrightarrow Cl^{\bullet}(g) + O_2(g)$ of radicals)

Evidence to support this mechanism is that the increase in the concentration of chlorine monoxide in the stratosphere over the Antarctic has been shown to mirror the decrease in the ozone concentration.

Nitrogen oxides also catalytically decompose ozone by a radical mechanism. The overall mechanism is complex. Essentially oxygen radicals are generated by the breakdown of NO_2 in ultraviolet light.

$NO_2(g) \longrightarrow NO(g) + O^{\bullet}(g)$

The oxygen radicals then reacts with ozone

$O^{\bullet}(g) + O_3(g) \longrightarrow 2O_2(g)$

The nitrogen oxide can also react with ozone to regenerate the catalyst

$NO(g) + O_3(g) \longrightarrow NO_2(g) + O_2(g)$

The overall reaction can be simplified as:

$$2O_3(g) \xrightarrow{\;NO_2(g)\;} 3O_2(g)$$

REASONS FOR GREATER OZONE DEPLETION IN THE POLAR REGIONS

During the winter the temperatures get very low in the stratosphere above the poles. At these low temperatures the small amount of water vapour in the air freezes to form crystals of ice. The crystals also contain small amounts of molecules, such as HCl and $ClONO_2$. It is believed that catalytic reactions occur on the surface of the ice crystals to produce species such as hypochlorous acid (HClO) and chlorine (Cl_2). With the advent of spring the Sun causes these molecules to break down to produce $Cl^{\bullet}$ radicals which catalyze the destruction of ozone. The largest holes occur during early spring. As the Sun warms the air the ice crystals disperse, warmer winds blow into the region, and the ozone concentration gradually increases again.

SUN-SCREENING COMPOUNDS

Sunlight can be harmful to the skin. At longer wavelengths it causes temporary reddening but at shorter wavelengths it can cause damage to proteins making the skin look leathery and wrinkled. It can also break bonds in DNA resulting in damage to genes and lead to skin cancer. Sunscreens block this damaging radiation. The atmosphere itself is an effective sunscreen but as the ozone concentration decreases more harmful ultraviolet light is reaching the Earth's surface. Glass is also an effective sunscreen so sunburn does not occur indoors behind a glass window. The body produces its own sunscreen – melanin. This is a natural dark brown pigment in the skin and its concentration is increased by the action of sunlight on the skin. Commercial sunscreens, such as 4-aminobenzoic acid (maximum absorption at 265 nm), are substances that are able to absorb light of a particular frequency. These contain conjugated double bonds (alternate $C{=}C$ and C—C bonds) with delocalized π electrons. The ultraviolet light is absorbed by exciting the π electrons in these bonds to higher energy levels. Some people now prefer to use sunblocks. These contain white pigments, such as zinc oxide or titanium(IV) oxide, which form a barrier to sunlight by reflecting and scattering the light.

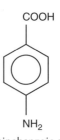

4-aminobenzoic acid
(also known as PABA,
para-aminobenzoic acid)

 # Smog

CONDITIONS FOR THE FORMATION OF SMOGS

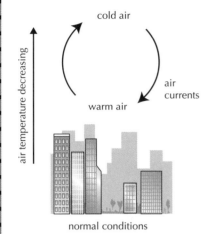

normal conditions

Smog is a poisonous mixture of smoke, fog, air, and other chemicals. It tends to form in large cities and is favoured by lack of wind. It also tends to occur more in cities which are 'bowl shaped', i.e. surrounded by higher ground in all directions. Both of these factors tend to prevent the movement of air. Smogs are most likely to occur when there is a **temperature inversion**. Normally the temperature decreases with altitude. Warm air rises taking the pollutants with it and is replaced by cleaner cooler air. However, sometimes atmospheric conditions cause a layer of still warm air to blanket a layer of cooler air. The trapped pollutants cannot rise and if the conditions persists the amount of pollutants in the warm air near the ground can increase to dangerous levels.

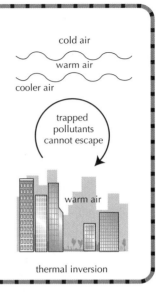

thermal inversion

PHOTOCHEMICAL SMOG

The classic London 'pea soup' smog that occurred before the introduction of clean air controls in the mid 1950s was a reducing smog, principally due to the combustion of coal and oil. This produced sulfur dioxide mixed with soot, fly ash, and partially oxidized organic material. This type of smog is now less common. Most smogs which now occur in cities, such as Los Angeles, are photochemical smogs. The primary pollutants are oxides of nitrogen and volatile organic compounds, VOCs, which are emitted from internal combustion engines and concentrate in the atmosphere to give the air a characteristic yellow/brown colour. In sunlight these are converted into secondary pollutants.

FORMATION OF SECONDARY POLLUTANTS IN PHOTOCHEMICAL SMOGS

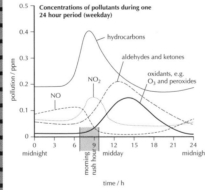

During the early morning rush hour there is a build up of hydrocarbons and nitrogen oxides from car exhausts. As the Sun comes out the nitrogen dioxide absorbs sunlight and breaks down to form radicals.

$$NO_2(g) \rightarrow NO(g) + O^{\bullet}(g)$$

These oxygen radicals can react with oxygen to form ozone and with water to form hydroxyl radicals. These secondary photochemical oxidants can react with a variety of molecules including nitrogen oxides to form nitric acid and VOCs to form peroxides ROOR, aldehydes RCHO, and ketones RCOR. Peroxides are extremely reactive and aldehydes and ketones reduce visibility by condensing to form aerosols.

$O^{\bullet}(g) + H_2O(l) \rightarrow 2OH^{\bullet}(g)$	(formation of hydroxyl radicals)
$OH^{\bullet}(g) + NO_2(g) \rightarrow HNO_3(aq)$	(nitric acid formation)
$OH^{\bullet}(g) + RH(g) \rightarrow R^{\bullet}(g) + H_2O(l)$	(radical propagation)
$R^{\bullet}(g) + O_2(g) \rightarrow ROO^{\bullet}$	(peroxide radical formation)

Chain termination can occur when peroxide radicals react with nitrogen dioxide to form **peroxyacylnitrates**, PANs. These compounds are eye irritants and are toxic to plants.

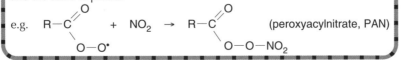

PRECIPITATION OF HEAVY METAL IONS AND PHOSPHATES FROM WATER

Even 'insoluble' salts are still very slightly soluble in water. For a salt formed from a metal M with a non metal X:

$MX(s) \leftrightarrows M^+(aq) + X^-(aq)$

The equilibrium expression will be $K_{sp} = [M^+(aq)] \times [X^-(aq)]$ where the equilibrium constant, K_{sp}, is known as the **solubility product**. (Note that MX(s) does not appear in the expression as a solid has no concentration.)

Many metal sulfides have very low solubility products so one effective way of precipitating heavy metal ions is by bubbling hydrogen sulfide through polluted water. The solubility product can be used to calculate the amount of a metal ion that will remain in solution after it has been precipitated. For example, the solubility product of lead sulfide, PbS is 1.30×10^{-28}.

$PbS(s) \leftrightarrows Pb^{2+}(aq) + S^{2-}(aq)$

$K_{sp} = [Pb^{2+}(aq)] \times [S^{2-}(aq)]$ but $[Pb^{2+}(aq)] = [S^{2-}(aq)]$

Therefore $K_{sp} = [Pb^{2+}(aq)]^2 = 1.30 \times 10^{-28}$ at 298 K

and $[Pb^{2+}(aq)] = (1.30 \times 10^{-28})^{1/2} = 1.14 \times 10^{-14}$ mol dm^{-3}

Hence the concentration of lead ions in the aqueous solution is 1.14×10^{-14} mol dm^{-3}. The relative atomic mass of lead is 207.19, the mass of lead ions that dissolves in one litre of water at 298 K is therefore $207.19 \times 1.14 \times 10^{-14}$ g which is equal to only 2.36×10^{-12} g. Hence the precipitation as the sulfide is very efficient. Even though this is very small it can reduced even further by adding more sulfide ions to the solution so that now the concentration of the lead ions is not the same as the concentration of the sulfide ions. This is known as the **common ion effect**. If the concentration of the sulfide ions is made to be 1.00 mol dm^{-3} then since

$K_{sp} = [Pb^{2+}(aq)] \times [S^{2-}(aq)]$ and $[S^{2-}(aq)] = 1.00$ mol dm^{-3}

$K_{sp} = [Pb^{2+}(aq)] = 1.30 \times 10^{-28}$ at 298 K.

The concentration of lead ions remaining in the solution at 298 K is now only 1.30×10^{-28} mol dm^{-3} and the mass of lead ions remaining in one litre is just 2.69×10^{-26} g.

Phosphate can be precipitated as aluminium phosphate, $AlPO_4$ and similar calculations can be performed using the K_{sp} value of 1.40×10^{-21}. For many heavy metal ions found in waste water it is slightly more complicated as the salts formed are not binary but the principle is the same. For example, nickel ions in waste water can be precipitated by adding hydroxide ions.

$Ni^{2+}(aq) + 2OH^-(aq) \leftrightarrows Ni(OH)_2(s)$

The solubility product for nickel(II) hydroxide at 298 K is 6.50×10^{-18}.

Hence $K_{sp} = [Ni^{2+}(aq)] \times [OH^-(aq)]^2$

Assuming that all the hydroxide ions in the solution come just from the dissolved nickel(II) hydroxide then $[OH^-(aq)] = 2[Ni^{2+}(aq)]$.

Therefore $K_{sp} = 6.50 \times 10^{-18} = [Ni^{2+}(aq)] \times (2[Ni^{2+}(aq)])^2 = 4[Ni^{2+}(aq)]^3$

Hence $[Ni^{2+}(aq)] = (6.50/4 \times 10^{-18})^{1/3} = (1.63 \times 10^{-18})^{1/3} = 1.18 \times 10^{-6}$ mol dm^{-3}.

Since the relative atomic mass of chromium is 58.71, the mass of nickel ions remaining in one litre of water at 298 K is 6.93×10^{-5} g. This small amount can be reduced even more by using the common ion effect, i.e. by adding more hydroxide ions. However care must be taken to ensure that soluble complex ions are not formed. For example, zinc hydroxide redissolves in excess hydroxide ions to form $[Zn(OH)_4]^{2-}(aq)$.

CATION EXCHANGE CAPACITY (CEC)

Both the soil organic matter and the clay particles in soil have a negative charge and will attract and bond to positive ions (cations). These ions are classified either as basic ions such as calcium, Ca^{2+}, magnesium, Mg^{2+}, sodium, Na^+, and potassium, K^+, or acidic ions such as hydrogen ions, H^+ or aluminium ions, Al^{3+}. The amount of positively charged cations that a soil can hold is described as the **cation exchange capacity, CEC**. The larger the CEC the more cations the soil can hold. These cations are exchanged with cations such as hydrogen ions on the root hairs of plants and thus provide nutrients to the plant. When soil is analyzed the total concentration of basic cations compared to the total concentration of acidic cations can be obtained. The more acidic the soil is (i.e. the lower the pH) the higher the percentage of acidic cations. Soil pH is important because acid cations such as aluminium ions are harmful to plants. This is the problem caused by acid rain as it increases the amount of aluminium ions which lowers the soil pH. Above pH 5 aluminium ions are virtually all precipitated out of the soil solution. Soil has a buffering capacity, but it is still sometimes necessary to add lime to soil to raise the pH and increase the concentration of basic cations held by the clay and soil organic material.

In addition to the nutrient cations required by the plants the soil organic matter, SOM can also bind to organic and inorganic compounds in the soil which helps to reduce the negative environmental effects of contaminants such as pesticides, heavy metal ions and other pollutants.

Acid deposition (2)

IB QUESTIONS – OPTION E – ENVIRONMENTAL CHEMISTRY

1. When the pH of rain water falls below about 5.6 it is known as *acid rain*.

   (a) What is the ratio of the hydrogen ion concentration in acid rain with a pH of 4 compared to water with a pH of 7? **[1]**

   (b) One of the two major acids present in acid rain originates mainly from the burning of coal. **Name** this acid and give equations to show how it is formed. **[3]**

   (c) The second major acid responsible for acid rain originates mainly from internal combustion engines. **Name** this acid and state **two** different ways in which its production can be reduced. **[3]**

   (d) Acid rain has caused considerable damage to buildings and statues made of marble ($CaCO_3$). Write an equation to represent the reaction of acid rain with marble. **[1]**

2. (a) In order to survive, fish require water containing dissolved oxygen. Discuss briefly how an increase in each of the following factors affects the amount of dissolved oxygen in a lake. **[3]**

   (i) Temperature
   (ii) Organic pollutants
   (iii) Nitrates and phosphates

   (b) Define *Biological Oxygen Demand (BOD)*. **[1]**

   (c) In a method to find the concentration of dissolved oxygen, manganese(IV) oxide is formed. This is then used to release iodine which is titrated with standard thiosulfate solution. The equations for these three steps are:

   $$2Mn^{2+}(aq) + 4OH^-(aq) + O_2(g) \rightarrow 2MnO_2(s) + 2H_2O(l)$$
   $$MnO_2(s) + 2I(aq) + 4H^+(aq) \rightarrow$$
   $$Mn^{2+}(aq) + I_2(aq) + 2H_2O(l)$$
   $$I_2(aq) + 2S_2O_3^{2-}(aq) \rightarrow S_4O_6^{2-}(aq) + 2I^-(aq)$$

   1000 cm^3 of a sample of water was processed by this method. It was found that 10.0 cm^3 of 0.100 mol dm^{-3} $Na_2S_2O_3$ solution were required to react with the iodine produced. Calculate the concentration of dissolved oxygen in **g dm^{-3}** in the water sample. **[3]**

3. Global warming is one of the major dilemmas confronting humans today.

   (a) Identify one substance produced by human activity that has been suggested to contribute to global warming. Give its source and explain briefly why the human contribution is so significant. **[3]**

   (b) Describe on the molecular level how the substance in **(a)** exerts its effect. **[2]**

   (c) Name one air pollutant that can counteract the effect of the substances in **(a)** and describe how it does so. **[2]**

4. **(a)** CO, NO, SO_2 and hydrocarbons are primary air pollutants.

   (i) The levels of CO and NO produced by automobiles can be lowered by a catalytic converter. Write a balanced chemical equation for the reaction that takes place between these two primary pollutants in a catalytic converter. **[2]**

   (ii) SO_2 is produced from the burning of coal. It can be removed from the exhaust gases of coal-burning power plants by alkaline scrubbing. Write a balanced chemical equation for the reaction that takes place in the scrubber. **[2]**

   (b) Write chemical equations to show the formation of acid rain from one of the primary pollutants above. **[2]**

   (c) State **one** adverse health effect of hydrocarbons. **[1]**

5. The demand for drinking water continues to be a problem for the world. About 97 % of all the water on the planet is present in the seas and oceans and most of the rest is in ice caps or glaciers.

   One method used to provide drinking water from sea water is reverse osmosis, which uses a partially permeable (semipermeable) membrane.

   (i) Outline what is meant by the terms *osmosis* and *partially permeable membrane*. **[2]**

   (ii) Explain the technique of reverse osmosis used to produce drinking water from sea water. **[3]**

   (iii) Suggest **one** way in which a householder could reduce the amount of water used. **[1]**

6. Waste water and sewage undergo primary, secondary and tertiary stages of treatment.

   (a) State **two** features of the activated sludge process that allow for the removal of impurities. **[2]**

   (b) Identify one major source of phosphate in waste water **[1]**

   (c) State the type of reaction used to remove Pb^{2+} and PO_4^{3-} ions from waste water. For each ion, give an equation to show its removal. **[3]**

7. **(a)** Identify **one** pollutant that contributes to the lowering of the ozone concentration in the upper atmosphere. State a source of the pollutant identified. **[2]**

   (b) Fluorocarbons and hydrofluorocarbons are now considered as alternatives to some ozone depleting pollutants. Outline **one** advantage and **one** disadvantage of the use of these alternatives. **[2]**

8. **(a)** Use the value for the average bond enthalpy for oxygen (in kJ mol^{-1}) given in Table 10 and information from Tables 1 and 2 of the IB data booklet to calculate the maximum wavelength of light that is able to decompose oxygen. Give your answer to three significant figures. **[3]**

   (b) Ozone absorbs radiation with a wavelength shorter than about 320 nm. Use Lewis structures to explain why ozone can be decomposed by light with a longer wavelength than that required to decompose oxygen. **[2]**

   (c) **(i)** Explain, with equations, how a CFC, such as dichlorodifluoromethane, is able to deplete ozone in the ozone layer. **[3]**

   (ii) Suggest an explanation to account for the fact that the depletion of ozone in the ozone layer is greatest as the winter months end over polar regions. **[2]**

9. **(a)** Photochemical smogs occur in many cities. List the two main primary pollutants responsible for photochemical smog and state their source. **[3]**

   (b) State the conditions which favour the build up of photochemical smog. **[3]**

   (c) A class of secondary pollutants present in photochemical smog is peroxyacylnitrates, PANs. Explain with the aid of equations how these secondary pollutants are formed from primary pollutants. **[5]**

10. Phosphate ions, PO_4^{3-} can be precipitated from polluted water by adding aluminium ions. The solubility product of aluminium phosphate at 298K is 1.40×10^{-21}.

    (a) Define the term *solubility product* using aluminium phosphate as an example. **[1]**

    (b) Calculate the concentration of phosphate ions in a saturated solution of aluminium phosphate at 298K. **[2]**

    (c) State what is meant by the *common ion effect* and explain how it could be used to lower the concentration of phosphate ions in a saturated solution of aluminium phosphate. **[2]**

Food groups

FOOD AND NUTRIENTS

The definition of food is very broad. Essentially it is any substance, whether processed, partially processed or raw that is intended for human consumption. This definition includes beverages, chewing gum and any substance used in the preparation or manufacture of 'food' such as colorants but it does not include cosmetics, tobacco or substances used only as drugs. For example, aspirin is taken by mouth but it is not classed as a food. The definition of nutrients is more specific. Nutrients are substances that are obtained from food. They are required to keep the body functioning and healthy. Nutrients provide energy, regulate growth and replenish chemicals for the maintenance and repair of the body's tissues. For the body to remain healthy it requires a balanced diet that includes all the essential nutrients taken from as wide a variety of foods as possible. Nutrients can be divided into six main groups: proteins, carbohydrates, lipids, vitamins, minerals and water. The amount of each required depends on several factors such as age, weight, gender and occupation. A well-balanced diet consists of about 60% carbohydrate, 20-30% protein, and 10-20% fats. Foods containing these three components will also provide the essential vitamins and the fifteen essential minerals which include calcium, magnesium, sodium, iron and sulfur along with trace elements like iodine and chromium. In addition a daily intake of about 2 dm^3 of water is required. Malnutrition can occur when either too little or too much of these essential components are taken.

CARBOHYDRATES

Simple carbohydrates are known as monosaccharides. All monosaccharides have the empirical formula CH_2O. In addition they contain a carbonyl group (C=O) and at least two –OH groups. They have between three and six carbon atoms. Monosaccharides with the general formula $C_5H_{10}O_5$ are known as pentoses (e.g. ribose) and monosaccharides with the general formula $C_6H_{12}O_6$ are known as hexoses (e.g. glucose).

Glucose can exist as a straight chain isomer or a ring compound.

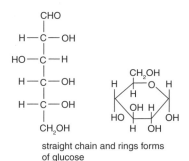

straight chain and rings forms
of glucose

Monosaccharides can undergo condensation reactions to form disaccharides and eventually polysaccharides. For example sucrose is a disaccharide formed from the condensation of glucose and fructose. The link between the two sugars is known as a glycosidic link.

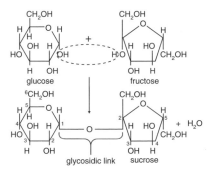

formation of sucrose by a condensation reaction

Other examples of disaccharides include lactose and maltose. Examples of polysaccharides include starch and cellulose.

LIPIDS (FATS AND OILS)

Fats and oils are triesters (triglycerides) formed from the condensation reaction of propane-1,2,3-triol (glycerol) with long chain carboxylic acids (fatty acids).

Fats are solid triglycerides, examples include butter, lard and tallow. Oils are liquid at room temperature and include castor oil, olive oil and linseed oil. The essential chemical difference between them is that fats contain saturated carboxylic acid groups (i.e. they do not contain C=C double bonds). Oils contain at least one C=C double bond and are said to be unsaturated. Most oils contain several C=C double bonds and are known as polyunsaturated.

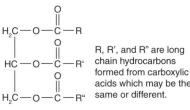

R, R', and R" are long chain hydrocarbons formed from carboxylic acids which may be the same or different.

general formula of a fat or oil.

PROTEINS

Proteins are large macromolecules made up of chains of 2-amino acids. About twenty 2-amino acids, which have the general formula $H_2NCHRCOOH$, occur naturally. The amino acids bond to each other through condensation reactions resulting in the formation of a polypeptide in which the amino acid residues are joined to each other by an amide link (peptide bond).

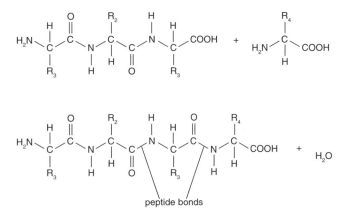

peptide bonds

Each protein contains a fixed number of amino acid residues connected to each other in strict sequence. This sequence e.g. gly-his-ala-ala-leu- ..... is known as the primary structure of proteins. The secondary structure describes the way in which the chain of amino acids folds itself due to intramolecular hydrogen bonding. The tertiary structure describes the overall folding of the chains by interactions between distant amino acids to give the protein its three-dimensional shape. Separate polypeptide chains can interact together to give a more complex structure – this is known as the quaternary structure. For example, hemoglobin has a quaternary structure that includes four protein chains grouped together around four heme groups.

Fats and oils

STRUCTURES OF SATURATED AND UNSATURATED FATTY ACIDS

Most naturally occurring fats contain a mixture of saturated, mono-unsaturated and poly-unsaturated fatty acids. Common examples of fatty acids include;

Name	Number of C atoms per molecule	Number of C=C bonds	Melting point/ °C
saturated fatty acids			
lauric acid $CH_3(CH_2)_{10}COOH$	12	0	44.2
myristic acid $CH_3(CH_2)_{12}COOH$	14	0	54.1
palmitic acid $CH_3(CH_2)_{14}COOH$	16	0	62.7
stearic acid $CH_3(CH_2)_{16}COOH$	18	0	69.6
unsaturated fatty acids			
oleic acid $CH_3(CH_2)_7CH=CH(CH_2)_7COOH$	18	1	10.5
linoleic acid $CH_3(CH_2)_4CH=CHCH_2CH=CH(CH_2)_7COOH$	18	2	−5.0

Unsaturated fatty acid may be either in the *cis-* form in which both hydrogen atoms are on the same side of the C=C double bond or in the *trans-* form in which both hydrogen atoms are on opposite sides of the C=C double bond.

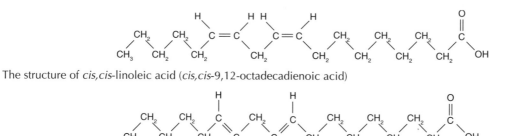

The structure of *cis,cis*-linoleic acid (*cis,cis*-9,12-octadecadienoic acid)

The structure of the *trans,trans-* form of linoleic acid

PROPERTIES OF FATS AND OILS

The melting point of fatty acids increases with increasing relative molecular mass. The melting points are also affected by the degree of unsaturation. Unsaturated fatty acids have lower melting points and tend to be less crystalline. The regular tetrahedral arrangement of the carbon 'backbone' of saturated acids means that they can pack together closely so the van der Waals' forces holding molecules together are stronger as the surface area between them is greater. As the bond angle at the C=C double bonds changes from 109.5° to 120° in unsaturated fatty acids it produces a 'kink' in the chain. They are unable to pack so closely and the van der Waals' forces between the molecules become weaker which results in lower melting points. The way in which unsaturated fatty acids can pack also depends on the geometrical isomerism of the double bonds. *Trans*-fatty acids can pack more closely so have higher melting points than *cis*-fatty acids where both hydrogen atoms are on the same side of the double bond. These packing arrangements in fatty acids are similar in fats and oils and explain why unsaturated fats (oils) have lower melting points than saturated fats. Fats and oils are chosen for cooking on the basis of their melting temperature. For example, cocoa butter melts at close to body temperature whereas fats used for cake making melt over a wide range of temperature.

Examples of fats and oils

Mainly saturated fats (solids at room temperature)	Mainly mono-unsaturated oils (liquids at room temperature)	Mainly poly-unsaturated oils (liquids at room temperature)
Palm, coconut, lard, butter, shortening, tallow	Olive, canola, peanut	Safflower, sunflower, linoleic, linolenic

Unsaturated fats are less stable chemically than saturated fats. This is because the carbon to carbon double bonds in unsaturated fats can react with water and undergo hydrolysis in the presence of heat or enzymes. They can also react with oxygen (auto-oxidation), light (photo-oxidation) and hydrogen (hydrogenation).

HYDROGENATION OF UNSATURATED FATS

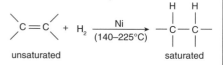

The addition of hydrogen in the presence of heat (140–225 °C) and a finely divided nickel (or zinc or copper) catalyst increases the amount of saturation. This increases the melting point and hardness and is used to manufacture margarine. Hydrogenation also increases the chemical stability of fats by making them less susceptible to oxidation. However there are also some disadvantages to hydrogenation. Oils naturally only contain *cis*-fatty acids but under partial hydrogenation oils containing *trans*-fatty acids can form. Unlike natural mono- and poly-unsaturated oils which increase HDL (high density lipoprotein) cholesterol and are healthier for the heart, oils containing *trans*-fatty acids behave more like saturated fats and increase levels of LDL (low density lipoprotein) cholesterol leading to an increased risk of heart disease. Fats containing *trans*-fatty acids are harder to metabolize and excrete so they accumulate in fatty tissue. They are also a lower quality energy source.

Shelf life

FACTORS AFFECTING SHELF LIFE

Packaged food on sale in many parts of the world has one or more dates stamped on it. These may be labeled as *best before, use by* or *display until*. These dates are detailing the **shelf life** of the food which is when the food no longer maintains the expected quality desired by the consumer because of changes in flavour, smell, texture and appearance (such as colour) and because of microbial spoilage. A food that has passed its shelf life may still be safe to consume but optimal quality is no longer guaranteed.

Chemical factors that cause a decrease in shelf life include:

- Water content   A change in the water content can cause loss of nutrients, browning and rancidity. Dry foods become more vulnerable to microbial spoilage if they absorb water.
- Change in pH   Can cause changes in flavour, colour, browning and loss of nutrients.
- Light   Can lead to rancidity, loss of vitamins and fading of natural colours.
- Temperature   A higher temperature increases the rate of other forms of spoilage.
- Exposure to air   Can increase the rate of oxidation causing browning, changes in flavour, colour and loss of nutrients.

RANCIDITY

Lipids (fats and oils) become rancid when our senses perceive them to have 'gone off' due to a disagreeable smell, taste, texture or appearance. Rancidity is caused either by the hydrolysis of the triesters or by the oxidation of the fatty acid chains.

Hydrolytic rancidity is caused by the breaking down of a lipid into its component fatty acids and propane-1,2,3-triol (glycerol).

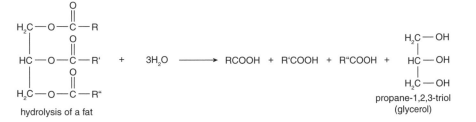

hydrolysis of a fat

This process of hydrolysis is the reverse of esterification. It occurs more rapidly in the presence of enzymes such as lipase and with heat and moisture. In deep frying the water present in the food and the high temperatures increases the rate of hydrolysis to fatty acids. Examples of fatty acids responsible for rancidity include:

- butanoic acid, C_3H_7COOH, hexanoic acid, $C_5H_{11}COOH$, and octanoic acid, $C_7H_{15}COOH$, in milk
- palmitic acid, $C_{15}H_{31}COOH$, stearic acid, $C_{17}H_{35}COOH$ and oleic acid, $C_{17}H_{33}COOH$ which give chocolate an oily or fatty flavour
- lauric acid, $C_{11}H_{23}COOH$ which gives palm and coconut oil, in cocoa butter substitutes, a soapy flavour
- butanoic acid, C_3H_7COOH, in butter.

Oxidative rancidity is due to the oxidation of the fatty acid chains, typically by the addition of oxygen across the C=C double bond in unsaturated fatty acids. Oily fishes, such as mackerel, contain a high proportion of unsaturated fatty acids and are prone to oxidative rancidity. The process proceeds by a free radical mechanism catalyzed by light in the presence of enzymes or metal ions. This photo-oxidation process leads to the formation of hydroperoxides. Hydroperoxides have the formula R-O-O-H and break down to form free radicals due to the weak O-O single bond they contain.

PROLONGING SHELF LIFE AND MINIMIZING THE RATE OF RANCIDITY

Methods of prolonging the shelf life of food are aimed at counteracting microbacterial growth in some way. Moulds and bacteria thrive in conditions of moderate temperature, high humidity and low salt concentrations. Traditional methods of preserving food include: fermentation (with alcohol), preserving with sugar, pickling with vinegar (ethanoic acid), salting (with sodium chloride), drying and smoking.

Other methods which slow down the rate of deterioration of food include processing, packaging and the use of additives.

Processing	Packaging	Use of additives
Refrigeration and freezing – storing dairy products at low temperature slows lipase hydrolysis. Reducing light levels by using coloured glass or keeping in a dark place. Radiation. Using gamma radiation or X-rays to destroy microorganisms	Using an inert gas to minimize contact with oxygen. Using hermetic sealing or low-gas permeability packaging film. Keeping jars full to minimize the amount of air in the headspace above oil. Sealing in tin cans.	Sodium sulfite, sodium hydrogensulfite and citric acid – delay the onset of enzymatic browning. Sodium and potassium nitrite and nitrate for curing meat, fixing colour and inhibiting micro-organisms. Sodium benzoate and benzoic acid as antimicrobial agents in fruit juices, carbonated drinks, pickles and sauerkraut. Sorbic acid, propanoic acid, calcium propanoate and sodium propanoate for delaying mould and bacterial growth in breads and cheeses. Ethanoic acid and benzoic acid for delaying mould and bacterial growth in pickled meats and fish products and to add flavour.

Antioxidants

NATURALLY OCCURRING ANTIOXIDANTS

Antioxidants delay the onset or slow down the rate at which oxidation occurs and hence extend the shelf life of food. Naturally occurring antioxidants include:

- vitamin C (ascorbic acid) found in citrus fruits, green peppers, broccoli, green leafy vegetables, strawberries, red currants, and potatoes

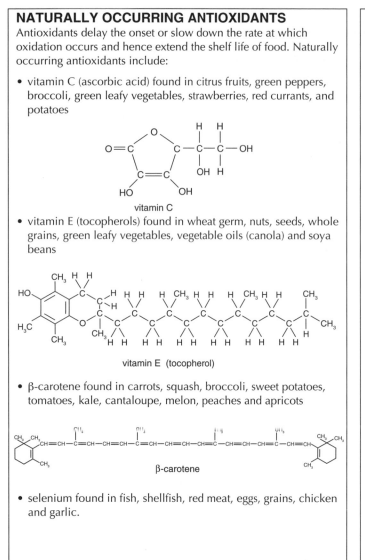

vitamin C

- vitamin E (tocopherols) found in wheat germ, nuts, seeds, whole grains, green leafy vegetables, vegetable oils (canola) and soya beans

vitamin E (tocopherol)

- β-carotene found in carrots, squash, broccoli, sweet potatoes, tomatoes, kale, cantaloupe, melon, peaches and apricots

β-carotene

- selenium found in fish, shellfish, red meat, eggs, grains, chicken and garlic.

THE ADVANTAGES AND DISADVANTAGES OF ANTIOXIDANTS IN FOOD

Advantages	Disadvantages
Naturally occurring vitamins C, E and carotenoids reduce the risk of cancer and heart disease by inhibiting the formation of free radicals.	Consumers perceive synthetic antioxidants to be less safe as they do not occur naturally.
Vitamin C is vital for the production of hormones and collagen.	Natural antioxidants are more expensive and less effective than synthetic antioxidants and can also add unwanted colour and leave an aftertaste to food.
β-carotene can be used as an additive in margarine to provide it with a yellow colour and act as a precursor for vitamin A synthesis.	Synthetic antioxidants are classed as food additives and need to be regulated by policies and legislation to ensure their safe use in food.
They are believed to enhance the health effects of other foods and boost overall health and resilience.	Policies regarding the safe use and labeling of food additives can be difficult to implement and monitor, especially in developing countries and across international borders.

SYNTHETIC ANTIOXIDANTS

Synthetic antioxidants include: butylated hydroxyanisole, BHA, butylated hydroxytoluene, BHT, propyl gallate, PG, trihydroxybutyrophenone, THBP, and *tert*-butylhydroquinone, TBHQ.

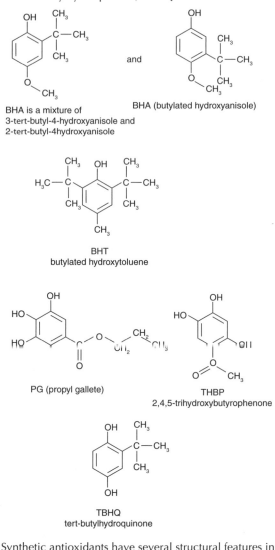

BHA is a mixture of 3-tert-butyl-4-hydroxyanisole and 2-tert-butyl-4hydroxyanisole

BHA (butylated hydroxyanisole)

BHT
butylated hydroxytoluene

PG (propyl gallete)

THBP
2,4,5-trihydroxybutyrophenone

TBHQ
tert-butylhydroquinone

Synthetic antioxidants have several structural features in common. All of the examples given contain a phenolic group (-OH joined directly to a benzene ring), and many contain a carbon atom bonded directly to three methyl groups which is known as a tertiary butyl group, $C(CH_3)_3$-. Both the phenolic group and the tertiary butyl group are free radical scavengers. They react with and remove the free radicals involved in the oxidation of the food and thus prolong the shelf life.

ANTIOXIDANTS IN TRADITIONAL FOOD

Many traditional foods found in different cultures contain natural antioxidants. Vitamin C and carotenoids found in many types of fruit and vegetables are obvious examples. Another class of natural antioxidants are flavonoids. These are found in all citrus fruits, green tea, red wine, oregano and dark chocolate (containing at least 70% cocoa). They have been linked to lowering levels of LDL (low density lipoprotein) cholesterol and blood sugar levels which reduce high blood pressure and to preventing the development of cancerous cells.

Colour

FOOD COLOUR

The colour of food is due to the ability of substances in the food to absorb light in the visible region of the electromagnetic spectrum and transmit the remaining light of the visible spectrum which has not been absorbed. The light that is transmitted is known as the complementary colour. Thus a substance that absorbs red light will transmit a blue-green colour.

The substances that cause food to be coloured may be natural or synthetic. Natural colorants which are found in the cells of plants and animals are called **pigments**. Synthetic food-grade water-soluble substances which are added to produce colour in food are known as **dyes**.

Complementary colours

When white light falls on food that contains a substance that absorbs in the blue-violet region of the electromagnetic spectrum then the food appears yellow as that is the colour of the remaining light that is transmitted.

Complementary colours are opposite each other.

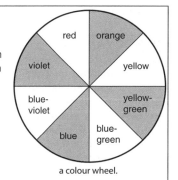

a colour wheel.

PIGMENTS

Pigments, which occur naturally, include anthocyanins, carotenoids, chlorophyll and haem.

Anthocyanins	Carotenoids	Chlorophyll
Most widely occurring pigments in plants. Responsible for the pink, red, purple and blue colours in fruits and vegetables, including cranberries, blueberries, strawberries and raspberries.	Most widespread pigments in nature (the large majority are produced by algae). Act as a precursor for vitamin A synthesis – those found in fruit and vegetables contribute 30–100% of the vitamin A requirement in humans. Colours range from yellow to red/orange. Found in bananas, carrots, tomatoes, watermelon, peppers and saffron. Red astaxanthin (complexed to a protein) is responsible for the blue or green colour of live lobsters and crabs and the pink colour of salmon and flamingos.	Major pigments necessary for photosynthesis found in green plants. **Heme** Myoglobin is responsible for the purple-red colour of fresh meat.

SYNTHETIC COLORANTS (DYES)

Many foods contain artificial ingredients to provide colour and flavour. Most countries now require the ingredients of all foods to be stated on the packaging. Food additives are given numbers so they can be identified. In some parts of the world these are called E numbers elsewhere an International Numbering System (INS) is used. Unfortunately many artificial dyes used in the past have been shown to be potentially carcinogenic and have been withdrawn.

One of the problems is that different countries have different regulations. Some dyes that are permitted in some countries are banned in others. With much food crossing international borders this is now of real concern and there is a need for international legislation on colorant legislation.

ANALYZING COLOUR FROM SPECTRA

The wavelength of visible light lies in the region of about 400 nm (4×10^{-7} m) to about 700 nm (7×10^{-7} m). The 'blue' light with the shorter wavelength of 400 nm has a greater energy than the 'red' light with a wavelength of 700 nm. In a visible spectrometer the amount of light absorbed is measured again the wavelength of the light passing through the sample to provide an absorption spectrum. This can be illustrated with the visible spectra of chlorophyll. Chlorophyll can exist in different forms and the precise absorption of each form is modified slightly in different species. The spectra of chlorophyll a and chlorophyll b are shown. Both forms of chlorophyll will appear green as the light that is not being absorbed, and is therefore being transmitted, lies in the region of 500–550 nm which is the green region of the spectrum. However they will not be exactly the same colour green as the precise values are different and it can be seen that the wavelengths where the maximum absorptions occur are also different. In chlorophyll a the maxima occur at about 420 and 680 nm whereas in chlorophyll b they occur at 480 and 630 nm.

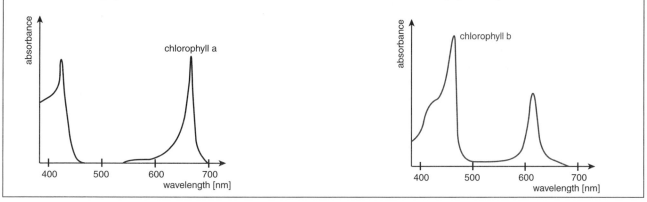

Factors affecting the colour stability of pigments

The colour stability of pigments is affected by many factors. Any factor which will change the structure of the molecule will affect the colour as this will affect the precise wavelength of visible light that the pigment absorbs and thus the complementary colours transmitted. These factors include: oxidation, temperature changes, pH changes and the presence of metal ions.

ANTHOCYANINS

Anthocyanins exist in different forms. In aqueous solution these different forms are in equilibrium with each other. Changing the pH and the temperature affects the position of equilibrium and thus the predominant species responsible for the colour. They are most highly coloured at low pH (in acidic solution) and at low temperatures. When exposed to heat the equilibrium moves to the right and the compounds are less thermodynamically stable. This causes a loss of colour and browning.

quinoid (**A**) ⇌ flavylium (**AH⁺**) ⇌ carbinol ⇌ chalcone

   (blue)   (red)   (colourless)   (colourless)

Anthocyanins also form complexes with metal ions such as aluminium ions, Al^{3+}, and iron(III) irons, Fe^{3+}. These ions are present in the metal of 'tin' cans. Food that is stored in cans may become discoloured due to this complex ion formation.

CHLOROPHYLL

Different forms of chlorophyll contain different groups attached in the R- position.

Chlorophyll contains a group with four nitrogen atoms which is called a porphin. The porphin ring forms a very stable complex with a magnesium ion. The stability of chlorophyll towards heat depends on the pH. In a basic solution with a pH of 9 it is thermodynamically stable but in acidic solution with a pH of 3 it is unstable. When heated, the cell membrane of the plant deteriorates releasing acids which decrease the pH. At this lower pH the magnesium ion is displaced by two hydrogen ions resulting in the formation of an olive-brown pheophytin complex. The breakdown of the cell during heating also increases the susceptibility of chlorophyll to decomposition by light.

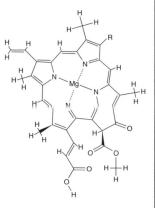

CAROTENOIDS

Carotenoids contain many alternate carbon to carbon single and double bonds which when bonded together account for the fact that they are coloured. An example is β-carotene. This is an antioxidant as well as being a pigment.

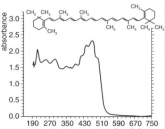

β-carotene is found in carrots and has a characteristic orange colour. It contains eleven conjugated double bonds and absorbs strongly in the violet-blue (400–510 nm) region.

Because of the unsaturation due to the C=C double bonds carotenoids are susceptible to oxidation. This oxidation process can be catalyzed by light, metals and hydroperoxides. This changes the type of bonding and results in the bleaching of colour, loss of vitamin A activity and is the cause of bad smells.

Carotenoids are stable up to 50 °C and in a pH range of 2–7 and are therefore not degraded by most forms of food processing. When heated the naturally occurring *trans-* isomer rearranges to the *cis-* isomer.

HAEM

The haem group also contains a porphin ring but it is complexed to an iron ion. During oxidation, oxygen binds to purple-red myoglobin (Mb), and red oxymyoglobin (MbO_2) forms. In both Mb and MbO_2 the iron in the haem group is in the form of iron(II), Fe^{2+}. Through auto-oxidation of Mb and MbO_2 the oxidation state of the iron is changed to iron(III), Fe^{3+}. In the Fe^{3+} state it is called metmyoglobin (MMb) and has an undesirable brown-red colour. Interconversion between these three forms occurs readily.

MbO_2 ⇌   Mb   ⇌   MMb
oxymyoglobin   myoglobin   metmyoglobin
(red, Fe^{2+})   (purple-red, Fe^{2+})   (brown, Fe^{3+})

In order to minimize the rate of formation of brown metmyoglobin from auto-oxidation meat can be stored free of oxygen. Packaging films with low gas permeability are used. Air is removed from the package and a storage gas, normally pure carbon dioxide, is injected.

NON-ENZYMATIC BROWNING OF FOOD

Most enzymatic browning, such as that caused when an apple is peeled, is normally undesirable. However natural browning through cooking both enhances the flavour and the appearance of food.

Foods high in carbohydrate content, especially sucrose and reducing sugars, e.g. glucose, lacking nitrogen-containing compounds can be **caramelized**. Although it can be achieved simply by heating, the chemical process is far from simple. Both sucrose and glucose when caramelized form many different products, among them acids, sweet and bitter derivatives, volatile molecules with a caramel aroma and brown-coloured polymers. Factors that increase the rate of caramelization include the pH and the temperature. At pH values below 3 acid catalysis occurs whereas base catalysis occurs at pH values greater than 9. Boiling will not cause caramelization as a temperature above 120 °C is required which occurs during the baking and the roasting of foods with a high sugar content. An example of caramelization is the browning on the top of baked egg dishes.

For foods that contain nitrogen the **Maillard reaction** involves the reaction of a carbohydrate, either a free sugar or one bound up in starch, with the amine group on an amino acid, which may also be free or part of a protein chain. Basically it involves a condensation reaction between the carbonyl group on the reducing sugar and the amine group. The presence of the amino acid lysine (which contains two amine groups) results in the most browning colour and cysteine the least.

Thus foods that contain lysine, for example milk, brown readily. Because moisture lowers the temperature, in order to make a good stew it is sensible to brown the meat, vegetables and flour well in hot oil to bring out the flavours before adding any liquid. Other examples of Maillard browning include making milk chocolate and heating sugar and cream to make fudge, toffees and caramels.

Genetically modified foods and texture

GENETICALLY MODIFIED FOODS
Genetic engineering involves the process of selecting a single gene for a single characteristic and transferring that sequence of DNA from one organism to another. Thus a genetically modified (GM) food can be defined as one derived or produced from a genetically modified organism. The GM food can be substantially different or essentially the same in composition, nutrition, taste, smell, texture and functional characteristics to the conventional food. An example of genetically modified food is the FlavrSavr tomato. In normal tomatoes a gene is triggered when they ripen to produce a substances that makes the fruit go soft and eventually rot. In the FlavrSavr tomato the gene has been inhibited to produce a tomato with a fuller taste and a longer shelf life.

BENEFITS OF GM FOODS
- With crops it can enhance the taste, flavour, texture and nutritional value and also increase the maturation time.
- Plants can be made more resistant to disease, herbicides and insect attack.
- With animals GM foods can increase resistance to disease, increase productivity and feed efficiency to give higher yields of milk and eggs.
- Anti-cancer substances and increased amounts of vitamins (such as vitamin A in rice) could be incorporated and exposure to less healthy fats reduced.
- Environmentally 'friendly' bio-herbicides and bio-insecticides can be formed. GM foods can lead to soil, water and energy conservation and improve natural waste management.

POTENTIAL CONCERNS OF GM FOODS
- The outcome of alterations is uncertain as not enough is known about how genes operate.
- They may cause disease as antibiotic-resistant genes could be passed to harmful microorganisms.
- Genetically engineered genes may escape to contaminate normal crops with unknown effects.
- They may alter the balance of delicate ecosystems as food chains become damaged.
- There are possible links to an increase in allergic reactions (particularly with those involved in the food processing).

THE TEXTURE OF FOOD
Food often appears homogeneous, that is, all the ingredients are in the same phase. However many ingredients in food are completely immiscible so will form separate phases within the food. The reason why the food often appears to be homogeneous is because the sizes of these phases can be very small. A **dispersed system** is a kinetically stable mixture of one phase within another largely immiscible phase. It follows that there are several types of dispersed systems as the dispersed particles must be in a different phase to the immiscible phase. Usually the continuous phase is a liquid. A solid dispersed in a liquid is known as a **suspension**, a liquid dispersed in a liquid is known as an **emulsion** and a gas dispersed in a liquid is known as a **foam**. An example of a suspension is molten chocolate, cream is an emulsion and beer is a foam.

EMULSIFIERS
There are essentially two types of food emulsion. One is water-in-oil emulsion such as found in butter. Water-in-oil emulsions essentially consist of the dispersion of water droplets in a continuous oil phase. The second type is an oil-in-water emulsion such as milk and salad dressing. This type of emulsion is more common in manufactured foods. Emulsifying agents (emulsifiers) are substances which aid the dispersal of these droplets and stabilize them to prevent them coalescing to form larger globules. Essentially they act as the interface between the solid, liquid or gas phases in the dispersed system so they are also used in making foams. Good emulsifiers will tend to be soluble in both fats (oils) and water. Common emulsifiers include lecithin (either pure or in egg yolk), milk protein and salts of fatty acids. To physically make an emulsion mechanical energy is also required in addition to adding an emulsifier. This aids dispersion which explains the need for beating, mixing and whisking in cooking. Stabilizers can also be added to prevent the emulsions and foams from separating out into the separate phases.

Oxidative rancidity and antioxidants

OXIDATIVE RANCIDITY

Before oxidative rancidity can occur the C-H bonds must be broken. Normally these are broken homolytically. The mechanism then proceeds via free radicals. The C-H bonds are strong with an average bond enthalpy of 412 kJ mol⁻¹ so that considerable energy is required to overcome this high activation energy to initiate radical formation.

$$R\text{-}H \rightarrow R^{\bullet} + H^{\bullet}$$

Where R represents any unsaturated fatty acid.

Once free radicals have been formed they can propagate in the presence of oxygen to form peroxide radicals which in turn can react with more of the unsaturated fatty acid molecules to form hydroperoxides.

$$R^{\bullet} + O_2 \rightarrow ROO^{\bullet}$$

$$ROO^{\bullet} + RH \rightarrow R^{\bullet} + ROOH$$

It is also possible that oxygen can react directly with the alkene of the hydrocarbon chain to form hydroperoxides directly and that it does not necessarily proceed by a free radical mechanism.

$$R\text{-}H + O_2 \rightarrow ROOH$$

Whichever way the hydroperoxide is formed the weak O-O peroxide bond (146 kJ mol⁻¹) then breaks either photochemically or through catalysis involving a metal or an enzyme to form two radicals which can lead to further radical propagation and other products from termination reactions. These other products include aldehydes, ketones and alcohols, all of which can be isolated from oxidizing lipid systems.

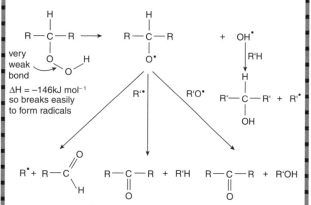

Break down of hydroperoxides leading to both radical and non-radical products.

Other termination steps can also occur when two free radicals combine to form non-radical products. These include:

$$R^{\bullet} + R^{\bullet} \rightarrow R\text{-}R$$

$$R^{\bullet} + ROO^{\bullet} \rightarrow ROOR$$

$$ROO^{\bullet} + ROO^{\bullet} \rightarrow ROOR + O_2$$

$$RO^{\bullet} + R^{\bullet} \rightarrow ROR$$

ANTIOXIDANTS

1. Free radical inhibitors

These work by either interrupting the formation of free radicals in the initiation step of auto-oxidation or by interrupting the propagation of the free radical chain. Examples of such antioxidants (with the simplified formula, AH) include BHA, BHT, TBHQ and vitamin E (tocopherols). These contain groups such as the phenolic group and the *tert*-butyl group which can scavenge (mop up) radicals. They do this by forming less stable and less reactive free radicals or non-radical products, e.g.

$$R^{\bullet} + AH \rightarrow R\text{-}H + A^{\bullet}$$

$$RO^{\bullet} + AH \rightarrow R\text{-}O\text{-}H + A^{\bullet}$$

$$ROO^{\bullet} + AH \rightarrow R\text{-}O\text{-}O\text{-}H + A^{\bullet}$$

$$R^{\bullet} + A^{\bullet} \rightarrow R\text{-}A$$

$$RO^{\bullet} + A^{\bullet} \rightarrow R\text{-}O\text{-}A$$

2. Complexing agents

These work by forming an irreversible complex with metals ions, such as the iron(II) ion, Fe^{2+}, to reduce the concentration of free metal ion in the solution so that the metal ion is unable to catalyze the oxidation reactions. This complexing is known as chelating and effective chelating agents include salts of EDTA, which is the shortened version of its old name, ethylenediaminetetracetic acid, and certain plant extracts such as rosemary, tea and ground mustard. EDTA has been shown to inhibit the Fe^{2+} catalyzed oxidation of raw beef but it is not currently approved for use in commercial meat products.

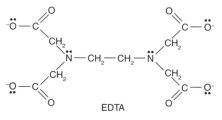

ethylenediaminetetraacetate ion (EDTA⁴⁻)

3. Reducing agents

The third group consists of reducing agents and includes vitamin C (ascorbic acid) and carotenoids. These are electron donors and they remove or reduce concentrations of oxygen. The half-equation for the oxidation of ascorbic acid is:

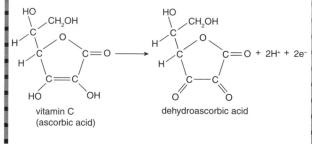

Stereochemistry in food

CONVENTIONS FOR NAMING ENANTIOMERS

Enantiomers occur when a pure compound contains an asymmetric or chiral carbon atom. They are non-superimposable mirror images.

Three different conventions exist for naming and hence differentiating bewteen the different enantiomers.

Two enantiomers

1. (+) or *d*– and (–) or *l*–isomers

The ability of enantiomers to rotate the plane of plane-polarized light has been discussed in Topic 10 – Organic chemistry. The enantiomer or optical isomer that rotates the direction of plane polarized light clockwise, known as dextrorotatory is identified by either by (+) or *d* (from the Latin *dextro* meaning right), whereas the one that rotates the plane-polarized light anticlockwise, known as laevorotatory (from the Latin *laevo* meaning left), is identified either by (–) or *l*. This form of nomenclature is useful as it relates directly to a physical property but it is impossible to state which enantiomer is which simply by looking at their chemical structure.

2. D and L notation

This notation uses the capital letters D and L and is based on the absolute configuration (i.e. on their spatial distribution) of each of the two enantiomers rather than their ability to rotate the plane of plane-polarized light in a particular direction. The D and L notation is commonly used when naming the enantiomers of carbohydrates, such as sugars, and amino acids. For naming carbohydrates the basic reference is (+)-2,3-dihydroxypropanal, more commonly known as (+)-glyceraldehyde, which is assigned the letter D. All other carbohydrates are assigned either D or L with reference to this.

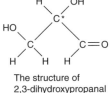

The structure of 2,3-dihydroxypropanal (glyceraldehyde) showing the chiral carbon atom marked by *.

When naming amino acids the 'CORN' rule is applied. This arranges the substituents: **CO**OH, **R**, **N**H$_2$ and H around the asymmetric carbon atom with the hydrogen atom pointing away from the viewer. If the 'CORN' groups are arranged clockwise then it is the D-enantiomer and if they are arranged anti-clockwise it is the L-enantiomer.

D-alanine   L-alanine

3. R and S notation

This way of naming enantiomers is mainly used by chemists when dealing with substances other than carbohydrates and amino acids. Each chiral, or asymmetric, carbon centre is labeled R or S according to the Cahn-Ingold-Prelog priority rules. These are based on atomic number. The lowest-priority of the four substituents is pointed away from the viewer. If the priority of the remaining three substituents decreases in a clockwise direction it is assigned the R-form, if it decreases in an anti-clockwise direction it is the S-form.

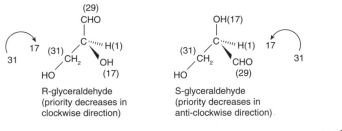

R-glyceraldehyde (priority decreases in clockwise direction)

S-glyceraldehyde (priority decreases in anti-clockwise direction)

PROPERTIES OF ENANTIOMERS IN FOOD

Enantiomers may have very different biological effects. One tragic example of this from the last century is the drug thalidomide. One of the enantiomers of thalidomide relieved morning sickness in pregnant women, the other enantiomer caused severe defects in the limbs of the unborn child (see Option D – Medicines and drugs). Enantiomers can have different effects in food too. For example, unlike the L- form, the D- form of vitamin C has no biological activity. Different enantiomeric forms may vary in their tastes, odour and toxicity. Most naturally occurring sugars exist in the D-form, whereas most naturally occurring amino acids are in the L-form. D-amino acids tend to taste sweet whereas L-forms often have no taste. Both caraway and dill seeds and spearmint contain carvone but the L-(+)-form of carvone in caraway and dill seeds tastes very different to the D-(-)form of carvone in spearmint.

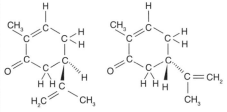

The two enantiomeric forms of carvone (2-methyl-5-(prop-1-en-2-yl)cyclohex-2-enone)

Smells and tastes can appear different to different people as the olfactory receptors also contain chiral receptor molecules which can interact differently with the enantiomeric molecules in food. The natural flavour of raspberries is due to R-α-ionone whereas synthetic raspberry flavourings contain both R- and S-isomers. In a similar way the +(*d*)- form of limonene smells of oranges whereas the –(*l*)- form of limonene smells of lemons. Other synthetically made foods often contain a racemic mixture of both enantiomers.

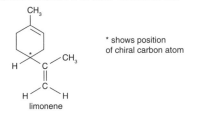

* shows position of chiral carbon atom

limonene

ULTRAVIOLET AND VISIBLE ABSORPTION IN ORGANIC MOLECULES

Organic compounds containing unsaturated groups such as C=C, C=O, -N=N-, $-NO_2$ and the benzene ring can absorb in the ultraviolet or visible part of the spectrum. Such groups are known as chromophores and the precise energy of absorption is affected by the other groups attached to the chromophore. The absorption is due to electrons in the bond being excited to an empty orbital of higher energy, usually an anti-bonding orbital. The energy involved in this process is relatively high and most organic compounds absorb in the ultraviolet region and thus appear colourless. For example ethene absorbs at 185 nm. However if there is extensive conjugation of double bonds (i.e many alternate C-C single bonds and C=C double bonds) in the molecule involving the delocalisation of pi electrons then less energy is required to excite the electrons and the absorption occurs in the visible region. Good examples include anthocyanins, carotenoids, chlorophyll and haem.

Anthocyanins

Anthocyanins contain the flavonoid $C_6C_3C_6$ skeleton.

It is the conjugation of the pi electrons contained in this structure which accounts for the colour of anthocyanins. The more extensive the conjugation, the lower the energy (longer the wavelength) of the light absorbed. This can be exemplified using cyanidin. In acidic solution it forms a positive ion and there is less conjugation than in alkaline solution where the pi electrons in the extra double bond between the carbon and oxygen atom are also delocalized.

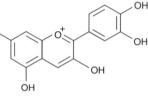

The flavonoid $C_6C_3C_6$ backbone

This difference in colour depending on pH explains why poppies which have acidic sap are red whereas cornflowers, which also contain cyanidin but have alkaline sap, are blue. Other anthocyanins differ in the number and types of other groups such as hydroxyl or methoxy groups which affect the precise wavelength of the light absorbed and hence the colour transmitted. The addition of other groups also affects other properties of anthocyanins. The basic flavonoid $C_6C_3C_6$ backbone is essentially non-polar. As more polar hydroxyl groups are added the potential for them to form hydrogen bonds with water molecules increases and many anthocyanins, such as cyanidin with several –OH groups, are appreciably soluble in water for this reason.

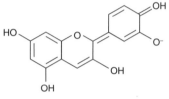

Structure of cyanidin in acidic solution. Less conjugation so absorbs in blue-green region and transmits red light.

Structure of cyanidin in alkaline solution. More conjugation so absorbs in the orange region of the spectrum and transmits blue light.

Carotenoids

In carotenoids the conjugation is mainly due to a long hydrocarbon chain (as opposed to the ring system in anthocyanins) consisting of alternate single and double carbon to carbon bonds.

The majority are derived from a (poly)ene chain containing forty carbon atoms which may be terminated by cyclic end groups and may also be complemented with oxygen-containing functional groups. The hydrocarbon carotenoids are known as xanthophylls. Examples include α-carotene, β-carotene and vitamin A.

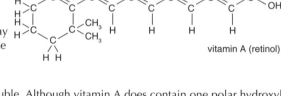

vitamin A (retinol)

α- and β-carotene and vitamin A are all fat soluble and not water soluble. Although vitamin A does contain one polar hydroxyl group the rest of the molecule is a large non-polar hydrocarbon

Chlorophyll and haem

Chlorophyll and haem both contain a planar heterocyclic unit with the general name of a porphin. Porphins contain a cyclic system in which all the carbon atoms are sp^2 hybridized. This results in a planar structure with extensive pi conjugation.

Porphin groups which contain substituents in the 1 to 8 positions are known more specifically as porphyrins. Porphins contain four nitrogen atoms. The non-bonding pairs of electrons on the nitrogen atoms enable the porphin to form coordinate bonds with metal ions. The general structure of chlorophyll, which contains magnesium has already been given. In chlorophyll the porphyrin complex with the original double bond between positions 7 and 8 is now saturated and there is an -R group on the third carbon atom, C_3. Chlorophyll is found in two closely related forms. In chlorophyll a, the -R group is a methyl group, $-CH_3$, and in chlorophyll b the -R group is an aldehyde group, -CHO. The haem group is present in both myoglobin and haemoglobin. Haemoglobin is the oxygen carrier found in mammalian blood. Myoglobin is the primary pigment in muscle tissue and is a complex of a protein, globin, together with a haem group which is a porphyrin ring containing a central iron(II) ion.

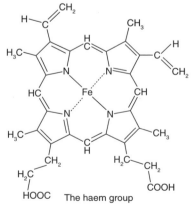

The structure of the free porphin group.

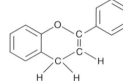

The haem group

IB QUESTIONS – OPTION F – FOOD CHEMISTRY

1. A molecule of a fat has the following structure.

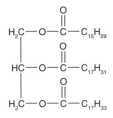

(a) State and explain whether the fat shown is a solid or a liquid at room temperature. **[2]**

(b) The fat is hydrogenated. In the process two molecules of hydrogen are added to one molecule of the fat.

  (i) State the conditions necessary for hydrogenation to occur. **[2]**

  (ii) Draw one possible structure for the molecule of hydrogenated fat. **[1]**

(c) Apart from the addition of hydrogen, describe what other change may occur to the structure of the molecule during the hydrogenation reaction. State one reason why such hydrogenated fats are considered less healthy. **[2]**

2. (a) Distinguish between a dye and a pigment in food chemistry. **[1]**

(b) One class of compounds which give food colour is the anthocynanins.

  (i) Explain why anthocynanins are coloured. **[2]**

  (ii) Explain why changing the pH may change the colour of an anthocyanin **[2]**

(c) Food is sometimes canned to preserve it. Explain why canned food that has been stored for some time may become discoloured. **[2]**

3. Two antioxidants are TBHQ and β-carotene.

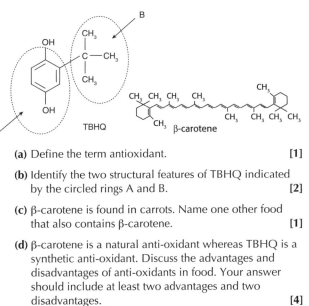

(a) Define the term antioxidant. **[1]**

(b) Identify the two structural features of TBHQ indicated by the circled rings A and B. **[2]**

(c) β-carotene is found in carrots. Name one other food that also contains β-carotene. **[1]**

(d) β-carotene is a natural anti-oxidant whereas TBHQ is a synthetic anti-oxidant. Discuss the advantages and disadvantages of anti-oxidants in food. Your answer should include at least two advantages and two disadvantages. **[4]**

4. Sugars are carbohydrates. The simplest carbohydrates are monosaccharides.

(a) Define the terms *carbohydrate* and *monosaccharide*. **[3]**

(b) Describe the type of reaction taking place when two monosaccharides react to produce a disaccharide. **[1]**

(c) Food with a high carbohydrate content can undergo caramelization which causes the food to brown and produce a desirable smell and flavour. State three factors which will increase the rate of caramelization. **[3]**

(d) Some other types of food can be browned by using the Maillard reaction. State the chemical composition of the foods affected by this reaction and describe the reaction taking place in terms of the functional groups of the reacting substances. **[4]**

(HL)

5. Cyanadin is a plant pigment. Its structure is pH dependent.

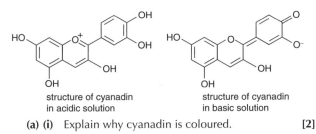

structure of cyanadin in acidic solution   structure of cyanadin in basic solution

(a) (i) Explain why cyanadin is coloured. **[2]**

  (ii) Cyanadin is red in acidic solution and blue in basic solution. Explain this observation in terms of the molecular structure of the compound. **[3]**

(b) Cyanadin can be classified as a flavonoid. Identify (by drawing) the part of the molecule which represents the flavonoid skeleton. **[1]**

(c) Explain why cyanadin is water-soluble and not fat-soluble. **[2]**

6. The structures of the 2-amino acids cysteine and valine are given.

$$H_2N-CH-COOH \qquad H_2N-CH-COOH$$
$$\quad\quad | \qquad\qquad\qquad\quad |$$
$$CH_2-SH \qquad\quad H_3C-CH-CH_3$$
cysteine   valine

(a) (i) Explain the significance of the 2 in 2-amino acids. **[1]**

  (ii) Explain why both acids can exist in two different enantiomeric forms. **[1]**

  (iii) Draw the structure of the D- enantiomer of valine and explain the D- and L- convention using the 'CORN' rule. **[3]**

  (iv) Describe how the (+) and (-) enantiomers of cysteine can be distinguished. **[2]**

(b) (i) Using structural formulas give the equation for a condensation reaction between lysine and valine. **[2]**

  (ii) Explain the significance of the above reaction in the primary structure of proteins. **[1]**

(c) (i) Myoglobin consists of a protein bound to a porphryin ring containing a central iron atom. Explain how the porphyrin ring bonds to the iron atom. **[2]**

Electrophilic addition reactions (1)

ELECTROPHILIC ADDITION TO SYMMETRIC ALKENES

Ethene readily undergoes addition reactions. With hydrogen bromide it forms bromoethane.

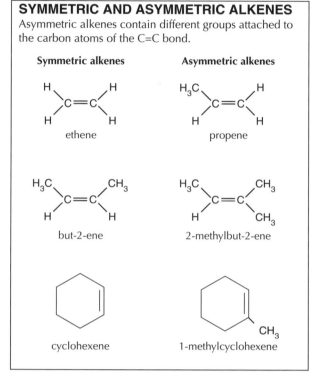

The reaction can occur in the dark which suggests that a free radical mechanism is not involved. The double bond in the ethene molecule has a region of high electron density above and below the plane of the molecule. Hydrogen bromide is a polar molecule due to the greater electronegativity of bromine compared with hydrogen. The hydrogen atom (which contains a charge of δ+) from the H–Br is attracted to the double bond and the H–Br bond breaks, forming a bromide ion. At the same time the hydrogen atom adds to one of the ethene carbon atoms leaving the other carbon atom with a positive charge. A carbon atom with a positive charge is known as a **carbocation**. The carbocation then combines with the bromide ion to form bromoethane. Because the hydrogen bromide molecule is attracted to a region of electron density it is described as an **electrophile** and the mechanism is described as **electrophilic addition**.

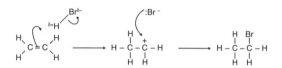

Electrophilic addition also takes place when bromine adds to ethene in a non-polar solvent to give 1,2-dibromoethane. Bromine itself is non-polar but as it approaches the double bond of the ethene an induced dipole is formed by the electron cloud.

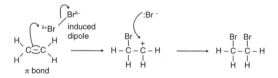

Evidence for this mechanism is that when bromine water is reacted with ethene the main product is 2-bromoethanol not 1,2-dibromoethane. This suggests that hydroxide ions from the water add to the carbocation in preference to bromide ions.

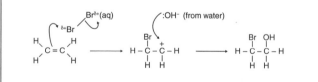

SYMMETRIC AND ASYMMETRIC ALKENES

Asymmetric alkenes contain different groups attached to the carbon atoms of the C=C bond.

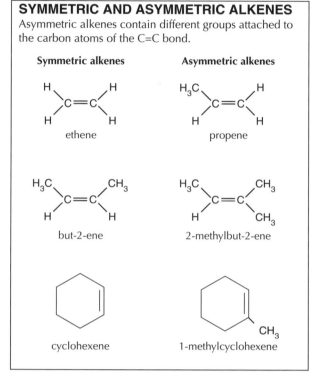

Electrophilic addition reactions (2)

MARKOVNIKOV'S RULE

When hydrogen halides add to asymmetric alkenes two products are possible depending upon which carbon atom the hydrogen atom bonds to. For example, the addition of hydrogen bromide to propene could produce 1-bromopropane or 2-bromopropane.

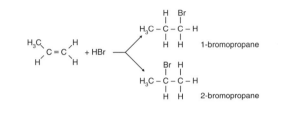

Markovnikov's rule enables you to predict which isomer will be the major product. It states that the hydrogen halide will add to the carbon atom that already contains the most hydrogen atoms bonded to it. Thus in the above example 2-bromopropane will be the major product.

EXPLANATION OF MARKOVNIKOV'S RULE

Markovnikov's rule enables the product to be predicted but it does not explain why. It can be explained by considering the nature of the possible intermediate carbocations formed during the reaction.

When hydrogen ions react with propene two different carbocation intermediates can be formed.

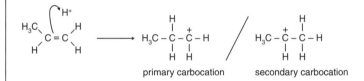

The first one has the general formula RCH_2^+ and is known as a **primary carbocation**. The second one has two R– groups attached to the positive carbon ion R_2CH^+ and is known as a **secondary carbocation**. A **tertiary carbocation** has the general formula R_3C^+. The R-groups (alkyl groups) tend to push electrons towards the carbon atom they are attached to which tends to stabilize the positive charge on the carbocation. This is known as a **positive inductive effect**. This effect will be greatest with tertiary carbocations and smallest with primary carbocations.

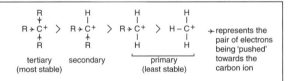

Thus in the above reaction the secondary carbocation will be preferred as it is more stable than the primary carbocation. This secondary carbocation intermediate leads to the major product, 2-bromopropane.

Understanding this mechanism enables you to predict what will happen when an interhalogen adds to an asymmetric alkene even though no hydrogen atoms are involved. Consider the reaction of iodine chloride ICl with but-1-ene. Since iodine is less electronegative than chlorine the iodine atom will act as the electrophile and add first to the alkene.

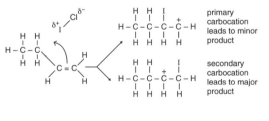

The major product will thus be 2-chloro-1-iodobutane.

Nucleophilic addition reactions and elimination reactions

ADDITION OF HYDROGEN CYANIDE TO ALDEHYDES AND KETONES

Aldehydes and ketones are both carbonyl compounds as they contain the $>C=O$ group. The double bond in carbonyl compounds is similar to the $C=C$ double bond in alkenes and the region around the bond is planar with bond angles of 120°.

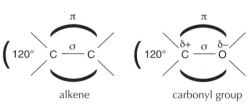

alkene   carbonyl group

The essential difference is that oxygen is much more electronegative than carbon so that the bond is polar. Whereas alkenes undergo electrophilic addition, carbonyl compounds undergo **nucleophilic addition reactions**.

A typical example is the reaction of ethanal with hydrogen cyanide. Hydrogen cyanide is weakly acidic and dissociates to form the cyanide ion which acts as the nucleophile. The intermediate anion is then protonated to form the product, 2-hydroxypropanenitrile. Hydroxynitriles are also known as cyanohydrins.

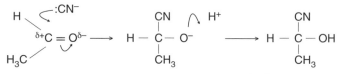

2-hydroxypropanenitrile

This reaction is useful in synthesis as it is a method by which an extra carbon atom can be introduced into a carbon chain. Nitriles can be hydrolysed in the presence of acid to form carboxylic acids so this product can easily be converted into 2-hydroxypropanoic acid (lactic acid).

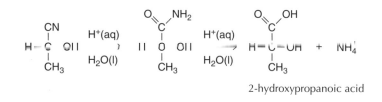

2-hydroxypropanoic acid

ELIMINATION REACTIONS OF ALCOHOLS

Alcohols can be dehydrated (elimination of water) by heating with concentrated sulfuric acid at 180 °C. In practice phosphoric acid is often used in place of sulfuric acid as fewer side reactions take place.

$$-\underset{|}{\overset{H}{\underset{|}{C}}}-\underset{|}{\overset{OH}{\underset{|}{C}}}-\quad\xrightarrow[180\,°C]{H_2SO_4 \text{ or } H_3PO_4}\quad >C=C< + H_2O$$

The strong acid acts as a catalyst by protonating the oxygen atom in the alcohol. Water is then lost to form a carbocation. The carbocation then donates a proton to form the double bond and regenerate the catalyst.

Tertiary alcohols are more readily dehydrated than primary and secondary alcohols because the intermediate tertiary carbocation is more stable.

Addition–elimination reactions

REACTION OF 2,4-DINITROPHENYLHYDRAZINE WITH ALDEHYDES AND KETONES

With nucleophiles containing the $-NH_2$ group add to aldehydes and ketones the addition can be followed by the elimination of water. For example, the reaction of hydrazine with propanone. The initial mode of attack is still nucleophilic addition but then the carbon atom forms a double bond with the nitrogen atom and a molecule of water is lost. The overall reaction is known either as an **addition–elimination reaction** or more simply as a **condensation reaction**.

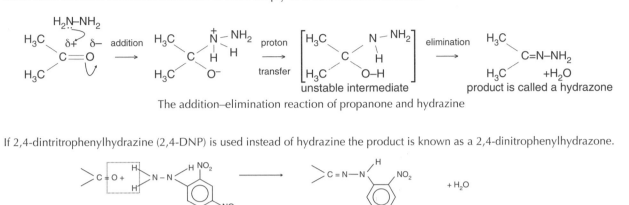

The addition–elimination reaction of propanone and hydrazine

If 2,4-dintritrophenylhydrazine (2,4-DNP) is used instead of hydrazine the product is known as a 2,4-dinitrophenylhydrazone.

aldehyde or ketone   2,4-DNP

2,4-dinitrophenylhydrazone

All aldehydes and ketones form red or orange crystalline solid 2,4-dinitrophenylhydrazones each of which has a characteristic melting point. These derivatives of 2,4-DNP are easy to make and purify so provide a convenient way of distinguishing between different aldehydes and ketones, although in a well-equipped laboratory chemists are now more likely to use a combination of spectroscopic techniques (e.g. ^1H NMR and mass spectrometry).

Melting points of 2, 4-dinitrophenyl-hydrazones of aldehydes and ketones

M. pt / °C		M. pt / °C	
methanal	166	propanone	126
ethanal	168	butanone	116
propanal	155	pentan-3-one	156
butanal	126	pentan-2-one	144
benzaldehyde	237	cyclohexanone	162

Arenes

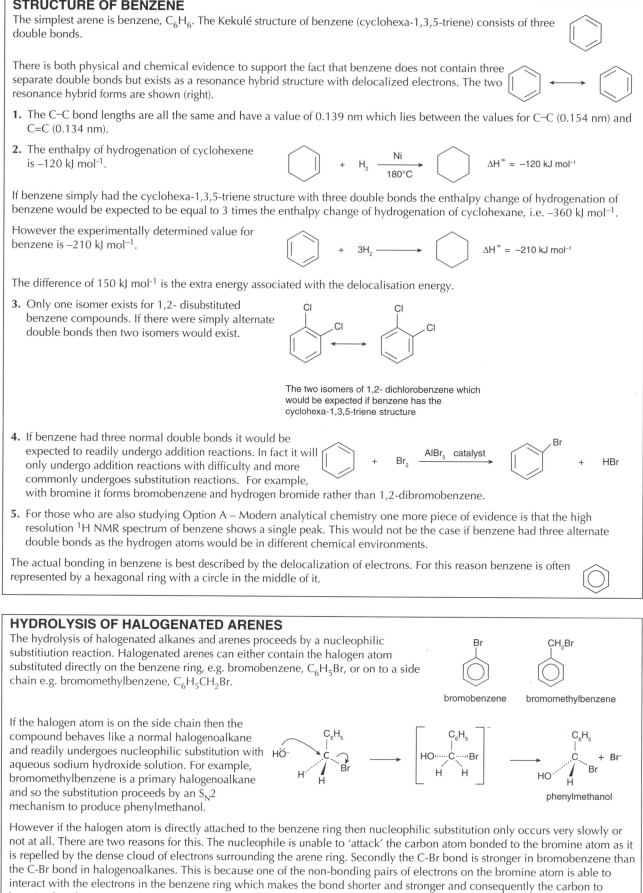

STRUCTURE OF BENZENE

The simplest arene is benzene, C_6H_6. The Kekulé structure of benzene (cyclohexa-1,3,5-triene) consists of three double bonds.

There is both physical and chemical evidence to support the fact that benzene does not contain three separate double bonds but exists as a resonance hybrid structure with delocalized electrons. The two resonance hybrid forms are shown (right).

1. The C–C bond lengths are all the same and have a value of 0.139 nm which lies between the values for C–C (0.154 nm) and C=C (0.134 nm).

2. The enthalpy of hydrogenation of cyclohexene is –120 kJ mol⁻¹.

$$\Delta H^\circ = -120 \text{ kJ mol}^{-1}$$

If benzene simply had the cyclohexa-1,3,5-triene structure with three double bonds the enthalpy change of hydrogenation of benzene would be expected to be equal to 3 times the enthalpy change of hydrogenation of cyclohexane, i.e. –360 kJ mol⁻¹.

However the experimentally determined value for benzene is –210 kJ mol⁻¹.

$$\Delta H^\circ = -210 \text{ kJ mol}^{-1}$$

The difference of 150 kJ mol⁻¹ is the extra energy associated with the delocalisation energy.

3. Only one isomer exists for 1,2- disubstituted benzene compounds. If there were simply alternate double bonds then two isomers would exist.

The two isomers of 1,2- dichlorobenzene which would be expected if benzene has the cyclohexa-1,3,5-triene structure

4. If benzene had three normal double bonds it would be expected to readily undergo addition reactions. In fact it will only undergo addition reactions with difficulty and more commonly undergoes substitution reactions. For example, with bromine it forms bromobenzene and hydrogen bromide rather than 1,2-dibromobenzene.

5. For those who are also studying Option A – Modern analytical chemistry one more piece of evidence is that the high resolution ¹H NMR spectrum of benzene shows a single peak. This would not be the case if benzene had three alternate double bonds as the hydrogen atoms would be in different chemical environments.

The actual bonding in benzene is best described by the delocalization of electrons. For this reason benzene is often represented by a hexagonal ring with a circle in the middle of it.

HYDROLYSIS OF HALOGENATED ARENES

The hydrolysis of halogenated alkanes and arenes proceeds by a nucleophilic substitiution reaction. Halogenated arenes can either contain the halogen atom substituted directly on the benzene ring, e.g. bromobenzene, C_6H_5Br, or on to a side chain e.g. bromomethylbenzene, $C_6H_5CH_2Br$.

bromobenzene   bromomethylbenzene

If the halogen atom is on the side chain then the compound behaves like a normal halogenoalkane and readily undergoes nucleophilic substitution with aqueous sodium hydroxide solution. For example, bromomethylbenzene is a primary halogenoalkane and so the substitution proceeds by an S_N2 mechanism to produce phenylmethanol.

phenylmethanol

However if the halogen atom is directly attached to the benzene ring then nucleophilic substitution only occurs very slowly or not at all. There are two reasons for this. The nucleophile is unable to 'attack' the carbon atom bonded to the bromine atom as it is repelled by the dense cloud of electrons surrounding the arene ring. Secondly the C-Br bond is stronger in bromobenzene than the C-Br bond in halogenoalkanes. This is because one of the non-bonding pairs of electrons on the bromine atom is able to interact with the electrons in the benzene ring which makes the bond shorter and stronger and consequently the carbon to bromine bond stronger and more difficult to break.

Organometallic chemistry and reaction pathways

FORMATION OF GRIGNARD REAGENTS

Organometallic compounds contain a carbon atom covalently bonded directly to a metal. One important group of organometallic compounds which are used in organic synthesis is known as Grignard reagents. They consist of an alkyl group bonded directly to magnesium. They are usually prepared *in situ* by reacting magnesium metal with a halogenoalkane using dry ethoxyethane (ether) as the solvent. Sometimes a small amount of iodine is added to initiate the reaction.

$$RX + Mg \longrightarrow R—Mg—X$$

They must be prepared in a non-polar solvent such as ether as Grignard reagents react with water to form alkanes, e.g.

$$C_2H_5—Mg—Br \xrightarrow{H_2O} C_2H_6 + Mg(OH)Br$$
$$\text{ethane}$$

REACTIONS OF GRIGNARD REAGENTS

The magnesium to carbon bond is polar with the negative part of the dipole on the carbon atom as carbon is more electronegative than magnesium. This enables the carbon atom to act as a good nucleophile. Grignard reagents can add to the partially positive carbon atom in aldehydes and ketones and then be hydrolyzed in acid solution to form alcohols with a longer carbon chain. For example with methanal they form a primary alcohol, with other aldehydes they form secondary alcohols and with ketones they form tertiary alcohols.

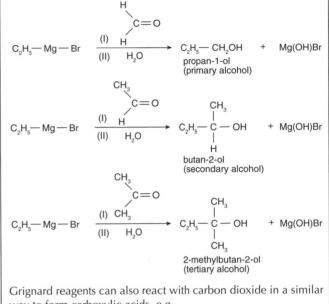

Grignard reagents can also react with carbon dioxide in a similar way to form carboxylic acids, e.g.

$$C_2H_5—Mg—Br \xrightarrow[\text{(II) } H_2O]{\text{(I) } CO_2} C_2H_5COOH + Mg(OH)Br$$
$$\text{propanoic acid}$$

REACTION PATHWAYS

The reactions studied in this part of the option can be summarized in the following two diagrams which include the required mechanisms.

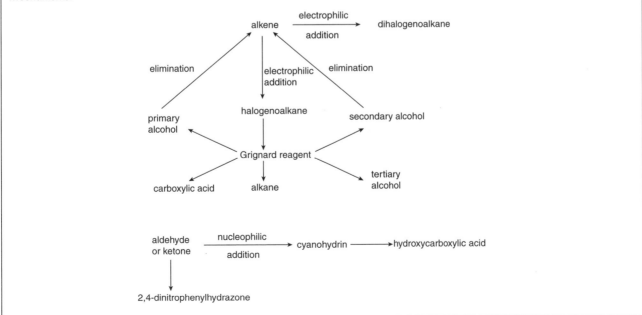

Acid–base reactions

STRENGTHS OF ORGANIC ACIDS AND BASES

The acid strength of organic acids can be compared using pK_a values. These are the values given in the IB data booklet. They work in a similar way to pH in that the lower the pKa value the stronger the acid. For organic bases pK_b values are used – the lower the pK_b value the stronger the base.

ACIDIC PROPERTIES OF PHENOLS

Phenol is weakly acidic because the negative charge in the conjugate phenoxide ion can be delocalized over the benzene ring. This makes it less likely to attract a proton than the ethoxide ion or hydroxide ion where delocalization cannot occur.

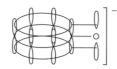

Phenol has a pK_a value of 10.00 making it a weaker acid than carboxylic acids. It is too weak to react with carbonates to give carbon dioxide but it will react with sodium to give hydrogen and form salts with sodium hydroxide.

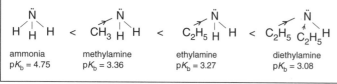

Substituted phenols containing electron withdrawing groups are much more acidic as the negative charge can be further delocalized. For example, 2-nitrophenol has a pK_a value of 7.21. In 2,4,6-trinitrophenol the effect is so great that the substituted phenol is nearly a strong acid with a pK_a value of 0.42.

2,4,6-trinitrophenol

ACIDIC PROPERTIES OF SUBSTITUTED CARBOXYLIC ACIDS

Although alcohols and carboxylic acids both contain an –OH group carboxylic acids are weak acids whereas alcohols are not. The electron withdrawing carbonyl group $\diagup C{=}O$ adjacent to the –OH group weakens the normally strong O–H bond so that the carboxyl group can lose a proton.

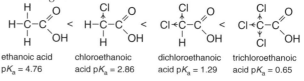

The acidity of carboxylic acids can also be explained by considering the conjugate base formed. In the carboxylate anion the negative charge is delocalized over three atoms as both C–O bond lengths are identical. This stabilizes the ion and makes it less likely to attract a proton as the charge density of the negative charge is reduced. Compare this with the strongly basic ethoxide ion where the negative charge is localized on the oxygen atom.

$$C_2H_5{-}O{-}H + H_2O \rightleftharpoons C_2H_5{-}O^- + H_3O^+ \quad pK_a = \text{approx. } 16$$

Electron withdrawing atoms such as chlorine can further delocalize the negative charge on the anion increasing the acid strength.

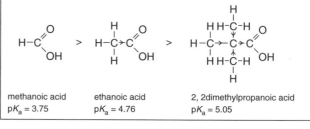

ethanoic acid	chloroethanoic	dichloroethanoic	trichloroethanoic
$pK_a = 4.76$	acid $pK_a = 2.86$	acid $pK_a = 1.29$	acid $pK_a = 0.65$

Groups with a positive inductive effect such as alkyl groups will decrease the ability of the charge to delocalize and the acid will be weaker.

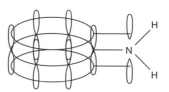

methanoic acid	ethanoic acid	2, 2dimethylpropanoic acid
$pK_a = 3.75$	$pK_a = 4.76$	$pK_a = 5.05$

RELATIVE BASICITIES OF AMMONIA AND AMINES

Ammonia is a weak base with a pK_b of 4.75 and forms salts with strong acids.

$$NH_3 + H_2O \rightleftharpoons NH_4^+ + OH^-$$

$$NH_3 + HCl \rightarrow NH_4^+Cl^-$$

Amines are more basic than ammonia because the positive inductive effect of the alkyl group 'pushes' electrons towards the nitrogen atom increasing the electron density of the non-bonding pair of electrons. Thus aminoethane (ethylamine) is more basic than aminomethane (methylamine) as the positive inductive effect of an ethyl group is greater than a methyl group. Secondary amines are more basic still, although tertiary amines are slightly less basic.

ammonia	methylamine	ethylamine	diethylamine
$pK_b = 4.75$	$pK_b = 3.36$	$pK_b = 3.27$	$pK_b = 3.08$

However, aminobenzene (phenylamine), although still basic with a pK_b of 9.48, is considerably less basic than ammonia. The non-bonding pair of electrons on the nitrogen atom is delocalized with the electrons in the ring and so is less available to donate to a proton.

All amines can be obtained from their salts by reacting with sodium hydroxide.

$$RNH_3^+ + OH^- \rightarrow RNH_2 + H_2O$$

Addition–elimination reactions of acid anhydrides

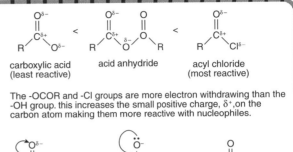

carboxylic acid (least reactive) < acid anhydride < acyl chloride (most reactive)

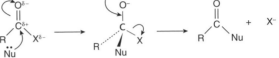

The -OCOR and -Cl groups are more electron withdrawing than the -OH group. this increases the small positive charge, δ^+, on the carbon atom making them more reactive with nucleophiles.

MECHANISM

When carboxylic acids react with alcohols to form esters the –OH group of the carboxylic acid is replaced by the –OR group of the alcohol. This is an example of an addition-elimination reaction as the alcohol acts as a nucleophile and the –OH group of the acid is eliminated as water. An acid anhydride is a carboxylic acid in which the hydroxyl group has been replaced by the -OCOR group from another molecule of the acid. Similarly in acyl chlorides the hydroxyl group has been replaced by a chlorine atom. Both -OCOR and -Cl are more easily replaced than -OH so acid anhydrides and acyl chlorides are much more reactive with nucleophiles than carboxylic acids.

Nu   represents a nucleophile e.g. H_2O, NH_3, ROH and amines
X   represents Cl or OCOR

ACID ANHYDRIDES

1. Water. Acid anhydrides react with water to form carboxylic acids, e.g.

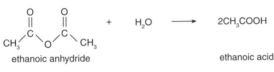

ethanoic anhydride   + H_2O $\longrightarrow$ $2CH_3COOH$
ethanoic anhydride   ethanoic acid

2. Alcohols. Acid anhydrides react with alcohols to form esters, e.g.

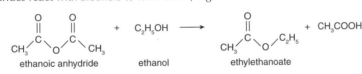

ethanoic anhydride   + C_2H_5OH $\longrightarrow$ + CH_3COOH
ethanoic anhydride   ethanol   ethylethanoate

The reaction of ethanoic anhydride with 2-hydroxybenzoic acid (salicylic acid), is used to prepare aspirin. In this reaction it is the –OH group on the benzene ring which is functioning as the alcohol.

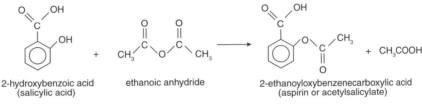

2-hydroxybenzoic acid   ethanoic anhydride   2-ethanoyloxybenzenecarboxylic acid
(salicylic acid)   (aspirin or acetylsalicylate)   + CH_3COOH

A similar reaction is used to turn morphine into heroin (see Option D – Medicines and drugs).

3. Ammonia and amines. Acid anhydrides react readily with concentrated aqueous ammonia to form amides, e.g.

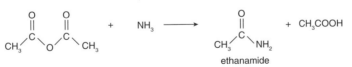

          + NH_3 $\longrightarrow$ + CH_3COOH
ethanoic anhydride   ethanamide

Secondary amines also contain a non-bonded pair of electrons on the nitrogen atom so can also function as nucleophiles. When 4-aminophenol is reacted with ethanoic anhydride in the presence of water at room temperature the product is N-(4-hydroxylphenyl)ethanamide. This has the common name of paracetamol in the U.K and acetaminophen in the U.S.A. and like aspirin it is also a pain killer and fever reducer.

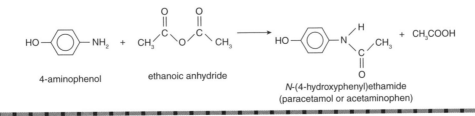

4-aminophenol   ethanoic anhydride   N-(4-hydroxyphenyl)ethamide   + CH_3COOH
                                                   (paracetamol or acetaminophen)

ACYL CHLORIDES

Acyl chlorides react in a very similar way to acid anhydrides except that now the other product is hydrochloric acid rather than the carboxylic acid. The reactions are more vigorous. This explains why acyl chlorides fume in moist air to form droplets of hydrochloric acid as soon as the stopper is taken off the bottle.

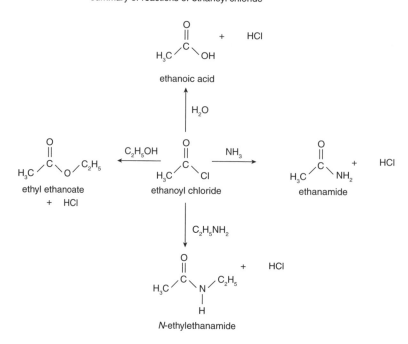

summary of reactions of ethanoyl chloride

The reaction of acyl chlorides with secondary amines is used to make nylon. For example nylon 6,6 can be prepared by reacting hexane-1,6-dioyl dichloride with 1,6-diaminohexane.

preparation of nylon 6,6

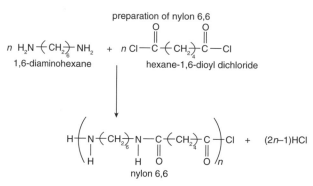

Electrophilic substitution reactions (1)

Benzene does not readily undergo addition reactions since an additional 150 kJ mol^{-1} of energy would be required to overcome the energy of delocalization. Instead it undergoes electrophilic substitution reactions. Electrophiles are attracted to the region of high electron density above and below the plane of the molecule due to the delocalized π bond.

NITRATION OF BENZENE

Benzene reacts with a mixture of concentrated nitric acid and concentrated sulfuric acid when warmed at 50 °C to give nitrobenzene and water. Note that the temperature should not be raised above 50 °C otherwise further nitration to dinitrobenzene will occur.

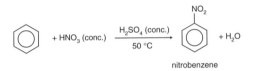

The electrophile is the **nitryl cation NO_2^+** (also called the nitronium ion). The concentrated sulfuric acid acts as a catalyst. Its function is to protonate the nitric acid which then loses water to form the electrophile. In this reaction nitric acid is acting as a base in the presence of the more acidic sulfuric acid.

$$H_2SO_4 + HNO_3 \rightleftharpoons H_2NO_3^+ + HSO_4^-$$
$$\downarrow$$
$$H_2O + NO_2^+$$

The NO_2^+ is attracted to the delocalized π bond and attaches to one of the carbon atoms. This requires considerable activation energy as the delocalized π bond is partially broken. The positive charge is distributed over the remains of the π bond in the intermediate. The intermediate then loses a proton and energy is evolved as the delocalized π bond is reformed. The proton can recombine with the hydrogensulfate ion to regenerate the catalyst.

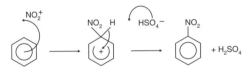

Although it is more correct to draw the intermediate as a partially delocalized π bond it can sometimes be convenient to show benzene as if it does contain alternate single and double carbon to carbon bonds. In this model the positive charge is located on a particular carbon atom.

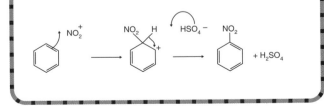

CHLORINATION, ALKYLATION AND ACYLATION OF BENZENE

Benzene can also undergo electrophilic substitution with chlorine. This only happens in the presence of a **halogen carrier**, such as anhydrous aluminium chloride or iron.

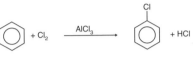

The halogen carrier is a catalyst (sometimes known as a Friedel–Crafts catalyst after its discoverers). It functions as a Lewis acid by accepting a pair of electrons from the chlorine. The electrophile is the resulting chlorine ion, Cl^+.

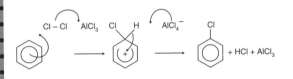

Halogen carriers can also be used to attract the electron pair in the C–Cl bond in halogenoalkanes R–Cl. The resulting electrophile will now be the positive alkyl ion R^+. The resulting substituted product will be an alkylbenzene. For example, methylbenzene can be prepared by warming benzene with chloromethane in the presence of anhydrous aluminium chloride.

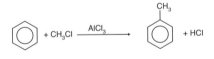

Mechanism:

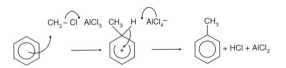

A similar reaction can occur with acyl chlorides to produce ketones – this is known as acylation. For example, benzene can react with ethanoyl chloride in the presence of a halogen carrier to form 1-phenylethanone (acetophenone). The electrophile is CH_3CO^+.

$$C_6H_6 + CH_3COCl \xrightarrow{AlCl_3} C_6H_5COCH_3 + HCl$$

Electrophilic substitution reactions (2)

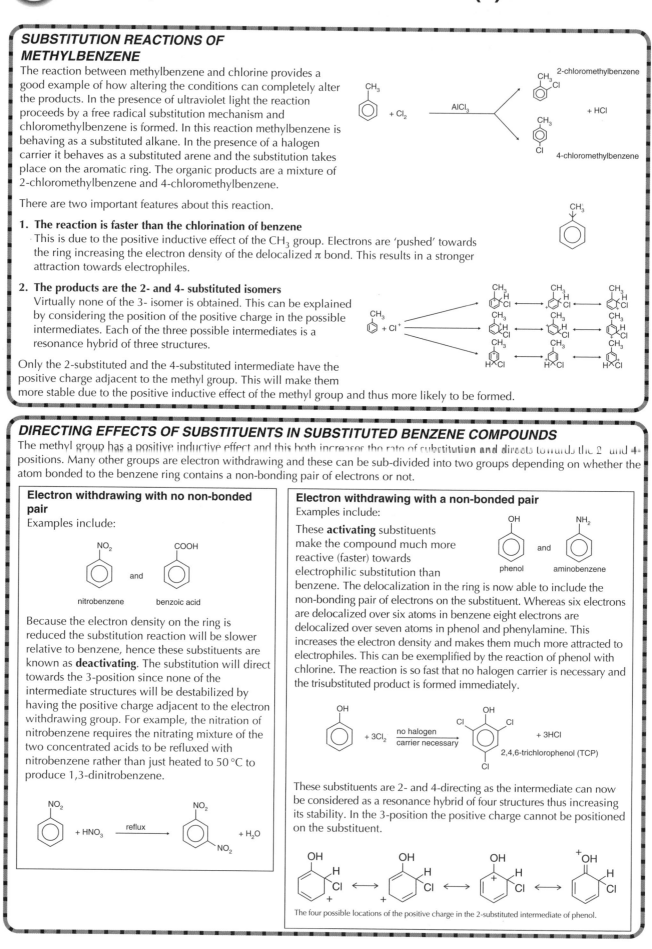

SUBSTITUTION REACTIONS OF METHYLBENZENE

The reaction between methylbenzene and chlorine provides a good example of how altering the conditions can completely alter the products. In the presence of ultraviolet light the reaction proceeds by a free radical substitution mechanism and chloromethylbenzene is formed. In this reaction methylbenzene is behaving as a substituted alkane. In the presence of a halogen carrier it behaves as a substituted arene and the substitution takes place on the aromatic ring. The organic products are a mixture of 2-chloromethylbenzene and 4-chloromethylbenzene.

There are two important features about this reaction.

1. **The reaction is faster than the chlorination of benzene**
   This is due to the positive inductive effect of the CH_3 group. Electrons are 'pushed' towards the ring increasing the electron density of the delocalized π bond. This results in a stronger attraction towards electrophiles.

2. **The products are the 2- and 4- substituted isomers**
   Virtually none of the 3- isomer is obtained. This can be explained by considering the position of the positive charge in the possible intermediates. Each of the three possible intermediates is a resonance hybrid of three structures.

Only the 2-substituted and the 4-substituted intermediate have the positive charge adjacent to the methyl group. This will make them more stable due to the positive inductive effect of the methyl group and thus more likely to be formed.

DIRECTING EFFECTS OF SUBSTITUENTS IN SUBSTITUTED BENZENE COMPOUNDS

The methyl group has a positive inductive effect and this both increases the rate of substitution and directs towards the 2- and 4- positions. Many other groups are electron withdrawing and these can be sub-divided into two groups depending on whether the atom bonded to the benzene ring contains a non-bonding pair of electrons or not.

Electron withdrawing with no non-bonded pair

Examples include:

nitrobenzene   benzoic acid

Because the electron density on the ring is reduced the substitution reaction will be slower relative to benzene, hence these substituents are known as **deactivating**. The substitution will direct towards the 3-position since none of the intermediate structures will be destabilized by having the positive charge adjacent to the electron withdrawing group. For example, the nitration of nitrobenzene requires the nitrating mixture of the two concentrated acids to be refluxed with nitrobenzene rather than just heated to 50 °C to produce 1,3-dinitrobenzene.

Electron withdrawing with a non-bonded pair

Examples include:

These **activating** substituents make the compound much more reactive (faster) towards electrophilic substitution than benzene. The delocalization in the ring is now able to include the non-bonding pair of electrons on the substituent. Whereas six electrons are delocalized over six atoms in benzene eight electrons are delocalized over seven atoms in phenol and phenylamine. This increases the electron density and makes them much more attracted to electrophiles. This can be exemplified by the reaction of phenol with chlorine. The reaction is so fast that no halogen carrier is necessary and the trisubstituted product is formed immediately.

phenol   aminobenzene

2,4,6-trichlorophenol (TCP)

These substituents are 2- and 4-directing as the intermediate can now be considered as a resonance hybrid of four structures thus increasing its stability. In the 3-position the positive charge cannot be positioned on the substituent.

The four possible locations of the positive charge in the 2-substituted intermediate of phenol.

Summary of reaction mechanisms

To assist in devising reaction pathways a summary of all the mechanisms covered in IB organic chemistry, with one example for each, is provided.

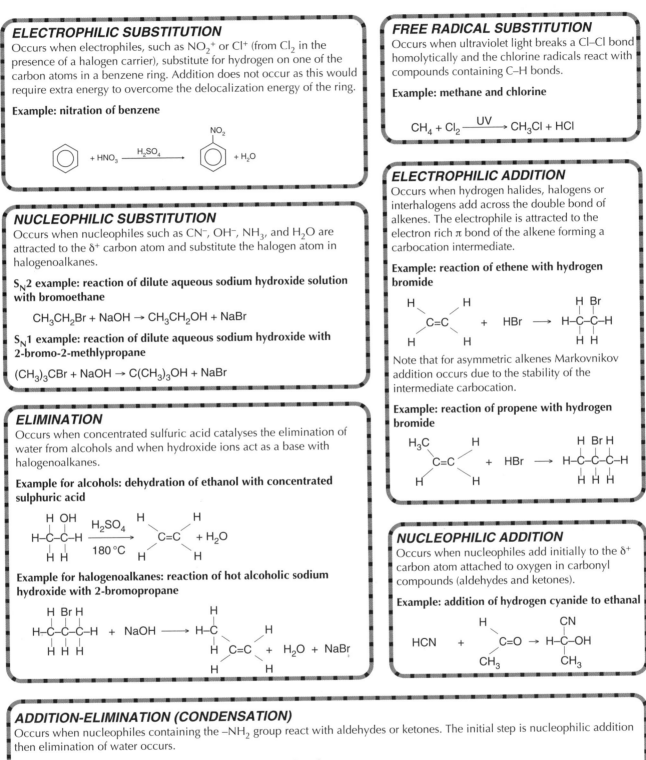

ELECTROPHILIC SUBSTITUTION

Occurs when electrophiles, such as NO_2^+ or Cl^+ (from Cl_2 in the presence of a halogen carrier), substitute for hydrogen on one of the carbon atoms in a benzene ring. Addition does not occur as this would require extra energy to overcome the delocalization energy of the ring.

Example: nitration of benzene

NUCLEOPHILIC SUBSTITUTION

Occurs when nucleophiles such as CN^-, OH^-, NH_3, and H_2O are attracted to the $\delta+$ carbon atom and substitute the halogen atom in halogenoalkanes.

S_N2 **example: reaction of dilute aqueous sodium hydroxide solution with bromoethane**

$$CH_3CH_2Br + NaOH \rightarrow CH_3CH_2OH + NaBr$$

S_N1 **example: reaction of dilute aqueous sodium hydroxide with 2-bromo-2-methlypropane**

$$(CH_3)_3CBr + NaOH \rightarrow C(CH_3)_3OH + NaBr$$

ELIMINATION

Occurs when concentrated sulfuric acid catalyses the elimination of water from alcohols and when hydroxide ions act as a base with halogenoalkanes.

Example for alcohols: dehydration of ethanol with concentrated sulphuric acid

Example for halogenoalkanes: reaction of hot alcoholic sodium hydroxide with 2-bromopropane

FREE RADICAL SUBSTITUTION

Occurs when ultraviolet light breaks a Cl–Cl bond homolytically and the chlorine radicals react with compounds containing C–H bonds.

Example: methane and chlorine

$$CH_4 + Cl_2 \xrightarrow{UV} CH_3Cl + HCl$$

ELECTROPHILIC ADDITION

Occurs when hydrogen halides, halogens or interhalogens add across the double bond of alkenes. The electrophile is attracted to the electron rich π bond of the alkene forming a carbocation intermediate.

Example: reaction of ethene with hydrogen bromide

Note that for asymmetric alkenes Markovnikov addition occurs due to the stability of the intermediate carbocation.

Example: reaction of propene with hydrogen bromide

NUCLEOPHILIC ADDITION

Occurs when nucleophiles add initially to the $\delta+$ carbon atom attached to oxygen in carbonyl compounds (aldehydes and ketones).

Example: addition of hydrogen cyanide to ethanal

ADDITION-ELIMINATION (CONDENSATION)

Occurs when nucleophiles containing the $-NH_2$ group react with aldehydes or ketones. The initial step is nucleophilic addition then elimination of water occurs.

Example: reaction of 2,4-dinitrophenylhydrazine with ethanal

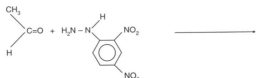
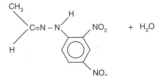

Addition-elimination also occurs when acid anhydrides and acyl chlorides react with H_2O, ROH, NH_3 and amines.

IB QUESTIONS – OPTION G – FURTHER ORGANIC CHEMISTRY

1. (a) Both pentan-2-one and pentan-3-one have a boiling point of 102 °C.

  (i) Describe how the two compounds can be distinguished using an addition-elimination reaction. **[2]**

  (ii) Draw the structure of the reagent and the organic product of the reaction in (a) (i) using pentan-2-one. **[2]**

(b) Pentan-2-one reacts with hydrogen cyanide. Describe the mechanism of the reaction, using curly arrows to represent the movement of electron pairs. **[4]**

2. (a) By referring to their structures, explain the difference in the acid strengths of ethanol and phenol. **[2]**

(b) Discuss how the acid strength of 2,2-dimethylpropanoic acid and trichloroethanoic acid compare with ethanoic acid. **[4]**

(c) Discuss how the basic strength of ethylamine and phenylamine compare with ammonia. **[4]**

3. When hydrogen bromide is added to 2-methylpropene two possible isomeric products could be formed.

(a) Give the structural formulas and the names of the two possible products. **[2]**

(b) State Markovnikov's rule for the addition of hydrogen halides to asymmetric alkenes. **[1]**

(c) According to the rule which isomer will be the major product in the above reaction? **[1]**

(d) Outline the mechanism for the reaction. **[2]**

(e) Explain clearly **why** only one of the possible products is formed. **[3]**

(f) Give the structural formula of the major product formed when iodine chloride, ICl, is added to 2-methylpropene. **[1]**

4. Suggest two-step syntheses for the following: Give details or any inorganic reagents and conditions.

(a) butan-2-ol from bromoethane and ethanal **[3]**

(b) 1,2-dibromopropane from propan-1-ol **[3]**

(c) 2-methyl-2-hydroxypropanoic acid from propanone **[3]**

HL

---

5. Benzene can be converted into phenylethene (styrene) by the following series of reactions:

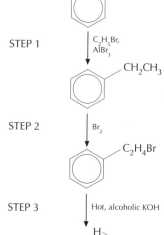

STEP 1 C_2H_5Br, $AlBr_3$

STEP 2 Br_2

STEP 3   Hot, alcoholic KOH

(a) Step 1: (i) Name the mechanism for this reaction. **[1]**

  (ii) Give mechanistic equations for the reaction clearly indicating the role of the aluminium bromide. **[4]**

(b) Step 2: (i) What reaction condition is necessary for Step 2? **[1]**

  (ii) State the name of the reaction mechanism and outline the steps involved. **[4]**

(c) Step 3: (i) Name and explain the mechanism of the reaction described in Step 3. **[3]**

  (ii) In Step 3 a different reaction takes place if a warm dilute aqueous solution of potassium hydroxide is used instead of a hot, alcoholic one. Give the structural formula of the major organic product formed under these conditions, and state the name of the mechanism. **[2]**

Study methods

This book has been written to provide you with all the information you need to gain the highest grade in Chemistry whether at SL or at HL. It is not intended as a 'teach yourself' book and is not a substitute for a good teacher nor for the practical work to support the theory. There is no magic solution which will compensate for a lack of knowledge or understanding but there are some pieces of advice that should ensure that you achieve to the best of your ability.

DURING THE COURSE

The IB course for both SL and HL is scheduled to last for two years, although some schools do attempt to cover the whole course in one year. There is a tendency for some students to take it easy in the first year as the final exams seem a long way off. Don't be tempted to do this as it will be hard to catch up later. Equally do not try to simply learn all the information given about each topic. The exam does not particularly test recall, more how to apply your knowledge in different situations. Although there are some facts that must be learned, much of Chemistry is logical and knowledge about the subject tends to come much more from understanding than from 'rote learning'. During each lesson concentrate on trying to understand the content. A good teacher will encourage you to do this by challenging you to think. At the end of the lesson or in the evening go over your notes, add to them or rearrange them to ensure you have fully understood everything. Read what this book has to say on the subject and read around the topic in other books to increase your understanding. If there are parts you do not understand ask your teacher to explain them again. You can also benefit much by talking and working through problems with other students. You will only really know if you understand something if you have to explain it to someone else. You can test your understanding by attempting the problems at the end of each topic in this book.

Some of the early parts of the course involve basic calculations. Some students do find these hard initially. Persevere and see if you can identify exactly what the difficulty is and seek help. Most students find that as the course progresses and more examples are covered their confidence to handle numerical problems increases considerably. If you ensure that you do understand everything during the course then you will find that by the time it comes to the exam, learning the essential facts to support your understanding is much easier.

MATHEMATICAL SKILLS

One big advantage of the IB is that all students study maths so the mathematical skills required for Chemistry should not present a problem. Essentially they concern numeracy rather than complex mathematical techniques. Make sure that you are confident in the following areas.

- Perform basic functions: addition, subtraction, multiplication and division.
- Carry out calculations involving means, decimals, fractions, percentages, ratios, approximations and reciprocals.
- Use standard notation (e.g. 1.8×10^5).
- Use direct and indirect proportion.
- Solve simple algebraic equations.
- Plot graphs (with suitable scales and axes) and sketch graphs.
- Interpret graphs, including the significance of gradients, changes in gradient, intercepts, and areas.
- Interpret data presented in various forms (e.g. bar charts, histograms, pie charts etc.).

USING YOUR CALCULATOR

Most calculators are capable of performing functions far beyond the demands of the course. When simple numbers are involved try to solve problems without using your calculator (you will need to do this for real in Paper 1). Even when the numbers are more complex try to estimate approximately what the answer will be before using the calculator. This should help to ensure that you do not accept and use a wrong answer because you failed to realize that you pushed the wrong buttons. Don't just give the 'calculator answer' but record the answer to the correct number of significant figures.

For HL students make sure you know how to use your calculator to work out problems involving logarithms for pH and pK_a calculations. The examples given below are for a TI-84 plus.

To convert a hydrogen ion concentration of 1.8×10^{-5} mol dm^{-3} into a pH.

$pH = -\log_{10} 1.8 \times 10^{-5}$

To obtain the value press the following keys in sequence.

(−) (LOG 1.8 × 10 ^ (−) 5) ENTER

This will give a value of 4.74.

To convert a pK_a value of 3.75 into a K_a value

Press

2nd 10^x (−) 3.75 ENTER

to give a value of 1.78×10^{-4} (don't forget to include the units in your answer).

The final examinations

PREPARING FOR THE EXAMINATIONS

Hopefully for much of the course the emphasis has been on enjoying learning and understanding Chemistry rather than always worrying about grades. Towards the end of the course, however, it does make sense to prepare yourself for the final exam. Examiners are human and mark positively (i.e. they look to give credit rather than penalize mistakes). You have to help them by being clear in your answers and addressing the particular question(s) asked.

- **Know what it is you have to know** Ask your teacher for a copy of the current programme for the core and the two options you are taking. Higher Level students should also have a copy of the Additional Higher Level material. Go through the programme carefully and make sure you recognize and have covered all the points listed for each topic and sub-topic.
- **Understand what depth you have to cover** Each sub-topic on the programme has an objective next to it. Objective 1 is the lowest and implies you just have to define or state the information. Objective 2 means you have to apply your knowledge of the topic in a straightforward situation. Objective 3 is the highest level and means that you will have to recognize the problem and select the appropriate method to solve it.
- **Be familiar with key command terms** Each statement in the programme and each question in the exam will contain a key command term. A list of all these verbs and their precise meaning can be found in the programme. If a question asks you to *describe* a reaction then a very different answer is required than if the question had asked you to *explain* a reaction. Examiners can only award marks for the correct answers to the question asked for. Not paying careful attention to the correct command term verb may cost you marks unnecessarily.
- **Practice with past papers** Most schools will give their students a 'mock' or 'trial' exam. This is helpful as it enables you to judge the correct amount of time to spend on each question. Make your mistakes in the mock exam and learn from them. Of course the IB questions are different each year but they do tend to follow a similar pattern. It helps to have seen similar questions before and know what level to expect.
- **Organize your notes** As you review your work it is often helpful to rewrite your notes. Concentrate on just the key points – they should trigger your memory. This book already contains the important points in a fairly condensed form. Condense them even more to make your own set of review notes. Each time you review each topic try to condense the notes even more. By the time you are ready to take the exam all you personal review notes should ideally fit onto a single page!
- **Be familiar with using the IB data booklet** You should get into the habit of using this throughout the course so that you are completely familiar with its contents and how to use them by the time of the exams.
- **Know the format of the exam papers** Both HL and SL students take three exam papers. Papers 1 and 2 examine the core (and additional HL) material. Paper 3 will be taken on the next day (or examination session if a weekend intervenes) and covers the options. Paper 1 is multiple choice and you are not allowed a calculator or the IB data booklet. A Periodic Table is provided. Paper 2 contains short answer questions in Section A that you must attempt and longer questions in Section B. SL students must choose one of the longer questions from a choice of three and HL students must choose two longer questions from a choice of four. Paper 3 contains questions on all the options. You are required to answer all the questions on two of the options.
- **Know the dates of the exams** Plan your review timetable carefully in advance. Remember that you will have exams in other subjects and that you may not have much time for a 'last minute' review.

TAKING THE EXAMINATIONS

- Try to ensure that the night before you are able to take some time to relax and get a good night's sleep.
- Take all you need with you to the examination room, i.e. pens, pencils, ruler, and a simple translating dictionary if English is not your first language. You will need your calculator for Paper 2 and Paper 3 – remember to include a spare battery.
- There is no reading time allowed for Paper 1. Work through the questions methodically. HL have 40 questions in 1 hour, SL have 30 questions in 45 minutes. If you get stuck on a question move on and then come back to it if you have time at the end. Make a note of those questions you are unsure about. You can then come back to these at the end rather than going through all of them again. Make sure you give one answer for each question. You are not penalized for wrong answers so if you run out of time make an educated guess rather than leave any questions unanswered.
- Use the five minutes reading time for Paper 2 wisely. You have to answer the questions on Section A so use this time to read through the questions on Section B thoroughly and decide which you will choose to answer.
- Read each question very carefully. Make a mental note of the key command term so that you give the required answer.
- Try to write your answers within the required space. Write as legibly as you can. For questions involving calculations do not round up too early but make sure your final answer is given to the correct number of significant figures. Always include the correct units. If you do need extra space continue in a separate answer booklet.
- Attempt all the required number of questions. If you do not attempt a question you can receive no marks. For sequential numerical questions even if you get the first part wrong continue as you will not be marked wrong twice for the same mistake. For this reason it is essential that you show your working. Do not answer more questions than required. The examiner will simply mark the required number in the order they are written, not necessarily the best ones.
- Leave yourself time to read through what you have written to correct any mistakes.
- Ensure that you have filled in the front of the paper correctly, including stating the number of the optional questions answered, before leaving the examination room.

Internal assessment (1)

INTRODUCTION

You are expected to spend 40 hours (SL) or 60 hours (HL) during the two years engaged in work that can be assessed internally. This is essentially time spent in the laboratory and comprises 24% of the final marks. Chemistry is an experimental science and practical work is an important component of the course. Your teacher should devise a suitable practical programme for you to follow. Practical work can have many different aims; for example, to improve your skills at different techniques, to reinforce the theoretical part of the course and to give you experience of planning your own investigations. Hopefully it will make studying Chemistry much more challenging and rewarding. Through the practical course you are expected to understand and implement safe practice and also to respect the environment.

INTERNAL ASSESSMENT – THE FACTS

Your practical work will be assessed continually throughout the two years of the course. The assessment is exactly the same for both SL and HL. It is assessed according to five different criteria. The first three criteria, Design (D), Data collection and processing (DCP) and Conclusion and evaluation (CE) carry a maximum of six marks each. The best two marks obtained for each of these three criteria over the two years will be used to give a maximum of 36 marks. The fourth criteria, Manipulative skills (MS) is assessed summatively throughout the course for a maximum of six marks and the fifth criteria Personal skills (PS) is assessed once only (during the group 4 project) for a maximum of six marks. This gives an overall total maximum mark of 48. This mark will then be halved to make up the 24% Internal Assessment component of the overall assessment mark. Some of your practical work may not be formally assessed at all and some may only be assessed for one or two of the criteria. However your teacher must assess the first three criteria at least twice during the two years. Most teachers will assess more than this and then submit the two best marks. Each criterion is broken down into three different aspects. Your teacher assesses whether you have covered each aspect **completely, partially** or **not at all** to arrive at the mark given for each criterion. Completely scores two marks, partially one mark and not at all zero marks. Each school will send the work of a few students for moderation so that everyone is graded uniformly.

GRADING OF INTERNAL ASSESSMENT

You will only gain good marks if you try to address each aspect of each criterion completely. It is important that you understand fully what is required for each of the different aspects. The description of each aspect to gain **complete** is given.

CRITERION	ASPECTS		
Design (D)	Formulates a focused problem/ research question and identifies the relevant variables	Designs a method for the effective control of the variables	Develops a method that allows for the collection of sufficient relevant data
Data collection and processing (DCP)	Records appropriate quantitative and associated qualitative raw data, including units and uncertainties where relevant	Processes the quantitative raw data correctly	Presents processed data appropriately and, where relevant, includes errors and uncertainties
Conclusion and evaluation (CE)	States a conclusion, with justification, based on a reasonable interpretation of the data	Evaluates weaknesses and limitations	Suggests realistic improvements in respect of identified weaknesses and limitations
Manipulative skills (MS)	Follows instructions accurately, adapting to new circumstances (seeking assistance when required)	Competent and methodical in the use of a range of techniques and equipment	Pays attention to safety issues
Personal skills (PS)	Approaches the group 4 project with self-motivation and follows it through to completion	Collaborates and communicates in a group situation and integrates the views of others	Shows a thorough awareness of their own strengths and weaknesses and gives thoughtful consideration to their learning experience

THE GROUP 4 PROJECT

The group 4 project is a collaborative activity whereby all the IB students in the school from the different group 4 subjects work together on a scientific or technological topic. You are required to spend about ten hours in total on the group 4 project. In the planning stage, which should last for about two hours, you should decide on an overall topic with your fellow students and then, in small groups, decide how you will investigate a particular aspect of the chosen topic. During the action stage, which lasts for about six hours, you should investigate your topic. The investigation may be practically or theoretically based and may be just in chemistry or across all the scientific disciplines. You should collaborate with other students and in any practical work pay attention to safety, ethical and environmental considerations. Finally there is the evaluation stage. This should last for about two hours and involves sharing your results, including your successes and failures, with all the other students. Unlike the remainder of your internally assessed work, the emphasis for the group 4 project is on the collaborative experience of working with other students. It is the **process** not the **product** that is important. To gain high marks for the personal skills criterion you need to show considerable self-motivation and perseverance. You need to be able to listen to others as well as put forward your own views and you need to be able to reflect on the learning experience. There are different ways in which this assessment may take place and each school will determine its own way. For example, you may be required to write a report or you may be asked to review your peers or you may be asked to complete a self-evaluation form.

Internal assessment (2)

HOW TO MAXIMIZE YOUR INTERNAL ASSESSMENT MARKS

General points

- Ascertain before you undertake the investigation which criteria (if any) are being assessed.
- Check that you are clear about all the aspects to be assessed.
- Record all your work as you proceed in your log book or laboratory notebook.
- Record the title of the experiment (or piece of work), the date and the name(s) of any partner(s) you worked with.
- Record precise details of all equipment used, e.g. a balance weighing to + or - 0.001 g, a thermometer measuring from -10 to + 110 °C to an accuracy of + or - 0.1°C, a 25.00 cm³ pipette measuring to + or -0.04 cm³ etc.
- Record precise details of any chemicals used, e.g. copper(II) sulfate pentahydrate $CuSO_4.5H_2O(s)$ and if it is a solution include the concentration, e.g. 0.100 mol dm⁻³ NaOH(aq).
- Record all measurements accurately to the correct number of significant figures and include all units.
- Record all observations. Include colour changes, solubility changes, whether heat was evolved or taken in etc.
- Draw up a checklist to cover each criterion being assessed. As you write the laboratory account check that each aspect is addressed fully. (Some students give each aspect a sub-heading. Although this is not strictly necessary it does help to draw the aspect to the attention of the teacher).
- Your work may be hand-written (in ink) or word-processed. Ensure that it is neat, correct and legible.
- Write clearly and succinctly.
- Hand your work in on time. Teachers are within their rights to refuse to mark work handed in late as you may benefit from using other students' marked assignments.
- Learn from your mistakes. In the early part of the course do not expect to get everything correct the first time you do it. Find out why you lost marks and improve your next presentation.
- Keep all your laboratory reports. At the end of the course some of them may need to be sent off for moderation.

Specific points for each criterion.

Design (D)

You will be given an instruction such as "Investigate an aspect of…." by your teacher but very little else in the way of instructions.

- Identify your own research question/problem and state it clearly.
- Identify all the variables. State clearly which variable are controlled, which one you will manipulate (the independent variable) and which one is the dependent variable that you will measure.
- Give accurate and concise details about the apparatus and materials used.
- Explain how the method chosen enables the controlled variables to be controlled and describe the method in sufficient detail so that it could be repeated by an independent researcher.
- Ensure that your method enables sufficient relevant data to be collected.

Data collection and processing (DCP)

- Ensure all raw data is recorded. Pay particular attention to significant figures and make sure all units are stated.
- Record the level of uncertainty for each quantitative reading.

- Include all qualitative data to describe what is observed during the experiment.
- Present your results clearly. Often it is better to use a table or a graph. If using a graph, ensure that the graph has a title and both axes are labelled clearly and that the correct scale is chosen to utilise most of the graph space.
- Draw the line or curve of best fit for graphical data
- When carrying out an acid-base titration ensure that the indicator is clearly stated and the change in colour recorded to signify the end-point.
- Ensure that you have used your data correctly to produce the required result.
- In quantitative experiments ensure that the limits of accuracy of each piece of apparatus have been stated and then summed to give the limits of accuracy with which you can state your result. Calculate it first in percentage terms then transform it into the + and - amount pertaining to your actual result.
- Include any other errors or uncertainties which may affect the validity of your result.

Conclusion and evaluation (CE)

- Include a valid conclusion. This should relate to the initial problem or hypothesis.
- Compare your result to the expected (Literature or IB data booklet) result.
- Calculate the percentage error from the expected value.
- Evaluate your method. Comment on random and systematic errors. State any assumptions that were made which may affect the result.
- Comment on the limitations of the method chosen by identifying any weaknesses and show an awareness of how significant the weaknesses are.
- Suggest how the method chosen could be realistically and specifically improved to obtain more accurate and precise results.

Manipulative skills (MS)

This criterion cannot be moderated from the written work you submit. It is assessed by your teacher on your performance throughout the two years when you are actually working in the laboratory. To gain high marks ensure:

- you follow instructions carefully and show initiative when necessary
- you ask when you are uncertain
- you show proficiency and competence in a wide range of different chemical techniques
- you are enthusiastic in your approach
- you show a high regard for safety in the laboratory
- you show respect for the environment in the way you conduct your experiments and dispose of any residues.

Personal skills (PS)

This is only assessed during the group 4 project. As you will only do the group 4 project once you will not be able to learn from your mistakes. To achieve good marks make sure:

- you show that you are highly motivated and involved
- you persevere throughout the whole of the group 4 project
- you collaborate well with others by listening to their views and incorporating them into your work as well as making your own suggestions
- you show an awareness of your own strengths and weaknesses
- you show that you have reflected well on the whole project and learned from the experience.

Extended Essays (1)

WHAT IS AN EXTENDED ESSAY?

In order to fulfil the requirements of the IB all Diploma candidates must submit an Extended Essay in an IB subject of their own choice. The Essay is an in-depth study of a limited topic within a subject. The purpose of the Essay is to provide you with an opportunity to engage in independent research. Approximately forty hours should be spent in total on the Essay. Each Essay must be supervised by a competent teacher. The length of the Essay is restricted to a maximum of 4000 words and it is assessed according to a carefully worded set of criteria. The marks awarded for the Extended Essay are combined with the marks for the Theory of Knowledge course to give a maximum of three bonus points.

EXTENDED ESSAYS IN CHEMISTRY

Although technically any IB Diploma student can choose to write their Essay in Chemistry it does help if you are actually studying Chemistry as one of your six subjects! Most Essays are from students taking Chemistry at Higher Level but there have been some excellent Essays submitted by Standard Level students. All Essays must have a sharply focused Research Question. Essays may be just library-based or also involve individual experimental work. Although it is possible to write a good Essay containing no experimental work it is much harder to show personal input and rarely do such Essays gain high marks. The experimental work is best done in a school laboratory although the word 'laboratory' can be interpreted in the widest sense and includes the local environment. It is usually much easier for you to control, modify, or redesign the simpler equipment found in schools than the more sophisticated (and expensive) equipment found in university or industrial research laboratories.

CHOOSING THE RESEARCH QUESTION

Choosing a suitable Research Question is really the key to the whole Essay. Some supervisors have a list of ready-made topics. The best Essays are usually where you, the student, identify a particular area or chemistry problem that you are interested in and together with the supervisor formulate a precise and sharply focused Research Question. It must be focused. A title such as 'A study of analysis by chromatography' is far too broad to complete in 4000 words. A focused title might be 'An analysis of the red dyes present in different brands of tomato ketchup by thin layer chromatography'. It is more usual to choose a topic then decide which technique(s) might be used to solve the problem. An alternative way is to look at what techniques are available and see what problems they could address.

SOME DIFFERENT TECHNIQUES (TOGETHER WITH A RESEARCH QUESTION EXAMPLE) THAT CAN BE USED FOR CHEMISTRY EXTENDED ESSAYS

The list below shows some examples of how standard techniques or equipment available in a school laboratory can be used to solve research questions. Although one example has been provided for each technique many research questions will, of course, involve two or more of these techniques.

Redox titration
Do different (specified) varieties of seaweed contain different amounts of iodine?

Extension of a standard practical
What gas is evolved when zinc is added to $CuSO_4$(aq) and what factors affect its formation?

Acid-base titration
How do storage time and temperature affect the vitamin C content of (specified) fruit juices?

Chromatography
Do strawberry jellies from (specified) different countries contain the same red dyes(s)?

Calorimetry
How efficient is dried cow dung as a fuel compared to fossil fuels?

pH meter
Can (specified) different types of chewing gum affect the pH of the mouth and prevent tooth decay?

Steam distillation
What is the amount of aromatic oil that can be extracted from (a specified) plant species?

Electrochemistry
What is the relationship between concentration and the ratio of O_2:Cl_2 evolved during the electrolysis of NaCl(aq)?

Refinement of a standard practical
How can the yield be increased in the laboratory preparation of 1,3-dinitrobenzene?

Microwave oven
What is the relationship between temperature increase and dipole moment for (specified substances)?

Polarimetry
Is it possible to prepare the different enantiomers of butan-2-ol in a school laboratory?

Data logging probes
What is the rate expression for (a specified reaction)?

Visible spectrometry
What is the percentage of copper in different ores found in (specified area)?

Gravimetric analysis
Do (specified) 'healthy' pizzas contain less salt than normal pizzas?

Inorganic reactions
An investigation into the oxidation states of manganese – does Mn(V) exist?

Microscale/Small Scale
How can the residues from a typical IB school practical programme be reduced?

Extended Essays (2)

RESEARCHING THE TOPIC

Once the topic is chosen research the background to the topic thoroughly before planning the experimental work. Information can be obtained from a wide variety of sources: a library, the internet, personal contacts, questionnaires, newspapers, etc. Make sure that each time you record some information you make an accurate note of the source as you will need to refer to this in the bibliography. Treat information from the internet with care. If possible try to determine the original source. Articles in journals are more reliable as they have been vetted by experts in the field. Together with your supervisor plan your laboratory investigation carefully. Your supervisor should ensure that your investigation is safe, capable of producing results (even if they are not the expected ones) and lends itself to a full evaluation.

THE LABORATORY INVESTIGATION

Make sure you understand the chemistry that lies behind any practical technique before you begin. Keep a careful record of everything you do at the time that you are doing it. If the technique 'works' then try to expand it to cover new areas of investigation. If it does not 'work' (and most do not the first time) try to analyse what the problem is. Try changing some of the variables, such as increasing the concentration of reactants, changing the temperature or altering the pH. It may be that the equipment itself is faulty or unsuitable. Try to modify it. Use your imagination to design new equipment in order to address your particular problem (modern packaging materials from supermarkets can often be used imaginatively to great effect). Because of the time limitations it is often not possible to get reliable repeatable results but attempt to if you can. Remember that the written Essay is all that the external examiner sees so leave yourself plenty of time to write the Essay.

WRITING THE ESSAY

Before starting to write the Essay make sure you have read and understood the assessment criteria. Your school or supervisor will provide you with a copy. It may be useful to look at some past Extended Essays to see how they were set out. Almost all Essays are word-processed and this makes it easier to alter draft versions but they may be written by hand. You will not be penalized for poor English but you will be penalized for bad chemistry so make sure that you do not make simple word-processing errors when writing formulas, etc.

Start the Essay with a clear introduction and make sure that you set out the Research Question clearly and put it into context. The rest of the Essay should then be very much focused on addressing the Research Question. Some of the marks are gained simply for fulfilling the criteria (e.g. numbering the pages, including a list of contents, etc.). These may be mechanical but you will lose marks if you do not do them. Give precise details of any experimental techniques and set out the results clearly and with the correct units and correct number of significant figures. If you have many similar calculations then show the method clearly for one and set the rest out in tabular form. Numerical results should give the limits of accuracy and a suitable analysis of uncertainties should be included. Relate your results to the Research Question in your discussion and compare them with any expected results and with any secondary sources of information you can locate. State any assumptions you have made and evaluate the experimental method fully. Suggest possible ways in which the research could be extended if more time were available. Throughout the whole Essay show that you understand what it is that you are doing and demonstrate personal input and initiative.

When you have completed the essay write an abstract. This should state clearly the Research Question, how the investigation was undertaken and the conclusion in less than 300 words.

Give your supervisor at least one draft version for comment before completing the final version. Before handing in the final version go through the following checklist very carefully. If you can honestly tick 'yes' to every box then your Essay will be at least satisfactory and hopefully much better than this. After you have handed in your Essay your supervisor is likely to give you a short interview (viva voce). This is to determine that the Essay is in fact all your own work and it also provide you with an opportunity to reflect on the successes and difficulties involved in the research process and an opportunity to reflect on what has been learned. It also helps your supervisor to write his or her report which will accompany your Essay when it is sent for external marking.

Extended Essays (3)

EXTENDED ESSAY CHECK LIST

In order to gain the maximum credit possible for your essay it is crucial that you can answer YES to the following questions *before* you finally submit your essay to your supervisor.

The maximum number of marks available for each criterion is given in brackets.

A. Research question [2]
Is the research question sharply focused? [ ]
Is the research question clearly and precisely stated in the introduction? [ ]
(N.B. the abstract and the title do NOT count as part of the essay)

B. Introduction [2]
Have you included all the necessary background knowledge? [ ]
Is the context of the research question clearly demonstrated? [ ]
Is it clear why the topic is worthy of investigation? [ ]

C. Investigation [4]
Has an imaginative range of information/data/sources been consulted/gathered? [ ]
Has the relevant material been carefully selected and the investigation well planned? [ ]
Have you shown how you personally have devised/adapted/modified the methods? [ ]
Has sufficient information been given so that all data gathered by you capable of replication by others? [ ]

D. Knowledge and understanding of the topic studied [4]
Have you shown excellent knowledge and understanding of the topic studied? [ ]
Does your essay clearly and precisely locate the investigation in an academic context? [ ]
Have you shown that you understand the underlying chemistry of all techniques and apparatus you have used? [ ]

E. Reasoned argument [4]
Are your ideas developed in a logical and coherent manner? [ ]
Does your essay develop a reasoned and convincing argument in relation to the research question? [ ]
Have you compared different approaches and methods that are relevant to the research question? [ ]

F. Application of analytical and evaluative skills appropriate to chemistry [4]
Have you shown that the whole emphasis of your essay is on chemistry? [ ]
Have you shown effective and appropriate analytical and evaluative skills relating to chemistry? [ ]
Have you examined all underlying assumptions and checked the quality of sources and data? [ ]

G. Use of language appropriate to chemistry [4]
Have you communicated clearly and precisely using the correct chemical terminology? [ ]
Have you used chemical terminology accurately and with skill and understanding? [ ]
Have you given all relevant formulas, equations and chemical structures and checked they are correct? [ ]
Are all units correctly given and has attention been paid to the correct use of significant figures? [ ]

H. Conclusion [2]
Is your conclusion stated clearly? [ ]
Is your conclusion relevant to the research question and consistent with the evidence? [ ]
Does your conclusion contain unresolved questions and suggest areas for further investigation? [ ]

I. Formal presentation [4]
Is the essay less than 4000 words? [ ]
Is there a title page and a list of contents which is clearly set out? [ ]
Is illustrative material appropriate, well set out and used effectively? [ ]
Are all the pages numbered? [ ]
Does the bibliography include all, and only, those works which have been consulted? [ ]
Are the references set out in a consistent standard format which specifies: author(s), title, date of publication and publisher? [ ]
Is all the work of others clearly acknowledged? [ ]
If there is an appendix, does it contain only information that is required in support of the text? [ ]

J. Abstract [2]
Is your abstract less than 300 words? [ ]
Does your abstract include the research question, how the investigation was undertaken and the conclusion? [ ]

K. Holistic judgement [4]
Have you demonstrated to the best of your ability the following qualities in your essay:
Intellectual initiative; Personal input; Inventiveness; Insight; Depth of understanding and Originality? [ ]

Answers to questions

QUANTITATIVE CHEMISTRY
1. C   2. C   3. B   4. A   5. D   6. D   7. A   8. B   9. A
10. A   11. B   12. C   13. D
14. (a) 2-hydroxybenzoic acid,   (b) 19.6 g,   (c) 69.9%.
15. (a) 4.00×10^{-2} moles,   (b) 4.00×10^{-2} moles,
(c) 362,   (d) $A_r = 133$, M = Cs.

ATOMIC STRUCTURE
1. A   2. D   3. A   4. D   5. B   6. B   7. D   8. A   9. D
10. C   11. B   12. B   13. B   14. C   15. A

PERIODICITY
1. A   2. A   3. D   4. C   5. C   6. C   7. B   8. C   9. D
10. A   11. D   12. D   13. D   14. A   15. A   16. A

BONDING
1. A   2. B   3. A   4. C   5. D   6. A   7. B   8. A   9. D
10. D   11. C   12. A   13. B   14. D   15. B   16. A

ENERGETICS
1. D   2. D   3. A   4. C   5. C   6. B   7. C   8. D   9. D
10. A   11. B   12. B   13. C   14. A   15. B

KINETICS
1. B   2. B   3. D   4. D   5. D   6. B   7. D   8. B   9. C
10. C   11. A   12. B   13. A   14. D   15. A   16. C   17. D

EQUILIBRIUM
1. A   2. B   3. D   4. A   5. D   6. A   7. B   8. B   9. B
10. C   11. D   12. D   13. D

ACIDS AND BASES
1. A   2. B   3. D   4. C   5. C   6. A   7. B   8. D   9. D
10. B   11. C   12. C   13. B   14. C   15. B   16. D

OXIDATION AND REDUCTION
1. C   2. D   3. A   4. B   5. B   6. A   7. B   8. A   9. D
10. A   11. A   12. C   13. A   14. C

ORGANIC CHEMISTRY
1. D   2. B   3. C   4. C   5. D   6. D   7. A   8. B   9. A
10. A   11. B   12. D   13. A   14. C   15. D   16. B   17. B

MEASUREMENT AND DATA PROCESSING
1. B   2. C   3. B   4. A   5. D   6. D   7. C   8. A   9. B
10. C

Option A – Modern analytical chemistry
1. (a) bonds vibrate (bend/stretch) [1], change in dipole moment required [1], different bonds (functional groups) absorb in different regions [1], precise absorption affected by neighbouring atoms [1].   (b)(i) ethanoic acid: C=O 1680-1750 cm^{-1}, O-H 2500-3300/3580-3650 cm^{-1}, C-H 2840-3095 cm^{-1} [2], methyl methanoate: C=O 1680-1750 cm^{-1}, C-H 2840-3095 cm^{-1}, C-O 1000-1300 cm^{-1} [2].   (c) O-H in ethanoic acid could be used or C-O absorption in ester [1], other absorptions cannot be used as they occur in both spectra [1].   (d) O-H [1], 3.03×10^{-4} cm [1].

2. (b)(i) R_f of P1 = 0.16 (answers accepted in range 0.14 – 0.20) [2]. (ii) Mixture, as more than one spot [1].

3. (a) It contains a C=O bond or aldehyde or ketone [1].
(b) C-H bond [1].   (c) It does not contain an O-H bond [1].

4. (a) A is ultraviolet: electronic transitions [2]; B is infrared : molecular vibrations [2]; A is higher in energy than B [1].
(b) Both A and B. A from electronic transitions when the bond breaks or is formed, B from molecular vibrations which depend on bond strength [1].

5. Pentan-2-one: 4 peaks in ratio of 3:2:2:3 [2]; pentan-3-one: 2 peaks in ratio of 3:2 [2].

6. (a) Many conjugated double bonds [1] leading to extended delocalized π bond [1].   (b) 1.54×10^{-5} mol dm^{-3} [2].
(c) Gas phase chromatography allows separation of the components of the mixture [1], mass spectrometry allows accurate identification of each component separated [1].

7. (a) SiC_4H_{12} [1] and any two from: sharp single peak, strong peak as all 12 Hs in same chemical environment, volatile so can be easily removed from sample, all other peaks shifted downfield from it [2].   (b) They are in the ratio 3:2:5 [1].
(c) It contains a –C_2H_5 group or a CH_2 group next to a CH_3 group [2].   (d) It contains a phenyl/aromatic/benzene group [1].

Option B – Human biochemistry
1. (a) Ketone (carbonyl) [1], alkene [1], testosterone would be expected to decolourise bromine water (or will give a crystalline solid with 2,4-DNPH) [1].   (b) Alkyne [1], alcohol [1], amine [1] phenyl [1]. Norethynodrel 'fools' the female reproductive system by mimicking the action of progesterone in true pregnancy [1], prevent the release of an egg [1], inhibits sperm from reaching the egg/makes the uterus unreceptive to any fertilised egg [1].

2. (a) Fats (or lipids) [1].   (b) Vegetable [1], all three fatty acid residues are unsaturated [1].   (c) Oil [1], the structure around the double bonds prevents close packing [1]. (d) Any two from: energy source/insulation/cell membrane [2].

3. (a)(i) Alkene [1], alcohol [1].   (ii) Retinol is fat soluble [1], as it has a long non-polar hydrocarbon chain [1], vitamin C is water soluble as it contains many -OH groups which can hydrogen bond with water molecules [1].   (iii) Vitamin A: night blindness/xerophthalmia [1], vitamin C: bleeding of gums etc. [1], scurvy [1].   (b) 0.0559 mol [1], each mole of oil contains four C=C double bonds [2].

4. (a) Form coloured compounds [1], form complex ions [1].
(b)(i) 55% - 25% = 30% [1].   (ii) Carbon dioxide and lactic acid are acidic so lower the pH [1], haemoglobin is less able to carry oxygen at lower pH [1].

Option C – Chemistry in industry and technology
1. (a)(i) Cryolite lowers the melting point of the alumina [1].
(ii) $2O^{2-} \rightarrow O_2 + 4e^-$ [1]. (iii) The oxygen produced reacts with the carbon electrode $C + O_2 \rightarrow CO_2$ [1].   (b) Alumina acts as an acidic oxide and reacts with NaOH [1], the basic impurities do not react with the NaOH [1].   (c) Low density/aviation industry; resistance to corrosion/window frames, electrical conductor/transmission lines etc. [2].   (d) Save energy, less damage to the environment [2].

2. (a) Crude oil was formed from marine organisms which contained S (in their amino acids) [1].   (b) S can poison catalysts used in refining (or when oil is burned the SO_2 produced leads to acid rain) [1].   (c) $C_{10}H_{22} \rightarrow C_8H_{18} + C_2H_4$ [1], larger molecules are broken down into smaller, more useful molecules [1].   (d) Include answers such fossil fuel so once it is used up there will be no more; as supplies deplete it will be more expensive to produce; will need to find an alternative source as a feedstock if it is all burned; causes increased global warming/climate change/marine pollution [4].

3. (a) Research and development of compounds in the 1-100 nm range [1].   (b) Catalyst is in a different phase to reactants [1].   (c) Large surface area/ confined space so reactants brought close together/ ability to adsorb other molecules [2].

(d) Unknown toxicity of small particles, humane immune system may be defenceless **[2]**.

4. HDPE higher melting point **[1]** as little branching **[1]** so the long chains can fit together closely **[1]** so stronger attractive forces between them **[1]**. Examples might include PVC adding plasticizer to make it more flexible **[2]** and Polystyrene where adding volatile hydrocarbons produces a lightweight packaging/insulation material **[2]**.　**(c) (i)** Not biodegradable so remain for a long time **[2]**;　**(ii)** May produce HCl and HCN as pollutants and if temperature not high enough may produce dioxins **[2]**,　**(iii)** Not broken down easily by light may get ingested into the stomach of marine animals **[2]**.

4. (a)

$$H_2C=C(CN)H \quad HOOC-\bigcirc-COOH \quad HO-CH_2-CH_2-OH \quad [3]$$

(b) Moderate pressure, 60°C, Ziegler catalyst (e.g. $TiCl_4$) **[2]**, high density polythene proceeds by an ionic mechanism whereas low density polythene involves a radical mechanism **[1]**.

5. (a) Shows liquid-crystal behaviour over a range of temperature between the solid and liquid states **[2]**. **(b) (i)** CN- group makes the molecule polar so increases intermolecular forces of attraction making them strong enough to align in a common direction **[2]**; **(ii)** Biphenyl makes molecules rod-shaped and rigid **[2]**; **(iii)** Alkyl group ensures molecules cannot pack so closely together so helps maintain the liquid-crystal state **[2]**. Liquid-crystal sandwiched between two glass plates with scratches at 90° **[1]** molecules line up with scratches and form a twisted arrangement **[1]** plane polarized light rotated by liquid-crystal **[1]** voltage applied so molecules align with field, twisted structure lost so plane-polarized light no longer rotated **[1]**.

Option D – Medicines and drugs

1. (a) A nucleic acid containing genetic materials within a protein coating **[1]**. **(b)** Any two from: they replicate quickly giving a high viral population/they mutate/they work by entering a host cell and any treatment is likely to also affect the host cell **[2]**. **(c)** **[1]**

(d) Any two from: The nitrogen atoms become protonated by the acid as they are bases to form a soluble salt / the -OH group becomes protonated to form a soluble salt/hydrogen bonding occurs between the Acyclovir and water **[2]**.　**(e)** Some drugs are decomposed by stomach acid **[1]** water-soluble drugs are not easily absorbed through the fatty tissue of the stomach wall **[1]**.

2. (a) Increases resistance to the penicillinase enzyme or alters the effectiveness of the 'penicillin' **[1]**.　**(b)** Some bacteria may remain unaffected **[1]**.　**(c)** Bacteria can become resistant to penicillins **[1]**.　**(d)** Antibiotics that are effective against a wide variety of bacteria **[1]**.　**(e)** Penicillins are effective against some, but not all, bacteria **[1]**. Penicillins work by interfering with the chemicals needed by bacteria to form normal cell walls **[2]**.

3. (a) $C_7H_6O_3$ **[1]**.　**(b)** Either circle <u>ester</u> group, pH = 7 **or** circle <u>carboxylic acid</u> group pH = 2 – 6 **[3]**.
(c) $H_2N-\bigcirc-OH$ **[2]**.

(d) Aspirin: stomach bleeding/Reyes syndrome/allergic reactions **[1]** paracetamol: overdose can permanently damage liver and kidneys **[1]**.

4. (a)(i) H_3N ＼ .Cl ∕Pt Cl ＼ ＼NH$_3$ **[1]**　**(ii)** The non-bonding pairs of electrons on the N atoms can form co-ordinate bonds with the platinum **[1]**, Lewis acid/base **[1]**

(b) The different enantiomers can have different biological properties **[1]**, one form may be beneficial and the other form harmful **[1]**. **(c)** A chiral auxilliary is itself an enantiomer **[1]**. It is bonded to the reacting molecule to create the stereochemical conditions necessary to follow a certain geometric path **[1]**. Once the desired enantiomer is formed the auxilliary is removed **[1]**.
(d)

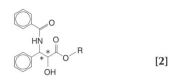 **[2]**

Option E – Environmental chemistry

1. (a) 1000:1 **[1]**.　**(b)** Sulfuric acid (or sulfurous acid) **[1]**, S + $O_2 \rightarrow SO_2$ (or $2SO_2 + O_2 \rightarrow SO_3$) **[1]**, $SO_2 + H_2O \rightarrow H_2SO_3$ (or $SO_3 + H_2O \rightarrow H_2SO_4$) **[1]**.　**(c)** Nitric acid **[1]**, any two from catalytic converter/richer petrol:air mixture/thermal exhaust system/switching to a different fuel e.g. solar power **[2]**.
(d) $CaCO_3 + 2H^+ \rightarrow Ca^{2+} + CO_2 + H_2O$ **[1]**.

2. (a)(i) As the temperature increases the solubility of oxygen decreases **[1]**, **(ii)** As organic pollutants decompose they use up available oxygen so the amount of dissolved oxygen decreases **[1]**, **(iii)** Nitrates and phosphates act as nutrients and increase the growth of algae. As the algae die they use up dissolved oxygen (eutrophication) **[1]**.　**(b)** The quantity of oxygen (in ppm) utilised when the organic matter in a fixed volume of water is decomposed biologically over a set time period (usually five days) **[1]**. **(c)** 8.00×10^{-3} g dm^{-3} (or 8 mg dm^{-3}) **[3]**.

3. (a) Carbon dioxide **[1]**, combustion of fossil fuels **[1]**, large scale combustion is causing an increase in atmospheric CO_2 levels **[1]**. **(b)** The molecules absorb infrared radiation emitted from the Earth **[1]**, this absorption prevents the heat from escaping into outer space and causes global warming **[1]**. **(c)** particulates **[1]**, counteracts the greenhouse effect by reflecting sunlight **[1]**.

4. (a) (i) $2CO+2NO \rightarrow 2CO_2+N_2$ **[2]**　**(ii)** $CaO+SO_2 \rightarrow CaSO_3$ (or $2CaO+2SO_2+O_2 \rightarrow 2CaSO_4$) **[2]** **(b)** $2NO+O_2 \rightarrow 2NO_2$ **[1]** $4NO_2+2H_2O+O_2 \rightarrow 4HNO_3$ **[1]** (or $2SO_2 + O_2 \rightarrow 2SO_3$ **[1]** $SO_2+H_2O \rightarrow H_2SO_4$ **[1]**) **(c)** Irritation of the mucous membranes or fatigue or cancer forming or respiratory problems **[1]**

5. (i) (Osmosis) – movement of solvent or water from dilute to concentrated solution **[1]**
(partially permeable membrane) – allows passage of solvent or water but not solute particles **[1]**;
(ii) Sea water is subjected to high pressure **[1]**; pressure must be greater than osmotic pressure **[1]**;
pure water passes through (partially permeable) membrane and dissolved solids left behind **[1]**;
salt / dissolved solids left behind; **(iii)** Shower rather than bath, use less water to flush toilet etc. **[1]**.

6. (a) Aeration with oxygen and bacteria **[2]**;　**(b)** Detergents **[1]**; precipitation **[1]** $Pb^{2+} + S^{2-} \rightarrow PbS$ **[1]** $Al^{3+} + PO_4^{3-} \rightarrow AlPO_4$ **[1]**.

7. (a) CFCs **[1]** from refrigerants etc. **[1]** or NO_x **[1]** from internal combustion engine etc. **[1]**; **(b)** *Advantage:* no weak C—Cl bonds so do not produce chlorine radicals **[1]** *disadvantage:* greenhouse gases so contribute to global warming or hydrofluorocarbons are flammable **[1]**.

8. (a) 2.41×10^{-7} m (241 nm) **[3]**, **(b)** From the Lewis structures **[1]** the bond is weaker in ozone as it is equivalent to 1.5 bonds rather than the double bond found in oxygen so requires less energy/longer wavelength to break **[1]**.　**(c)(i)** $CCl_2F_2 \rightarrow CClF_2 + Cl$ (uv light causes radical formation **[1]**, $Cl + O_3 \rightarrow ClO + O_2$ (chlorine radicals react with ozone) **[1]**, $ClO + O \rightarrow Cl + O_2$

(chlorine radicals regenerated)[1]. **(ii)** During the arctic winter small amounts of water vapour freeze into ice crystals [1], the surface of the ice crystals acts as a catalyst to produce species such as Cl_2, when the winter is over the sun converts this into chlorine radicals[1].

9. (a) NO_x and VOCs [2] from internal combustion engines [1]; **(b)** Bowl-shaped cities, lack of wind, temperature inversion [3]; **(c)** In sunlight NO_2 breaks down to form oxygen radicals $NO_2 \rightarrow NO + O^{\bullet}$ [1] these react to form hydroxyl radicals $O^{\bullet} + H_2O \rightarrow 2OH^{\bullet}$ [1] which react with hydrocarbons $OH^{\bullet} + RH \rightarrow R^{\bullet} + H_2O$ [1] the alkyl radicals react with oxygen to form peroxides $R^{\bullet} + O_2 \rightarrow ROO^{\bullet}$ [1] which react with NO_2 to form PANs $RCOOO^{\bullet} + NO_2 \rightarrow RCOOONO_2$ [1].

10. (a) $K_{sp} = [Al^{3+}][PO_4^{3-}]$ [1]; **(b)** $[PO_4^{3-}] = 3.74 \times 10^{-10}$ mol dm^{-3} [2]; **(c)** The addition of more of one of the ions, in this case Al^{3+} ions, so that $[PO_4^{3-}]$ must decrease to keep K_{sp} constant [2].

Option F – Food chemistry

1. (a) Liquid [1], It is polyunsaturated. **(b)(i)** Ni [1], 140-225 °C [1]. **(ii)** There are several possible answers e.g. –$C_{15}H_{31}$, –$C_{17}H_{33}$ –$C_{17}H_{33}$ or any other combination where four hydrogen atoms have been added but C_{15} can never exceed H_{31} and C_{17} can never exceed H_{35} [1]. **(c)** During the partial hydrogenation some of the *trans-* isomer may form [1], increases levels of LDL cholesterol so increased risk of heart disease [1].

2. (a) A dye is a synthetic colorant a pigment is a natural colorant [1]. **(b)(i)** They absorb and reflect [1] different wavelengths of visible light [1]. **(ii)** Changing the pH may alter the structure [1] so that they absorb and reflect in a different region of the visible spectrum [1]. **(c)** Cans contain metal ions (e.g. Fe^{2+} and Al^{3+}) [1] which can form coordination complexes with the food molecules responsible for colour [1].

3. (a) A substance that delays the onset or slows the rate of oxidation in food [1]. **(b)** A = phenol [1], B = tertiary butyl [1]. **(c)** Squash, broccoli, sweet potatoes, tomatoes, kale, cantaloupe, melon, peaches or apricots [1]. See list of advantages [2] and disadvantages [2] on p 148.

4. (a) Carbohydrate: contains C, H and O [1]; monosaccharide: empirical formula CH_2O with one carbonyl group and at least two –OH groups [2]. **(b)** Condensation [1]. **(c)** Lack of nitrogen containing compounds [1], pH [1] and temperature above 120 °C [1]. **(d)** Amino acid or protein with a reducing sugar [2], the amino group undergoes a condensation reaction with the carbonyl group on the sugar [2].

5. (a)(i) It contains conjugated double bonds [1] which absorb light in the visible region as electrons are promoted to higher energy levels [1]. **(ii)** In acidic solution it absorbs in the blue region and transmits red light [1]. In basic solution there is more conjugation (as all three rings are linked) [1] this means it absorbs at a lower wavelength (red) so the light transmitted is blue [1]. **(b)** 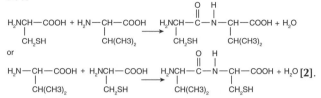 [1] **(c)** It contains several –OH groups which makes it very polar [1] so it is able to form hydrogen bonds with water molecules and it is not attracted to non polar fat molecules [1].

6. (a)(i) The amine group is on the second carbon atom (the first carbon atom is the –COOH group) [1]. **(ii)** They contain an asymmetric (or chiral) carbon atom [1].

(iii)
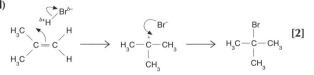
[1] the CORN rule states that for the D isomer the groups follow the clockwise order COOH, R, NH_2

when the hydrogen is behind [2]. **(iv)** The (+) enantiomers will rotate the plane of plane-polarized light clockwise and the (-) will rotate the plane of plane-polarized light by the same amount anticlockwise [2].

(b)(i)

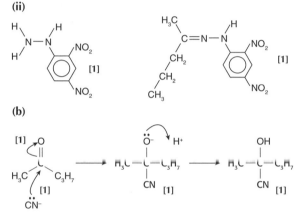

(ii) The primary sequence contains a chain of many amino acids condensed together by peptide (amide) linkages [1]. **(c)** The porphyrin ring contains four nitrogen atoms [1] which are able to form coordinate bonds with the iron atom [1].

Option G – Further organic chemistry

1. (a)(i) Add 2,4-dintrophenylhydrazine to make the 2,4-dinitrophenylhydrazone[1] The crystalline product has a characteristic melting point which can be compared with the data book value [1].

(ii)

H₃C...C=N-...NO₂ [1]

(b)

[1] ...CN⁻ ... H₃C—C—C₃H₇ CN [1] ... H₃C—C—C₃H₇ CN [1]

2. (a) Ethanol is an extremely weak acid/not acidic as the negative charge of the anion is localised on the oxygen atom so the anion is a very strong base and attracts a proton strongly [1] in phenol the negative charge of the anion can be delocalised over the whole ring so the charge density is less and the attraction for a proton is less so phenol is acidic [1]. **(b)** The acid strength depends on the ability of the negative charge on the anion to be localised or delocalised. In 2,2-dimethylpropanoic acid the positive inductive effect of the two alkyl groups makes the charge more localised [1] so it is a weaker acid than ethanoic acid [1] trichloroethanoic acid is a stronger acid than ethanoic acid [1] as the chlorine groups are electron withdrawing [1]. **(c)** Ethylamine is a stronger base than ammonia[1] as the positive inductive effect of the ethyl group increase the electron density of the non bonding pair of electrons on the nitrogen atom so it attracts a proton more strongly [1] phenylamine is a weaker base than ammonia [1] as the non bonding pair of electrons on the nitrogen atom delocalises with the electrons in the benzene ring so is less available to attract a proton[1].

3. (a) $CH_3C(CH_3)BrCH_3$ 2-bromo-2-methylpropane [1], $CH_3CH(CH_3)CH_2Br$ 1-bromo-2-methylpropane [1]. **(b)** When a hydrogen halide adds to the C=C double bond the hydrogen atom bonds to the carbon atom which is already bonded to the greatest number of hydrogen atoms [1]. **(c)** 2-bromo-2-methylpropane [1].

(d)

H₃C—C=C—H ... H₃C—C⁺—CH₃ CH₃ ... H₃C—C—CH₃ Br CH₃ [2]

(e) The 2-bromo isomer involves a tertiary carbocation intermediate **[1]**. This is much more stable than the primary carbocation intermediate which would be formed if the 1-bromo isomer were the product **[1]**. The inductive effect of the CH_3- groups 'pushes' electrons towards the carbocation. This disperses the positive charge and makes it more stable **[1]**. **(f)** $CH_3C(CH_3)ClCH_2I$ **[1]**.

4. (a) React bromoethane with magnesium in ether **[1]** to make ethyl magnesium bromide **[1]** then react the Grignard reagent with ethanal followed by hydrolysis with water to make butan-2-ol **[1]**. **(b)** Dehydrate propan-1-ol by heating with concentrated phosphoric or sulfuric acid **[1]** to make propene **[1]** then add bromine (ideally in a non polar solvent) to form 1,2-dibromopropane **[1]**. **(c)** Add HCN **[1]** to propanone to form the cyanohydrin **[1]** then hydrolyse with a dilute acid to form 2-methyl-2-hydroxypropanoic acid **[1]**.

5. (a)(i) Electrophilic substitution **[1]**.   **(ii)** $C_2H_5 - Br$ $AlBr_3 \rightarrow$ $C_2H_5^+ + AlBr_4^-$ **[2]**

[2]

(b)(i) Ultra-violet light **[1]**.   **(ii)** Free radical substitution **[1]**, $Br-Br \rightarrow 2Br$ **[1]** $Br + C_6H_5C_2H_5 \rightarrow C_6H_5C_2H_4 + HBr$ **[1]**, $C_6H_5C_2H_4 + Br_2 \rightarrow C_6H_5C_2H_4Br + Br$ **[1]**

(c)(i) Elimination **[1]**, The OH^- ion acts as a strong base **[1]**, removing H^+ from the bromoalkane **[1]**. **(ii)** $C_6H_5CH_2CH_2OH$ **[1]**. nucleophilic substitution **[1]**.

Origin of individual questions

The questions detailed below are all taken from past IB examination papers and are © IBO. The questions are from the May (M) or November (N) examinations, with the year given, paper 1 (P1), paper 2 (P2) or paper 3 (P3) with the question number in brackets. All other questions are IB style questions made by the author for this book.

QUANTITATIVE CHEMISTRY
11. N98HLP1(16);   **12.** M00HLP1(16);   **13.** M99HLP1(16)
14. N98SLP2(1).

ATOMIC STRUCTURE
3. M99HLP1(5);   **4.** M98HLP1(5);   **5.** M98HLP1(5);
6. N99HLP1(5);   **7.** N99HLP1(7);   **8.** M98HLP1(6);
9. N99SLP1(8);   **10.** M99SLP1(7);   **11.** M99HLP1(6),
12. M00HLP1(6);   **13.** N98HLP1(6); **14.** N98HLP1(7);
15. M98HLP1(8).

PERIODICITY
3. N98SLP1(9);   **4.** M00SLP1(10);   **6.** M99SLP1(8);
7. M99SLP1(10);   **8.** N98SLP1(8);   **9.** N98SLP1(10);
10. N99HLP1(9);   **11.** HLM99P1(10);   **12.** M00HLP1(8);
13. N99HLP1(10);   **14.** M00HLP1(10);   **15.** M99HLP1(9);
16. M98HLP1(10).

BONDING
1. M99HLP1(11);   **4.** N99SLP1(12);   **5.** M99SLP1(12);
6. N98SLP1(13);   **7.** M99SLP1(14);   **8.** M00SLP1(14);
10. N98SLP1(14);   **11.** M99HLP1(13);   **12.** N99HLP1(12);
13. N99HLP1(13);   **14.** N98HLP1(11);   **16.** N98HLP1(12).

ENERGETICS
1. N99SLP1(16);   **2.** M98SLP1(16);   **3.** M99SLP1(16);
4. M98SLP1(17);   **5.** M99SLP1(17);   **6.** N99SLP1(17);
7. M98SLP1(18);   **8.** M05SLP1;   **9.** M99HLP1(20);
10. N99HLP1(21);   **11.** N04HLP1(18);   **12.** N98HLP1(19);
13. M98HLP1(20);   **14.** N99HLP1(20);   **15.** M00HLP1(20).

KINETICS
1. M00SLP1(19);   **2.** M98SLP1(19);   **3.** M99SLP1(19);
4. N99SLP1(19);   **5.** M98SLP1(22);   **6.** N98SLP1(19);
7. N98SLP1(18);   **8.** M00SLP1(20);   **10.** M03SLP1(19);
11. M00HLP1(20);   **12.** M98HLP1(23);   **13.** N98HLP1(23);
14. N99SLP1(20);   **15.** N99HLP1(23).   **17.** M98SLP1(20)

EQUILIBRIUM
1. M00SLP1(21);   **2.** N99SLP1(21);   **3.** M00HLP1(26);
4. M99SLP1(21);   **5.** M00SLP1(22);   **6.** M98SLP1(21);
7. M99SLP1(22);   **8.** N99HLP1(27);   **9.** N98SLP1(21);
10. M99HLP1(26);   **11.** M99HLP1(27);   **12.** N98HLP1(27);
13. M98HLP1(27).

OPTION A – MODERN ANALYTICAL CHEMISTRY
1. M99HLP3(G2)   **2.** M03HLP3(G2);   **3.** N01HLP3(G1);
4. M05HLP3(G1);   **7.** N01HLP3(G1).

OPTION B – HUMAN BIOCHEMISTRY
1. N98HLP3(C3);   **2.** M98HLP3(C1);   **3.** N99HLP3(C1,C2b);
4. N98HLP3(C2).

OPTION C – CHEMISTRY IN INDUSTRY AND TECHNOLOGY
1. M99HLP3(E1);   **2.** N00HLP3(E2);   **4.** N06HLP3(E2);
5. N98HLP3(E2)

OPTION D – MEDICINES AND DRUGS
1. M98SL(S2);   **2.** M98SL(S1);   **3.** M99SL(S1).

OPTION E - ENVIRONMENTAL CHEMISTRY
1. N00HLP3(D1);   **2.** N00SLP3(D2);   **3.** M98HLP3(D2);
4. N01HLP3(D1);   **5.** M03HLP3(D1);   **6.** N06HLP3(D3);
7. M05HLP3(D2);   **8.** N00HLP3(D3).

OPTION G – FURTHER ORGANIC CHEMISTRY
1. N06HLP3(H2);   **2.** M03HLP3(H2(b) and N06HLP3(H3);
3. N98HLP3(H2);   **5.** N98HLP3(H1).

Index

Periodic Table for use with the IB

GROUP

| Atomic number |
| Element |
| Relative atomic mass |

1	2											3	4	5	6	7	0
1 **H** 1.01																	2 **He** 4.00
3 **Li** 6.94	4 **Be** 9.01											5 **B** 10.81	6 **C** 12.01	7 **N** 14.01	8 **O** 16.00	9 **F** 19.00	10 **Ne** 20.18
11 **Na** 22.99	12 **Mg** 24.31											13 **Al** 26.98	14 **Si** 28.09	15 **P** 30.97	16 **S** 32.06	17 **Cl** 35.45	18 **Ar** 39.95
19 **K** 39.10	20 **Ca** 40.08	21 **Sc** 44.96	22 **Ti** 47.90	23 **V** 50.94	24 **Cr** 52.00	25 **Mn** 54.94	26 **Fe** 55.85	27 **Co** 58.93	28 **Ni** 58.71	29 **Cu** 63.55	30 **Zn** 65.37	31 **Ga** 69.72	32 **Ge** 72.59	33 **As** 74.92	34 **Se** 78.96	35 **Br** 79.90	36 **Kr** 83.80
37 **Rb** 85.47	38 **Sr** 87.62	39 **Y** 88.91	40 **Zr** 91.22	41 **Nb** 92.91	42 **Mo** 95.94	43 **Tc** 98.91	44 **Ru** 101.07	45 **Rh** 102.91	46 **Pd** 106.42	47 **Ag** 107.87	48 **Cd** 112.40	49 **In** 114.82	50 **Sn** 118.69	51 **Sb** 121.75	52 **Te** 127.60	53 **I** 126.90	54 **Xe** 131.30
55 **Cs** 132.91	56 **Ba** 137.34	57+ **La** 138.91	72 **Hf** 178.49	73 **Ta** 180.95	74 **W** 183.85	75 **Re** 186.21	76 **Os** 190.21	77 **Ir** 192.22	78 **Pt** 195.09	79 **Au** 196.97	80 **Hg** 200.59	81 **Tl** 204.37	82 **Pb** 207.19	83 **Bi** 208.98	84 **Po** (210)	85 **At** (210)	86 **Rn** (222)
87 **Fr** (223)	88 **Ra** (226)	89‡ **Ac** (227)															

+

58 **Ce** 140.12	59 **Pr** 140.91	60 **Nd** 144.24	61 **Pm** 146.92	62 **Sm** 150.35	63 **Eu** 151.96	64 **Gd** 157.25	65 **Tb** 158.92	66 **Dy** 162.50	67 **Ho** 164.93	68 **Er** 167.26	69 **Tm** 168.93	70 **Yb** 173.04	71 **Lu** 174.97

‡

90 **Th** 232.04	91 **Pa** 231.04	92 **U** 238.03	93 **Np** (237)	94 **Pu** (242)	95 **Am** (243)	96 **Cm** (247)	97 **Bk** (247)	98 **Cf** (251)	99 **Es** (254)	100 **Fm** (257)	101 **Md** (258)	102 **No** (259)	103 **Lr** (260)